**Microreactors in Organic
Synthesis and Catalysis**

Edited by
Thomas Wirth

Related Titles

Hessel, V., Schouten, J. C., Renken, A., Wang, Y., Yoshida, J.-I. (eds.)

Handbook of Micro Reactors

Chemistry and Engineering

approx. 1500 pages in 3 volumes with approx. 1300 figures
2008
Hardcover
ISBN: 978-3-527-31550-5

Kiwi-Minsker, L., Renken, A., Hessel, V.

Structured Catalytic Microreactors and Catalysts

approx. 350 pages with approx. 250 figures
2007
Hardcover
ISBN: 978-3-527-31520-8

Sheldon, R. A., Arends, I., Hanefeld, U.

Green Chemistry and Catalysis

448 pages with 476 figures and 21 tables
2007
Hardcover
ISBN: 978-3-527-30715-9

Cornils, B., Herrmann, W. A., Muhler, M., Wong, C.-H. (eds.)

Catalysis from A to Z

A Concise Encyclopedia

approx. 1560 pages in 3 volumes with 2778 figures and 104 tables
2007
Hardcover
ISBN: 978-3-527-31438-6

Microreactors in Organic Synthesis and Catalysis

Edited by
Thomas Wirth

WILEY-VCH

WILEY-VCH Verlag GmbH & Co. KGaA

The Editor

Prof. Dr. Thomas Wirth
School of Chemistry
Cardiff University
Park Place Main Building
Cardiff CF10 3AT
United Kingdom

All books published by Wiley-VCH are carefully produced. Nevertheless, authors, editors, and publisher do not warrant the information contained in these books, including this book, to be free of errors. Readers are advised to keep in mind that statements, data, illustrations, procedural details or other items may inadvertently be inaccurate.

Library of Congress Card No.: applied for

British Library Cataloguing-in-Publication Data
A catalogue record for this book is available from the British Library.

Bibliographic information published by the Deutsche Nationalbibliothek
Die Deutsche Nationalbibliothek lists this publication in the Deutsche Nationalbibliografie; detailed bibliographic data are available in the Internet at <http://dnb.d-nb.de>.

© 2008 WILEY-VCH Verlag GmbH & Co. KGaA, Weinheim

Typesetting Thomson Digital, Noida, India
Printing Strauss GmbH, Mörlenbach
Binding Litges & Dopf Buchbinderei GmbH, Heppenheim

Printed in the Federal Republic of Germany
Printed on acid-free paper

ISBN: 978-3-527-31869-8

Contents

Microreactors in Organic Synthesis and Catalysis. Edited by Thomas Wirth
Copyright © 2008 WILEY-VCH Verlag GmbH & Co. KGaA. All rights reserved.
ISBN: 978-3-527-31869-8

Preface

Microreactor technology is no longer in its infancy and its applications in many areas of science are emerging. This technology offers advantages to classical approaches by allowing miniaturization of structural features up to the micrometer regime. This book compiles the state of the art in organic synthesis and catalysis performed with microreactor technology. The term 'microreactor' has been used in various contexts to describe different equipment, and some examples in this book might not justify this term at all. But most of the reactions and transformations highlighted in this book strongly benefit from the physical properties of microreactors, such as enhanced mass and heat transfer, because of a very large surface-to-volume ratio as well as regular flow profiles leading to improved yields with increased selectivities. Strict control over thermal or concentration gradients within the microreactor allows new methods to provide efficient chemical transformations with high space–time yields. The mixing of substrates and reagents can be performed under highly controlled conditions leading to improved protocols. The generation of hazardous intermediates *in situ* is safe as only small amounts are generated and directly react in a closed system. First reports that show the integration of appropriate analytical devices on the microreactor have appeared, which allow a rapid feedback for optimization.

Therefore, the current needs of organic chemistry can be addressed much more efficiently by providing new protocols for rapid reactions and, hence, fast access to novel compounds. Microreactor technology seems to provide an additional platform for efficient organic synthesis – but not all reactions benefit from this technology. Established chemistry in traditional flasks and vessels has other advantages, and most reactions involving solids are generally difficult to be handled in microreactors, though even the synthesis of solids has been described using microstructured devices.

In the first two chapters, the fabrication of microreactors useful for chemical synthesis is described and opportunities as well as problems arising from the manufacture process for chemical synthesis are highlighted. Chapter 1 deals with the fabrication of metal- and ceramic-based microdevices, and Brandner describes different techniques for their fabrication. In Chapter 2, Frank highlights the

Microreactors in Organic Synthesis and Catalysis. Edited by Thomas Wirth
Copyright © 2008 WILEY-VCH Verlag GmbH & Co. KGaA. All rights reserved.
ISBN: 978-3-527-31869-8

microreactors made from glass and silicon. These materials are more known to the organic chemists and have therefore been employed frequently in different laboratories. In Chapter 3, Barrow summarizes the use and properties of microreactors and also takes a wider view of what microreactors are and what their current and future uses can be.

The remaining chapters in this book deal with different aspects of organic synthesis and catalysis using the microreactor technology. A large number of homogeneous reactions performed in microreactors have been sorted and structured by Ryu *et al.* in Chapter 4.1, starting with very traditional, acid- and base-promoted reactions. They are followed by metal-catalyzed processes and photochemical transformations, which seem to be particularly well suited for microreactor applications. Heterogeneous reactions and the advantage of consecutive processes using reagents and catalysts on solid support are compiled by Ley *et al.* in Chapter 4.2. Flow chemistry is especially advantageous for such reactions, but certain limitations to supported reagents and catalysts still exist. Recent advances in stereoselective transformations and in multistep syntheses are explained in detail. Other biphasic reactions are dealt with in the following two chapters. In Chapter 4.3, we focus on liquid–liquid biphasic reactions and focus on the advantages that microreactors can offer for intense mixing of immiscible liquids. Organic reactions performed under liquid–liquid biphasic reaction conditions can be accelerated in microreactors, which is demonstrated using selected examples. The larger area of gas–liquid biphasic reactions is dealt with by Hessel *et al.* in Chapter 4.4. After introducing different contacting principles under continuous flow conditions, various examples show clearly the prospects of employing microreactors for such reactions. Aggressive and dangerous gases such as elemental fluorine can be handled and reacted safely in microreactors. The emergence of the bioorganic reactions is described by van Hest *et al.* in Chapter 4.5. Several of the reactions explained in this chapter are targeted toward diagnostic applications. Although on-chip analysis of biologic material is an important area, the results of initial research showing biocatalysis can also now be used efficiently in microreactors are summarized in this chapter. In Chapter 5, Hessel *et al.* explain that microreactor technology is already being used in the industry for the continuous production of chemicals on various scales. Although only few achievements have been published by industry, the insights of the authors into this area allowed a very good overview on current developments. Owing to the relatively easy numbering up of microreactor devices, the process development can be performed at the laboratory scale without major changes for larger production. Impressive examples of current production processes are given, and a rapid development in this area is expected over the next years. I am very grateful to all authors for their contributions and I hope that this compilation of organic chemistry and catalysis in microreactors will lead to new ideas and research efforts in this field.

August 2007

Thomas Wirth
Cardiff

List of Contributors

Batoul Ahmed-Omer
School of Chemistry
Cardiff University
Main Building
Park Place
Cardiff CF10 3AT
UK

Ian R. Baxendale
Department of Chemistry
University of Cambridge
Lensfield Road
Cambridge CB2 1EW
UK

Juergen J. Brandner
Forschungszentrum Karlsruhe GmbH
Institute for Micro Process Engineering
Hermann-von-Helmholtz-Platz 1
76344 Eggenstein-Leopoldshafen
Germany

Thomas Frank
Little Things Factory GmbH
Ehrenbergstraße 1
98693 Ilmenau
Germany

Takahide Fukuyama
Department of Chemistry
Faculty of Science
Osaka Prefecture University
Sakai
Osaka 599-8531
Japan

John J. Hayward
Department of Chemistry
University of Cambridge
Lensfield Road
Cambridge CB2 1EW
UK

Volker Hessel
Department of Chemical Engineering
and Chemistry
Eindhoven University of Technology
Eindhoven
The Netherlands
Chemical Process Technology
Institut für Mikrotechnik Mainz GmbH
Carl-Zeiss-Strasse 18-20
55129 Mainz
Germany

Microreactors in Organic Synthesis and Catalysis. Edited by Thomas Wirth
Copyright © 2008 WILEY-VCH Verlag GmbH & Co. KGaA. All rights reserved.
ISBN: 978-3-527-31869-8

Kaspar Koch
Organic Chemistry
Institute for Molecules and Materials
Radboud University Nijmegen
Toernooiveld 1
6525 ED Nijmegen
The Netherlands

Steve Lanners
Department of Chemistry
University of Cambridge
Lensfield Road
Cambridge CB2 1EW
UK

Steven V. Ley
Department of Chemistry
University of Cambridge
Lensfield Road
Cambridge CB2 1EW
UK

Patrick Löb
Chemical Process Technology
Institut für Mikrotechnik Mainz GmbH
Carl-Zeiss-Strasse 18-20
55129 Mainz
Germany

Holger Löwe
Department of Chemistry
Pharmaceutics and Earth Sciences
Johannes-Gutenberg-Universität Mainz
Germany
Chemical Process Technology
Institut für Mikrotechnik Mainz GmbH
Carl-Zeiss-Strasse 18-20
55129 Mainz
Germany

Md Taifur Rahman
Department of Chemistry
Faculty of Science
Osaka Prefecture University
Sakai
Osaka 599-8531
Japan

Floris P.J.T. Rutjes
Organic Chemistry
Institute for Molecules and Materials
Radboud University Nijmegen
Toernooiveld 1
6525 ED Nijmegen
The Netherlands

Ilhyong Ryu
Department of Chemistry
Faculty of Science
Osaka Prefecture University
Sakai
Osaka 599-8531
Japan

Christopher D. Smith
Department of Chemistry
University of Cambridge
Lensfield Road
Cambridge CB2 1EW
UK

Jan C.M. van Hest
Organic Chemistry
Institute for Molecules and Materials
Radboud University Nijmegen
Toernooiveld 1
6525 ED Nijmegen
The Netherlands

Thomas Wirth
School of Chemistry
Cardiff University
Main Building
Park Place
Cardiff CF10 3AT
UK

1
Fabrication of Microreactors Made from Metals and Ceramics
Juergen J. Brandner

The material used to manufacture microstructure devices is heavily dependent on the desired application. Factors such as the temperature and pressure range of the application, the corrosivity of the fluids used, the need to have catalyst integration or to avoid catalytic blind activities, thermal conductivity and temperature distribution, specific heat capacity, electrical properties as well as some other parameters have a large influence on the choice of material. Finally, the design of the microstructures itself is an important consideration. Very specific designs are achievable only with special materials because certain manufacturing techniques are needed for them. It might also be necessary to take care of a special surface quality, which is achievable only with certain manufacturing techniques and materials.

Moreover, depending on the number of the devices needed, some manufacturing techniques are considered suitable while others are not.

In this section, fabrication as well as bonding and packaging of microstructure components and devices made from metals and ceramics will be described briefly. The manufacturing processes of both metal microstructure components and ceramic microstructure devices will also be described setting the focus on some well-established technologies. The detailed description of all these techniques can, however, be found in Refs [1–5]. A very short bonding section in which the most common bonding and sealing techniques are briefly described will complete the description of metals and ceramics [4,6].

Two different principal manufacturing techniques, that is erosive and generative have been considered with the discussed materials. Following this, other techniques such as embossing or molding are included into the list of generative manufacturing techniques.

1.1
Manufacturing Techniques for Metals

Metals and metal alloys are the most often used materials for conventional devices in process engineering, and thus applied in microprocess or technology as well. The range of materials is spread from noble metals such as silver, rhodium, platinum or palladium via stainless steel to metals such as copper, titanium, aluminum or nickel-based alloys [1,4–6]. Most manufacturing technologies for metallic microstructures have their roots either in semiconductor (in most cases, silicon) device production or in conventional precision machining. Of these, the techniques that are well known have been used for microstructure dimensions. Further, they have been adapted and improved to reach the desired precision and surface quality. In some rare cases, it was possible to use the same manufacturing process for macroscale and microscale devices and to get the desired results. In most of the cases, substantial changes in the design of the device, the methodology of the process and the manufacturing process itself were more or less necessary to provide the accuracy and quality needed for microstructure devices suitable for process engineering. Almost all but one technique used for microstructures in metals are abrasive, and the exception (selective laser melting SLM) will be discussed later.

1.1.1
Etching

Dry and wet etching techniques based on silicon and other semiconductor technologies are well known. For many metals, etching is a relatively cheap and well-established technique to obtain freeform structures with dimensions in the submillimeter range. This technique is well described in the literature [1–5,7]. A photosensitive polymer mask material is applied on the metal to be etched. The mask is exposed to light via a primary mask with structural layers. Here, different technologies are applicable, and their details can be found in the literature on semiconductor processing or in Refs [1–3]. The polymer is then developed. This means that the non-exposed parts are polymerized in such a way that they cannot be diluted by a solvent that is used to remove the rest of the polymer covering the parts to be etched. Thus, a mask is formed, and the metal is etched through the openings of this mask. To generate the etching mask, other techniques such as direct mask writing with a laser are also possible and common.

When etching techniques are used, two main considerations have to be given. First, the aspect ratio (the ratio between the width and depth of a structure), for wet chemical etching, can only be <0.5 at the optimum. As a result of the isotropic etching of the wet solvents, the minimum width of a structure is two times the depth plus the width of the mask openings. Dry etching (e.g. laser) is not limited to this aspect ratio, but it shows other limitations and is rather expensive (see Ref. [1]). Second, wet chemical etching always results in semielliptic or semicircular structures, which is again due to the isotropic etching. Dry etching often leads to other channel geometries. Here, rectangular channels are also possible. In Figure 1.1, a stainless steel microchannel structure manufactured by wet chemical etching is shown. The microchannels are used to build a chemical reactor for heterogeneously catalyzed gas-phase reactions. They are about 360 μm wide and 130 μm deep. Figure 1.2 shows the entrance area of such a microchannel. The semicircular structure is clearly seen. Detailed descriptions of the etching processes and etching agents can be found in Refs [1,4,7,8].

Figure 1.1 Wet chemically etched microchannels in a stainless steel foil.

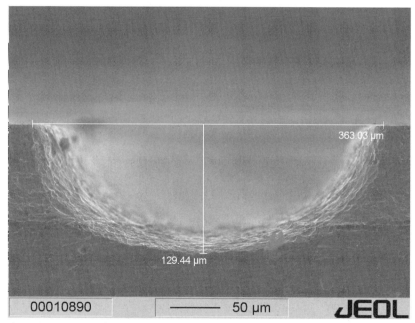

Figure 1.2 Structure of the microchannels from Figure 1.1. The semielliptic shape of the channels is clearly seen. The dimension of the microchannel is about 360 μm wide and 130 μm deep.

1.1.2
Machining

Not all materials can be etched in an easy and cheap way. Especially, noble metals or tantalum are stable against most of these corrosive structuring methods. Hence, precision machining may be used to generate microstructures from these metals as well as from standard metal alloys such as stainless steel or hastelloy. Depending on the material, precision machining can be performed by spark erosion (wire spark erosion and countersunk spark erosion), laser machining or mechanical precision machining. In this case, mechanical precision machining means milling, drilling, slotting and planning. Although the machining technology used is comparable to the techniques well known from conventional dimensions in the millimeter range or above, the tools used are much smaller. Whereas spark erosion and laser machining are suitable for any metal, the use of mechanical precision machining and the tools suitable for this type depend on the stability of the alloy. For brass and copper, natural diamond microtools are suitable and widely used, while for stainless steel and nickel-based alloys, hard metal tools are needed. Figure 1.3 shows a natural diamond cutter, whereas Figure 1.4 shows a hard metal drill. Figure 1.5 shows photos of a rhodium honeycomb microchannel catalyst system. The channels have been machined by wire spark erosion and therefore show a semicircular face area that is shown in detail in Figure 1.6.

The range of surface quality reached with the different techniques is widespread depending on the material as well as on the machining parameters. Spark erosion techniques lead to a considerably rough surface. The surface quality obtained with laser ablation heavily depends on the material to be structured and on the correct parameter settings. Values between some 10 µm and about 1 µm are common. In

Figure 1.3 Natural diamond cutter for micromachining of metals.

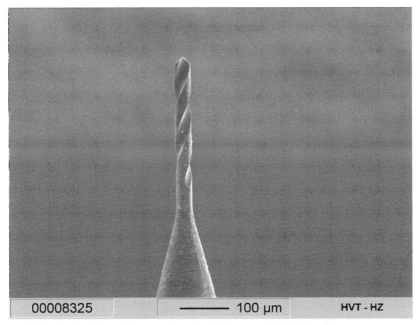

Figure 1.4 Microdrill made from hard metal. The diameter of the drill is about 30 μm.

Figure 1.5 Rhodium honeycomb catalyst microstructure device. The microchannels have been manufactured by wire erosion.

Figure 1.6 Detail of Figure 1.5. Clearly, the semicircular shape of the microchannels obtained by wire spark erosion can be seen.

Figure 1.7, results of laser ablation obtained in stainless steel by using incorrect parameters are shown.

By using brass or copper structural material, the best surface quality is achievable with mechanical precision machining. However, an electropolishing step must follow the micromechanical machining. A surface roughness ranging down to 30 nm can be reached. Figure 1.8 shows the surface of some microchannels machined into oxygen-free copper after the electropolishing step.

Details regarding all techniques can be found in Refs [1,4,9–15].

1.1.3
Generative Method: Selective Laser Melting (SLM)

A special method to manufacture metallic microstructures is SLM. It is one of the rare generative methods for metals and is normally taken into the list of rapid prototyping technologies. The technique is completely different than the abrasive techniques described so far. On a base platform made of the desired metal material, a thin layer of a metal powder is distributed. A focused laser beam is ducted along the structure lines given by a 3D CAD model, which is controlled by a computer. With the laser exposure, the metal powder is melted, forming a welding bead. The first layer of welding beads forming a copy of the 3D CAD structure is generated. After this, the platform is lowered by a certain value, new powder is distributed and the process is repeated. Thus, microstructures are generated layer by layer. In principle, any metal

00012395 ——— 300 µm JEOL

Figure 1.7 Surface quality of a stainless steel microchannel foil machined by laser ablation. The laser parameters have been incorrectly set.

powder can be used for SLM as long as the melting temperature can be reached with the help of the laser. For metal alloys, some problems might occur with dealloying by melting. Details of this relatively new technology can be found in Refs [16–18]. Figure 1.9 gives a schematic sketch of the working principle of this technique, whereas Figure 1.10 shows a picture of a microstructure stainless steel body manufactured by the process of SLM.

1.1.4
Metal-Forming Techniques

Almost all technologies described so far are suitable for prototyping or for small series production only. It simply takes a lot of time and is therefore costly to manufacture a large number of microstructures by laser ablation or wire erosion and by milling or SLM. This is not so in the case of the etching techniques. Here, a large number of microstructure devices can be very easily generated.

Another possibility to obtain a large number of microstructures is by embossing. As it was shown [19], even microstructures ranging down to a few 10 µm structure size can be easily realized with embossing technology. For embossing, a tool providing the negative structure design has to be cut into a hard metal. This negative is then pressed into the desired material using high mechanical forces, generating the positive of the structure design.

Figure 1.8 Surface quality obtained in oxygen-free copper by micromachining, followed by an electropolishing step. The mean roughness is about 30 nm.

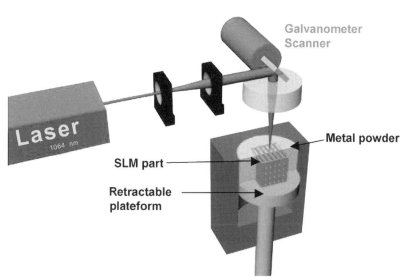

Figure 1.9 Schematic sketch of the SLM technology for metals.

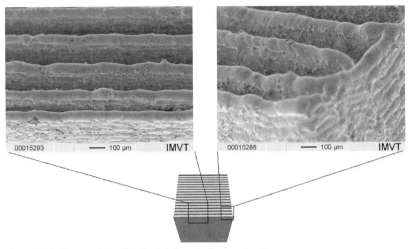

Figure 1.10 Photo and details of a stainless steel microstructure cube generated by SLM method. Clearly, the single welding beads that have been generated layer by layer to form the walls are shown.

1.1.5
Assembling and Bonding of Metal Microstructures

Although assembling of a number of device parts is not really a problem in the macroscale world, it needs to be delicately handled in the microscale world. The main point is the adjustment and alignment accuracy of the parts. Moreover, problems of sealing, fixation and bonding technology may also occur depending on the material and the parameters of the designated process of the device. Depending on the surface quality and the bonding technology applied, aligning errors may reach similar dimensions compared to the microstructure itself. An example of the same is shown in Figure 1.11. Here, a number of wet chemically etched microstructure foils have been aligned in a poor way to form elliptically shaped microchannels. Figure 1.12 shows two correctly aligned foils forming nearly circular microchannels. Misalignment will lead to non-regular channels and therefore may interfere with the bonding technique; in severe cases, it may lead to the destruction of the complete device. A correct alignment will lead to only small deviations from the desired elliptical shape, and the distortion while the bonding process takes place will be minimum. Alignment techniques used to avoid errors can be simple mechanical methods (e.g. use of alignment pins), edge catches in a specially designed assembling device or optical methods such as laser alignment. These methods are easily automated as shown in the semiconductor technology. In fact, most of the methods come from silicon processing technology where precise alignment of multiple mask layers is needed to guarantee the functionality of the manufactured devices [1,3].

Another problem of the microscale is the surface quality of the single parts of a device. Burr formation generated by mechanical micromachining or laser machining

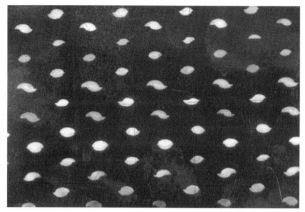

Figure 1.11 Photo of an arrangement of wet chemically etched microchannel foils. Owing to misalignment, in some layers the microchannels are not formed correctly to elliptically shaped channels.

may lead to significant problems while assembling of device parts as well as bonding is performed. Thus, special attention has to be paid to burr microstructures or to avoid burr formation. It might even be necessary to apply special techniques such as electropolishing to burr the single parts.

Bonding of metals can be done by numerous techniques. The common techniques for microstructures are welding (laser, e-beam, etc.), brazing, diffusion bonding and low-temperature as well as high-temperature soldering. Even clamping and sometimes, for very specific applications, gluing, including different sealing techniques, might be the other options. Details of the processes can be found in Refs [1,2,4,5,20–28].

For high-pressure applications and very secure run of chemical reactions, diffusion-bonded metal devices are the optimum choice. As a result of the process

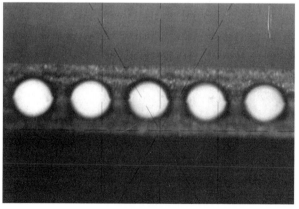

Figure 1.12 Photo of two wet chemically etched foils arranged and aligned correctly to form nearly circular microchannels.

of diffusion bonding (stacking, applying defined mechanical pressure force to the stack, heating in vacuum or inert atmosphere ranging to about 80% of melting temperature and cooling down while the mechanical pressure force is applied), a more or less monolithic block including microstructures is generated, which is extremely stable at high pressures. Owing to the diffusion of material from one foil to another, no borderline limitations between single foils in terms of heat transfer exist any more. Thus, the thermal behavior of diffusion-bonded devices is superior in comparison to that of the devices manufactured by other bonding techniques. In Figure 1.13, the diffusion bonding process chain is shown clockwise, starting with the single foils stack of a cross-flow stainless steel device. Figure 1.14 shows a cut through a diffusion-bonded stainless steel device. It is clearly visible that there was a crystal growth across the foil borderlines.

It is obvious that the choice of the bonding technique has to be made depending on the process parameters. It is not possible to run a device bonded by low-temperature soldering at some 100 °C. Thus, the most relevant parameters for the choice of

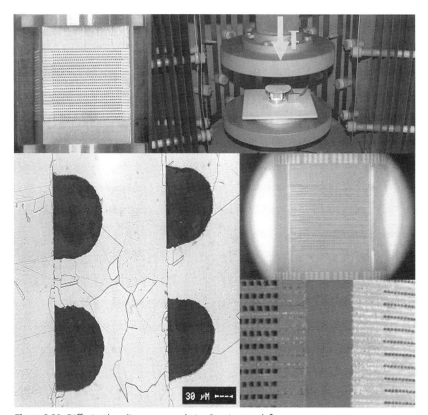

Figure 1.13 Diffusion bonding process chain. Starting top left: stacking, diffusion bonding furnace with mechanical pressure force, diffusion bonding and a cut through a microchannel system after diffusion bonding.

Figure 1.14 Cut through a diffusion-bonded cross-flow arrangement of stainless steel microchannel foils. The crystal growth over the borderlines between the single foils is clearly shown. Thus, diffusion bonding led to the formation of a monolithic stainless steel device.

bonding technique are process temperature, process pressure and corrosivity of the process chemicals.

1.2
Ceramic Devices

Microstructure devices made from ceramic and glass can be used for processes with reaction parameters that are reachable neither with metals nor with polymers. High temperatures measuring above 1000 °C, absence of catalytic blind activity and some easy ways to integrate catalytic active materials make ceramics a very interesting material. Glass is chemically resistant against almost all chemicals and also provides good resistivity at elevated temperatures. In addition, optical transparency of glass leads to some very interesting possibilities such as photochemistry or a closer look into several fluid dynamics and process parameters with online analytical methods using optical fibers. Nevertheless, microfabrication of components made from glass and ceramics is limited only to some known technologies and thus is not very cost efficient.

The conventional way to obtain ceramic microstructures is to prepare a feedstock or a slurry, fluid or plastic molding, injection molding or casting (CIM, HPIM and tape casting), demolding, debinding and sintering. Most ceramic materials will shrink during the sintering process, thus a certain tolerance to the dimensions have to be added. Solid free-form techniques such as printing, fused deposition or stereolithography are also possible with ceramic slurry. There are certain ceramic materials that can be machined mechanically. Details of these manufacturing processes can be found in Refs [6,29–40].

During the previous year, efforts were made to apply SLM with ceramics, and it proved to be successful. This is a new technology available now for ceramic materials. The principle of this technique has been described earlier. First preliminary experiments show promising results [18].

Independently of the manufacturing process, the grain size of the ceramic powder used to generate the precursor or the slurry has to be small enough to reproduce precisely all details of the desired microstructure. Even after sintering, which is normally accompanied by a coarsening of the grain size, the grains should be at least one order of magnitude smaller than the smallest dimension of the device. Additives also play an important role in the manufacturing process. Removing additives in a wrong way may lead to distortions and cracks, or even to debinding of microscopic parts of the desired microstructure device. Densification of the material is achieved by sintering; for example, for alumina, a temperature of about 1600 °C is needed, whereas for zirconia temperatures around 1500 °C only will be needed.

The most crucial point is the correct microstructure design. Owing to the specific properties of ceramics, it is not suitable simply to transfer the design of metallic or polymer devices to ceramic devices. Special needs for sealing, assembling and joining as well as interconnections to metal devices have to be considered. Moreover, guidelines for the micrometer design are still missing, and experiences obtained with macroscopic devices cannot be transferred directly down to microscale [6]. The interconnection between conventional process engineering equipment and ceramic devices is also critical because the thermal expansion of those materials is different. This may lead to thermal stress, weakening of the connections or ruptures.

Another possibility of ceramic material application is the use of coatings and foams inside, for example, metallic microstructure devices. Here, well-known technologies such as CVD processes, sputtering, electrophoretic deposition, sol–gel methods in combination with either spin coating or dip coating or wash coating methods or the use of anodic oxidation for aluminum-based devices will lead to either dense, protective ceramic coatings or porous layers used as a catalyst support. In Figure 1.15 an example of a sol–gel layer is given. Here, the dark sol–gel layer surrounds the rectangular microchannels completely. Figure 1.16 shows a porous layer obtained by anodic oxidation. The overview photo shows how the ceramic layer surrounds the microchannel, whereas the detailed picture shows the porous system within the ceramic layer. With anodic oxidation, the size and number of pores can be controlled to a certain amount by the choice of the electrolyte and the applied voltage and current density.

Ceramic foams can be inserted into microstructure devices made from metals and polymers to enhance the surface area, act as catalyst supports or even work as heaters. Details of these processes can be found in Refs [29–42].

Figure 1.15 Sol–gel generated catalyst support layer inside rectangular stainless steel microchannels. The support layer (dark line in photo) surrounds the microchannels completely, providing a porous system to be wet impregnated.

1.2.1
Joining and Sealing

Joining of ceramic materials should only involve materials with similar properties. Especially, the thermal expansion coefficient is a crucial point while either joining ceramic materials to each other or, even worse, joining ceramics to metals. The ideal joining of ceramics to each other is done in the green state before the firing process. When the firing process takes place, the ceramics are bound together tightly to form a single ceramic body from all parts. Another possibility is the soldering with, for example, glass–ceramic sealants. Here, the working temperature of the device is limited by the melting temperature of the sealant. Reversible assembling and sealing with clamping technologies or gluing are also possible. Conventional seals such as

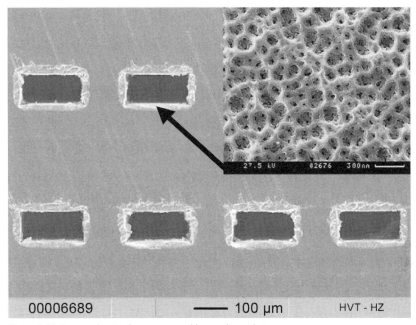

Figure 1.16 Porous alumina layer generated by anodic oxidation. With an Al alloy, a porous system can be generated inside the microchannels by anodic oxidation. The detailed photo shows some pores.

polymer o-rings or metal gaskets may be used in metal technology as well. The adaptation of ceramic microstructure devices to metallic process equipment should be done as far away from high temperatures as possible. Owing to the very different thermal expansion coefficients of both material classes, problems will most likely occur here. Then the sealing used should be designed to minimize tensile stresses as far as possible. For more details, see Refs [1,6,29–42].

References

1 Madou, M. (1997) *Fundamentals of Microfabrication*, CRC Press, London, UK.
2 Menz, W. and Mohr, J. (1997) *Mikrosystemtechnik fuer Ingenieure*, Wiley-VCH Verlag GmbH, Weinheim, Germany.
3 Eigler, H. and Beyer, W. (1996) *Moderne Produktionsprozesse der Elektrotechnik, Elektronik und Mikrosystemtechnik*, Expert Verlag, Renningen, Germany.
4 Brandner, J.J., Gietzelt, T., Henning, T., Kraut, M., Moritz, H. and Pfleging, W.

(2006) *Advanced Micro & Nanosystems: Micro Process Engineering*, vol. 5 (eds H. Baltes, O. Brand, G.K. Fedder, C. Hierold, J. Korvink and O. Tabata), Wiley-VCH Verlag GmbH, Weinheim, Germany, Chapter 10.
5 Brandner, J.J., Bohn, L., Schygulla, U., Wenka, A. and Schubert, K. (2003) *Microreactors: Epoch-Making Technology for Synthesis* (ed. J.I. Yoshida), MCPT, 2001, CMC Publishing Company, Tokyo, Japan, pp. 75–87, 213–223.

6 Knitter, R. and Dietrich, T. (2006) *Advanced Micro & Nanosystems: Micro Process Engineering*, vol. 5 (eds H. Baltes, O. Brand, G.K. Fedder, C. Hierold, J. Korvink and O. Tabata), Wiley-VCH Verlag GmbH, Weinheim, Germany, Chapter 12.

7 Petzow, G. (1994) *Metallographisches, Keramographisches und Plastographisches Ätzen*, Gebrüder Bornträger, Berlin, Germany.

8 Harris, T.W. (1976) *Chemical Milling*, Clarendon Press, Oxford, UK.

9 Slocum, A.H. (1992) Precision machine design: macromachine design philosophy and its applicability to the design of micromachines. Proceedings of IEEE MEMS 1992, Travemünde, Germany.

10 Boothroyd, G. and Knight, W.A. (1989) *Fundamentals of Machining and Machine Tools*, Marcel Dekker, New York.

11 Surjaprakash, M.V. (2004) *Precision Engineering: Copen 2003–2004*, Narosa Publishing House, India.

12 Walker, J.R. (2004) *Machining Fundamentals: From Basic to Advanced Techniques*, Goodheart-Wilcox Company, Inc., USA.

13 Shaw, M.C. (1984) *Metal Cutting Principles*, Clarendon Press, Oxford, UK.

14 DeVries, W.R. (1992) *Analysis of Material Removal Processes*, Springer, New York.

15 Chryssolouris, G. (1991) *Laser Machining*, Springer, New York.

16 Vansteenkiste, G., Boudeau, N., Leclerc, H., Barriere, T., Celin, J.C., Carmes, C., Roques, N., Millot, C., Benoit, C. and Boilat, C. (2004) Investigations in direct tooling for micro-technology with SLS. Proceedings of LANE2004, Erlangen, Germany, pp. 425–434.

17 Fischer, P., Blatter, A., Romano, V. and Weber, H.P. (2004) Highly precise pulsed selective laser sintering of metal powders. *Laser Physics Letters*, 620–628.

18 Brandner, J.J., Hansjosten, E., Anurjew, E., Pfleging, W. and Schubert, K. (2007) Microstructure devices generation by selective laser melting. Proceedings of SPIE Photonics West, January 25–27, San Jose, CA, USA.

19 Pfeifer, P. (2004) MicroMotive – development and fabrication of miniaturised components for gas generation in fuel cell systems. Proceedings of E&E China 2004: The 7th Biennial China International Environmental Protection and Energy Saving and Comprehensive Resource Utilization Exhibition, September 27–30, Beijing, China; see also Ref. [25].

20 Ehrfeld, W., Gärtner, C., Golbig, K., Hessel, V., Konrad, R., Löwe, H., Richter, T. and Schulz, C. (1997) Microreaction technology. Proceedings of the 1st International Conference on Microreaction Technology (ed. W. Ehrfeld), Springer, Berlin, pp. 72–90.

21 Kolb, G., Cominos, V., Drese, K., Hessel, V., Hofmann, C., Löwe, H., Wörz, O. and Zapf, R. (2002) Proceedings of the 6th International Conference on Microreaction Technology (eds P. Baselt, U. Eul, R.S. Wegeng, I. Rinard and B. Hoch), AIChE, March 10–14, New Orleans, LA, USA, pp. 61–69.

22 Ziogas, A., Löwe, H., Küpper, M. and Ehrfeld, W. (2000) Microreaction technology. Proceedings of the 3rd International Conference on Microreaction Technology (ed. W. Ehrfeld), Springer, Berlin, Germany, pp. 136–150.

23 Meyer, H., Crämer, K., Kurtz, O., Herber, R., Friz, W., Schwiekendick, C., Ringtunatus, O. and Madry, C. Patent Application DE 10251658 A1.

24 Pfeifer, P. Haas-Santo, K., Görke, O. and Schubert, K. (2004) Micromotive, unpublished results.

25 Pfeifer, P., Görke, O., Schubert, K., Martin, D., Herz, S., Horn, U. and Gräbener, T. (2005) Micromotive – development and fabrication of miniaturised components for gas generation in fuel cell systems. Proceedings of the 8th International Conference on Microreaction Technology IMRET 8, April 10–14, Atlanta, GA, USA.

26 Paul, B.K., Hasan, H., Dewey, T., Alman, D. and Wilson, R.D. (2002) Proceedings of the 6th International Conference on Microreaction Technology (eds P. Baselt, U. Eul, R.S. Wegeng, I. Rinard and B. Hoch), AICHE, March 10–14, New Orleans, LA, USA, pp. 202–211.

27 Bier, W., Keller, W., Linder, G., Seidel, D. and Schubert, K. (1990) Proceedings of the Symposium Volume, DSC-vol. 19, ASME, New York, pp. 189–197.

28 Pfleging, W. and Lambach, H. Unpublished results.

29 Heule, M., Vuillemin, S. and Gauckler, L.J. (2003) Powder-based ceramic meso- and microscale fabrication processes. *Advanced Engineering Materials*, **15**, 1237–1245.

30 Yu, Z.Y., Rakurjar, K.P. and Tandon, A. (2004) Study of 3D micro-ultrasonic machining. *Journal of Manufacturing Science Engineering, Transactions of the ASME*, **126**, 727–732.

31 Knitter, R., Günther, E., Maciejewski, U. and Odemer, C. (1994) Preparation of ceramic microstructures. *cfi/Ber. DKG*, **71**, 549–556.

32 Mutsuddy, B.C. and Ford, R.G. (1995) *Ceramic Injection Molding*, Chapman & Hall, London, UK.

33 Griffith, M.L. and Halloran, J.W. (1996) Freeform fabrication of ceramics via stereo lithography. *Journal of the American Ceramic Society*, **79**, 2601–2608.

34 Blazdell, P.F., Evans, J.R.G., Edirisinghe, M.J., Shaw, P. and Binstead, M.J. (1995) The computer aided manufacture of ceramics using multiplayer jet printing. *Journal of Materials Science Letters*, **14**, 1562–1565.

35 Agrarwala, M.K., Bandyopadhyay, A., van Weeren, R., Safari, A., Danforth, S.C., Langrana, N., Jamalabad, V.R. and Whalen, P.J. (1996) FDC, rapid fabrication of structural component. *American Ceramic Society Bulletin*, **75**, 60–65.

36 Evans, J.R.G. (1996) Injection moulding, in *Materials Science and Technology: Processing of Ceramics Part 1*, vol. 17a (ed. R.J. Brook), Wiley-VCH Verlag GmbH, Weinheim, Germany, Chapter 8.

37 Bauer, W. and Knitter, R. (2002) Development of a rapid prototyping process chain for the production of ceramic microcomponents. *Journal of Materials Science*, **37**, 3127–3140.

38 Mistler, R.E. (1995) The principles of tape casting and tape casting applications, in *Ceramic Processing* (eds R.A. Terpstra, P.P.A.C. Pex and A.H. de Vries), Chapman & Hall, London, UK, Chapter 5.

39 Ritzhaupt-Kleissl, H.-J., von Both, H., Dauscher, M. and Knitter, R. (2005) Further ceramic replication techniques, in *Advanced Micro and Nanosystems: Microengineering of Metals and Ceramics*, vol. 4 (eds H. Baltes, O. Brand, G.K. Fedder, C. Hierold, J. Korvink and O. Tabata), Wiley-VCH Verlag GmbH, Weinheim, Germany, Chapter 15.

40 Su, B., Button, T.W., Schneider, A., Singleton, L. and Prewett, P. (2002) Embossing of 3D ceramic micro-structures. *Microsystem Technologies*, **8**, 359–362.

41 Haas-Santo, K., Görke, O., Pfeifer, P. and Schubert, K. (2002) Catalyst coatings for microstructure reactors. *Chimia*, **56**, 605–610.

42 Hessel, V. and Löwe, H. (2002) Mikroverfahrenstechnik: Komponenten – Anlagenkonzeptionen – Anwenderakzeptanz. *Chemie Ingenieur Technik*, (74), **1–2**, 17–30, **3**, 185–207, **4**, 381–400.

2
Fabrication and Assembling of Microreactors Made from Glass and Silicon

Thomas Frank

2.1
How Microreactors are Constructed

In general, microreactors are systems consisting of tiny channels in which a number of fluidizable substances are combined under specific physical conditions. The temperature, pressure and dwell time are the most important parameters, which are either set or altered by peripheral equipment such as pumps, heaters/coolers and control systems.

Microreactors can be fabricated with a variety of materials available, such as metals, polymers, glass, ceramics and semiconductors, manufactured using microsystems engineering or conventional manufacturing methods. The processes involved are geared to create flat, planar components, similar to the wafers made for microelectronics – features that tend to dictate the basic construction of microreactors. The systems of channels in the microreactors are formed by hermetically sealed layers bearing cannular structures. Modern bonding technology enables the individual layers to be combined into a single functional component with inlets and outlets. Only two design elements are required: cavities and through-holes. The other vital engineering technique is that of achieving a hermetically sealed bond between the individual layers, impermeable to chemicals. To illustrate the construction of microreactors, the design of a simple microreactor, with a Y-junction to enable two fluid substances to be brought together, will be taken up here. In addition, two more functional layers to control the temperature in the microreactor or, alternatively, to operate it directly within a thermostat can be included. The model microreactor is defined by its individual layers, each layer bearing a structure in a two-dimensional pattern, with a given height as the third dimension, see Figure 2.1.

Microreactors in Organic Synthesis and Catalysis. Edited by Thomas Wirth
Copyright © 2008 WILEY-VCH Verlag GmbH & Co. KGaA. All rights reserved.
ISBN: 978-3-527-31869-8

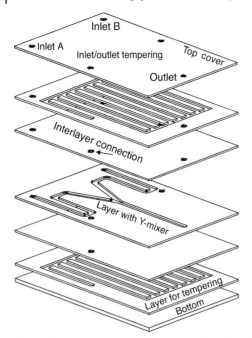

Figure 2.1 Setup of a simple microreactor with a Y-junction.

2.2
Glass as Material

Glass is much used as a material for technological purposes. One of its obvious applications is in optics; however, it has also had a huge influence on the progress in chemistry, pharmacology, electrical engineering and electronics. Its traditional use in the technical fields of laboratory experiment, electron tubes and lamps has led to the development of a huge variety of special types of glass for technical purposes because of the different chemical or physical properties required. The newer applications such as in microsystems engineering have continued to extend the shaping processes employed. Glass is obviously appreciated for its chemical resistance; moreover, its transparency is also a great advantage, allowing observation and even analysis from outside.

Glass is a noncrystalline solid substance with similarities to a liquid. To put it simply, glass is made up of an irregular network of particles (usually SiO_2) into the gaps between which the other components (the network converters such as Na_2O, K_2O) are woven, Figure 2.2.

The structuring possibilities depend on the mechanical, chemical and thermal properties. Glass and ceramics possess a high structural resistance ($\sigma_B > 10^4$ N/mm^2) but this is not of practical significance, because the resistance to breakage of glass articles depends on manufacturing defects in the surface of the glass. It is

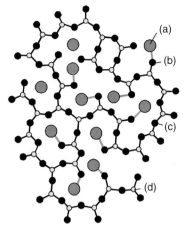

(a)

(b)

(c)

(d)

Figure 2.2 (a) Network converters (Na$_2$O): (b) oxygen, not make bridges; (c) oxygen, make bridges; and (d) make bridges SiO$_2$.

the slight surface damage in the form of fine notches and cracks that cause an article to break, because a large increase in tension develops at the ends of the cracks when they are subjected to mechanical loads. In ductile materials such as metals, this excess tension will be dissipated by a plastic type of flow. Glass and glass ceramics, on the contrary, behave as if they were brittle. At the temperatures and for the time spent under pressure in 'normal use', these two materials do not manifest any plastic flow so as to enable the peaks of stress at the tips of cracks and notches to dissipate.

On account of its chemical resistance, glass is generally excellent for water, saline solutions, acids, organic substances and even alkalis, so that in all these cases it is superior to most metals and plastics. Hydrofluoric acid, strong alkaline solutions and concentrated phosphoric acid are the only chemicals that have a noticeably adverse effect on glass, particularly at high temperatures.

As its heat conductivity is low (typically 0.9–1.2 W/[m K] at 90 °C), temperature changes within the glass will cause relatively steep temperature gradients. On heating, expansion tends to generate high mechanical tension. The resistance to temperature changes is the significant factor; the greater the resistance the lower the specific linear thermal expansion. Of the various types of glass, flint glass is the one with the lowest specific thermal expansion and the highest resistance to temperature changes. Glass would only be a second choice in situations where highly efficient heat transfer is required, as its thermal conductivity is low.

Despite its high chemical resistance and structural stability, its tendency to fracture under tension and the often inadequate resistance to temperature changes make glass more difficult to be structured using the classic structure-imposing methods. Wet and dry chemical etching is only possible within limits, but it produces good geometric resolution while the depth of the structure remains shallow. Inadequate resistance to temperature changes also complicates laser ablation, which can be used successfully only in the case of fused silica, the type with the lowest coefficient of thermal linear expansion. Of the machining methods, which rely on a shaving process, with few exceptions, only those can be used in which the cutter operates with indefinite geometry, such as lapping, ultrasound lapping, grinding and sandblasting on a miniature scale.

One exception is the special types of photostructurable glass. For the construction of microreactors, the borosilicate types of glass as well as fused silica are the most important ingredients.

- Fused silica is a single-component type, which consists of SiO_2. Its technical importance lies mainly in its low coefficient of thermal expansion, its excellent resistance to high temperatures (up to $1000\,^\circ$C), its very high resistance to temperature changes and its extremely good transparency to ultraviolet light.

- Borosilicate glass contains a higher proportion of SiO_2 than most other varieties of silicate glass, which together with varying amounts of B_2O_3 constitutes up to 13% of the mass. This type is characterized by high resistance to the influence of chemicals and to differences in temperature. It thus finds application mainly in the chemical and pharmaceutical industries and as domestic ovenproof glass. The chemical composition can vary widely. The actual properties mainly depend on the manner in which the boron compounds suited to the glass melt are combined with other metallic oxides. Borosilicate 33 is used often; the 33 stands for the thermal coefficient of expansion, $\alpha = 3.3 \times 10^{-6}\,K^{-1}$, and the trade names often met are Borofloat 33$^\circledR$, Duran$^\circledR$, Pyrex$^\circledR$.

- Photoetchable special glasses belong to the doped lithium-aluminosilicate group, which characteristically crystallizes on the areas exposed to light if, after masking, they are subjected to UV-irradiation and then heated. The crystallized parts are more easily dissolved in hydrofluoric acid, so that a geometric microstructure can be created on the glass. Some of the types that have made an impact in this field are FOTURAN (made by Schott, Germany), PEG3 (made by Hoya, Japan), among others. The Department of Inorganic Non-Metallic Materials at the Technische Universität Ilmenau is using a type of photostructurable glass (FS21) it has developed for its current research. Table 2.1 shows the most important properties of this glass [9].

2.3
Silicon as Material

Silicon as a material is very common in microsystems engineering, as it is not only suitable for micro switches but is also successfully used to create structures for mechanical purposes and fluids. The reasons lie not only in its useful mechanical properties but also in its ready availability and ease of structuring.

Silicon is used in its monocrystalline form. The single crystal is developed by a variety of processes as a cylindrical ingot or boule. This cylindrical shape is ground to

Table 2.1 Physical properties.

Physical properties	Fused silica	Borosilicate glass	Photoetchable special glasses
Maximum working temperature (°C)	1100	450	$450 \leq T_g \leq 465$
Coefficient of mean thermal expansion, $\alpha_{(20-300\,°C)}$ $(10^{-6}\,K)$	0.55	3.25	$8.4 \leq \alpha \leq 10.6$
Density, ρ (g/cm^{-3}) (25 °C)	2.2	2.2	$2.34 \leq \rho \leq 2.37$
Young's modulus (kN/mm^2)	66	64	$74 \leq E \leq 81$
Flexural strength, σ (MPa)	50	25	25
Hydrolytic class	1	1	5
Acid class	1	1	2
Alkali class	1	2	2
Specific heat capacity, C_p (20–100 °C) (KJ/(kg K))	1.05	0.83	0.87
Transformation temperature, T_g	1130	525	$450 \leq T_g \leq 465$
Thermal conductivity, λ (90 °C) (W/(m K))	1.38	1.2	1.35

It is available in a rectangular wafer form. The structuring methods most often used are described in the text.

a nominal diameter and then sliced. The resultant wafers are polished. Silicon crystallizes with a diamond structure. The wafers may have the (1 0 0), (1 1 0) or (1 1 1) orientation; the (1 0 0) orientation is most commonly used. Figure 2.3 shows the layers in a cubic crystal system, described by their Miller's indices [4]. The most important properties are given in Table 2.2.

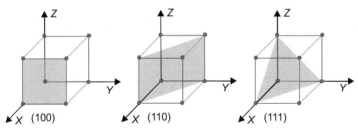

Figure 2.3 Layers in the cubic crystalline.

Table 2.2 Physical properties of silicon [9].

Coefficient of mean thermal expansion, $\alpha_{(20-300\,°C)}$ $(10^{-6}\,K)$	2.6
Density, ρ (g/cm^3) (25 °C)	2.329
Flexural strength, σ (MPa)	6000
Thermal conductivity, λ (90 °C) (W/(m K))	150
Young's modulus (GPa)	130–188
Melting point (°C)	1413

2.4
The Structuring of Glass and Silicon

In microsystems engineering, another important process, besides the structuring of thin layers, is the etching of the solid material. This will now be described in more detail.

From the viewpoint of the mechanical characteristics, glass and silicon resemble each other. They have a similar mechanical hardness, are brittle as they lack plasticity and are thus prone to fracture. Of the standard precision engineering procedures available for shaping, only those that do not use a geometrically defined cutter can be used, such as grinding and lapping. Microengineering techniques are much more efficient, but they do prove difficult for deeper structures.

2.5
Structuring by Means of Masked Etching in Microsystems Technology

On the whole, microsystems procedures rely on masking, whether the aim is to create the structure by adding or subtracting materials. The areas not intended to be affected by the procedure are shielded from it by a protective mask. In the cases described here, through-holes or cavities are produced by the subtraction of material. The masking acts as a shield, for instance, against an aggressive etching or the chemical changes associated with the photoresist technique (Figure 2.4). A distinction must be made between the mask for photolithography and the type of masking required for an ensuing process of substance removal. In the case of photolithography, a light pattern

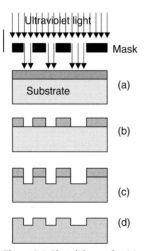

Figure 2.4 Photolithography, (a) photoresist lay on and exposed through a mask, (b) developing, (c) etching the substrate and (d) photoresist remove.

is directed onto the surface of the substrate after it has been coated with photosensitive material (photoresist) resulting in resolutions on the micrometer scale. For the required structures this is sufficiently accurate. The mask, or reticle, used in photolithography is a layer of glass or transparent polymer coated with certain absorber structures. For microelectronics, procedures that can achieve much higher resolution are available [3].

The photochemical processes in photoresist technique may work as a positive resist enhancing the solubility of the exposed areas or as a negative resist reducing the solubility. At a later stage of the manufacturing process, the easier-to-dissolve portions will be removed. This photoresist mask can be used for certain substance removal procedures, as in the case of the special resist for microsandblasting, or for plasma etching of silicon. In cases where this cannot be used, because it is not selective enough, two-stage masking is necessary, for instance, for deep wet chemical etching of glass and anisotropic etching of silicon. Here, a masking layer of several hundred nanometer thickness is painted onto the substrate; for structuring the glass, the mask is made of CrNi and polysilicon and for silicon it is a single SiO_2, Si_3N_4 layer. This coating is then etched with the aid of the photoresist and a suitable etching medium, see Figure 2.5. The selectivity achieved with this combined masking layer is much greater than that achieved with photoresist alone. In addition to photolithography, other, more direct, types of lithography are possible.

The means of masking will be described together with the structuring processes for which they are used.

To produce through-holes and cavities in a wafer, special forms of lithography are employed. If it is assumed that a roughly constant amount of material requires to be removed from all over the wafer, the creation in one and the same process of both cavities and holes is not feasible. The only chance of achieving this is to perform masking and lithography from both sides and then etch both sides simultaneously. The through-holes will be on both sides but the cavities only on one side. The

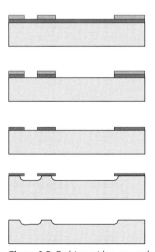

Figure 2.5 Etching with a second masking layer.

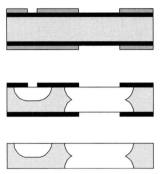

Figure 2.6 Two-stage (double-sided) lithography.

second option is a two-stage lithography. In the first stage, the cavities are made and then the wafer is masked again and the holes are created in the second stage (Figure 2.6).

2.6
Etching Technologies

As far as microtechnology is concerned, etching processes offer the highest geometrical resolution. Using these processes, structures accurate to less than a micrometer are possible. This is in contrast to the way in which machining or 'shaving' processes work because etching processes deal with the whole wafer at once, treating the entire surface of one or more wafers simultaneously. Two important parameters are the etch rate and selectivity.

The types of etching process are divided into wet or dry chemical etching and isotropic or anisotropic etching as shown in Figure 2.7.

If the material etched away is removed as a liquid, the process is called wet chemical etching and when the material is removed in the gaseous state, it is referred to as chemical etching. Each type can be isotropic, with the same etch rate operating in all directions, or anisotropic, with the etch rates reflecting direction, which may be a spatial direction or crystal orientation.

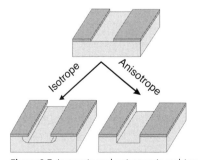

Figure 2.7 Isotropic and anisotropic etching.

Table 2.3 Different mask layers.

Material of mask layers	Chemical structuring	Fabrication	Use
SiO$_2$	Buffert hydrofluoric acid, RIE Dry etching, deep reactive ion etching: DRIE	Thermische thermal oxidation, low-pressure chemical vapor deposition	Crystallographic and wet chemical etching of silicon
Si$_3$N$_4$	RIE	Low-pressure chemical vapor deposition	Crystallographic and wet chemical etching of silicon
Poly Si	RIE	Low-pressure chemical vapor deposition	Glass isotrope
Cr−Ni Photoresist	Wet chemical etching Photolithography	Cathode sputtering Spincoater	Glass isotrope DRIE of Si RIE of Si
Chromium/gold	Wet chemical etching	Cathode sputtering	Glass isotrope

The glass and silicon materials can be subjected to every combination. To achieve structure sizes required for microreactors, the following processes are used:

(1) Anisotropic (crystallographic) wet chemical etching of silicon (KOH).
(2) Isotropic wet chemical etching of silicon (HF + HNO$_3$ + CH$_3$COOH + H$_2$O).
(3) Anisotropic dry chemical etching of silicon (reactive ion etching, RIE).
(4) Isotropic wet chemical etching of glass (buffered 10% hydrofluoric acid).

A number of commonly used masking layers are listed in Table 2.3.

2.6.1
Anisotropic (Crystallographic) Wet Chemical Etching of Silicon (KOH)

In anisotropic crystallographic wet chemical etching of silicon, the dependency of the etch rate on crystal orientation is exploited. Even along the main levels of the crystal, for example along the (1 1 1) and (1 1 0) levels, the etch rate can vary by a factor of 100. Aqueous solutions of alkaline hydroxides such as KOH and NaOH are anisotropic etch solutions for silicon. The etching speed in individual directions will depend on the temperature and the etch solution used. The relation of the etch rate to the crystal's direction is shown in Figure 2.8 for the widely produced wafers made of (1 0 0)-type silicon [3].

Figure 2.9 shows the shapes obtained when silicon is etched anisotropically. The four (1 1 1) levels cut the surface of a disk of (1 0 0) material along the (1 1 0) directions within it. The etchant, in turn, attacks each layer of the (1 0 0) levels and removes

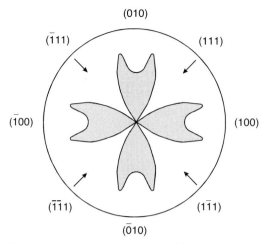

Figure 2.8 Schematic representation of the speed of etching of the lateral direction $(1\,0\,0)$ silicon (50% aq. KOH at 78 °C).

them; Figure 2.9 shows the process diagrammatically. Any push forward on the part of the $(1\,0\,0)$ base is prevented laterally by the $(1\,1\,1)$ levels slowly being etched. Apart from a slight undercutting of the mask, the $(1\,1\,1)$ levels seem stable. The angle β to the vertical is 54.75°. At that, the etching effectively ceases. The ratio of the width of the trough (w) to its depth (t) is

$$\frac{w}{t} = \sqrt{2}.$$

The masking material used is SiO_2 or Si_3N_4 (see Table 2.3). Rectangular mask openings oriented in the $(1\,1\,0)$ direction lead to the creation of rectangular troughs that are limited by the four $(1\,1\,1)$ levels. These have the effect of a natural etch barrier. A square opening implies that inverted pyramids will form. Openings shaped in any other ways will always produce V-shaped troughs if the etching is continued until the

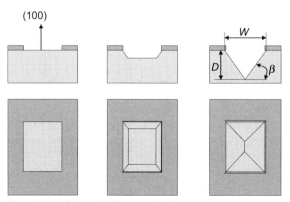

Figure 2.9 Etching until the natural etch stops.

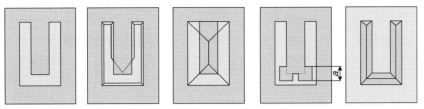

Figure 2.10 Different forms, simple corner compensation and design elements. The edge length a depends on the etching depth D; according to the empirical formula the complete compensation reached with $a = D$.

natural stop effect is reached. At the convex corners of the (1 1 1) levels, levels with higher indices and a faster etch rate are exposed. These are etched away in a lateral direction until new (1 1 1) levels form concave corners, which are the natural barriers and which prevent further etching (see Figure 2.10).

Waiting for the natural etch stop means that 'L', 'T' or 'U'-shaped V-troughs necessary for microreactors cannot be produced. To produce these troughs, therefore, etching is interrupted when the desired depth has been reached. These V-cross-sectional troughs have convex 90° corners at any bends, and at these points there is considerable undercutting of the mask. The remedy is the corner compensation. Sacrificial surfaces are created at the convex corners, calculated in such a way that the etch time in the direction of the convex 90° corner will produce the correct shape.

Figure 2.10 shows the effect of various corner compensation arrangements. Corner compensation, if given the right dimensions, can ensure the production of channels in a variety of depths and even etching right through the wafer. Cavities and holes can be achieved in one and the same process. If a two-sided process is employed, V-troughs can be created on each side and these can even be connected by holes etched right through. Such structures with holes are required for connection positioned on top or to permit fluid connection between the individual layers. If channels with a trapezoid cross section are needed, the depth of the channel is set by the etch time [5].

2.6.2
Isotropic Wet Chemical Etching of Silicon

Isotropic etching of silicon is done with an aqueous mixture of $HF + HNO_3 + CH_3COOH$ (hydrofluoric, nitric acid and acetic acid). Often the acetic acid is omitted. The etch rates are very high at room temperature: 940 μm/min with 20–46% HNO_3 (69% HNO_3, 31% H_2O) complemented by HF (49% HF, 51% H_2O). The quality of the silicon surface produced will depend on the solution used. Smoother surfaces are achieved with a higher proportion of nitric acid and a lower proportion of acetic acid.

The disadvantage, as far as silica is concerned, is the low degree of selectivity. For SiO_2 the etch rates are between 300 and 700 nm/min. Silicon nitride and noble metals demonstrate better resistance to etching. Because of the isotropy, the aspect

Gas inlet

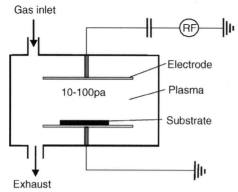

Exhaust
Figure 2.11 Apparatus for RIE.

ratio is only 0.5. Therefore, it is very seldom employed for the creation of structures to be used in microreactors [3].

2.6.3
Anisotropic Dry Etching of Silicon

In contrast to anisotropic wet etching, reactive dry etching offers the possibility of creating a relatively free selection of geometric patterns independent of the crystal orientation. For the use described here, the most suitable method is the reactive ion etching (RIE). RIE is a combination of etching by sputtering (a purely physical method) and plasma etching (a purely chemical method). It simultaneously uses energetic particles and reactive plasma particles to remove the material. The apparatus used is the same as used for sputter etching (see Figure 2.11. It differs from sputtering in that instead of the inert gas, a reactive gas (or mixture containing one) is fed into the reactor. The energetic particles always knock out surface particles with a sputter effect, whether or not there is a chemical reaction. But RIE means that there is also a chemical reaction between the reactive gas and the substrate – here, RIE is similar to plasma etching. The unwanted material is removed physically, but some chemical is reaction at the surface [5] also assists in the process. The physical part of the process is anisotropic and the chemical is isotropic. Different etching profiles can be set by judicious selection of parameters, see Figure 2.12.

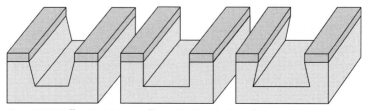

Figure 2.12 Different etching profiles setup by selection of parameters.

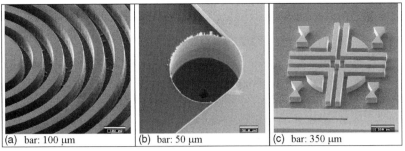

(a) bar: 100 μm   (b) bar: 50 μm   (c) bar: 350 μm

Figure 2.13 Examples of RIE etched silicon, with friendly permission of the Technical University of Ilmenau, Faculty of Mechanical Engineering, Department of Micromechanical Systems. (a) Circular channels, (b) channel with a through-hole and (c) adjustment mark.

Various procedures are used. One common dry chemical etching process employed in microreactor manufacturing is the deep reactive ion etching (DRIE). To increase the anisotropy, the RIE procedure is modified so that as the etching advances, some of the reaction products settle on the side wall, passivating it. Reaction products at the front, to which the etching has advanced, are removed by means of physical components. The resulting troughs are very deep, with vertical walls. The etch rate will depend on the pattern left open by the mask and on the depth to which etching is permitted. Troughs of different widths will etch to different depths in the same period of time [5,6].

The gas mixture used to etch the silicon consists of SF_6 and O_2; SF_6^+ ions and oxygen radicals are generated in the plasma. Photoresist is the masking layer employed in DRIE. Aspect ratios up to 50 are possible. The channels can be formed on any pattern and will have a rectangular cross section. A two-sided process or two-stage masking is used if holes are required in combination with cavities. Figure 2.13 shows examples.

2.6.4
Isotropic Wet Chemical Etching of Silicon Glass

In the case of glass, which has higher chemical resistance than silicon, the choice of structuring procedures is narrower for this reason. One possibility is isotropic wet chemical etching. The nature of the process means the aspect ratio is only 0.5. The choices of masking material are polysilicon, chromium-nickel or chromium-gold; for the application of the mask and the structuring, see Table 2.3. The structuring of the layer depends on lithography. As the masking layer is applied defects, known as pinholes, may develop. The number of pinholes may vary for a particular procedure, depending on the window in the manufacturing process. As the etching is isotropic, mask defects are magnified. The tiniest gap will lead to a hole with a diameter twice the depth of the etch. Layer combinations are often used to prevent pinholes.

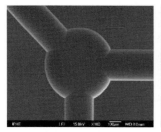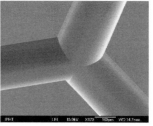

Figure 2.14 Isotropic wet chemical etching of silicon glass (with friendly permission of the Institute of Physical High Technology).

Whenever there is a chemical erosion of glass, water or its dissociation products, H^+ or OH^- ions will be involved. Because of this, a distinction is made between the resistance of glass to water (its hydrolytic resistance) and to alkali or acid. Under attack from water or acids, small numbers of cations, particularly monovalent and divalent, are released. On resistant types of glass, a very thin layer of silica gel forms in this way on the glass surface and usually inhibits further erosion. In contrast, hydrofluoric acid, alkaline solutions and, under certain circumstances, phosphoric acid will slowly remove the inhibitor layer and thus the entire surface. Nonaqueous solutions (organic solutions), however, are practically nonreactive with glass.

The pinhole problem, adhesion difficulty and low aspect ratio all restrict the depth of the etch for the structuring. Depths ranging down to 120 μm are achieved simply; in Ref. [1] there is a report of 300 μm achieved by using a 1 μm Cr/Au masking system.

Through-holes are only possible in very thin substrates. This process can be combined with other means of structuring glass; for example, microsandblasting or ultrasonic lapping (Figure 2.14).

2.6.5
Photostructuring of Special Glass

None of the etching processes described so far enables structures to be created in glass that have a high aspect ratio or that has the cross sections necessary for microreactors. To allow the manufacture of fine structures with a high aspect ratio in glass, a number of photostructurable types of glass have been developed in parallel. These types are based on the general threefold glass system, $Li_2O-Al_2O_3-SiO_2$, with tiny amounts of additives, making them photosensitive. By following procedures similar to that of photoresist technique, a photolithographic process imparts a structure to these types of glass. In Table 2.4 the approximate composition of photostructurable types of glass is given.

The process followed with glass is more complicated than described below, adequate enough to understand the sequence. The glass is melted under reducing conditions so that cerium is present in the form of Ce^{3+} and silver in the form of Ag^+. The Ce^{3+} will emit another electron if illuminated with UV light ($\lambda = 300–320$ nm), and this electron can be taken up by the Ag^+ to form Ag^0. The amount of energy

Table 2.4 Configuration of photostructurable special glass.

SiO_2	74–80%	ZnO	0–1%
Li_2O	9–12%	Ag_2O	0.01–0.2%
Na_2O	0–3%	SnO	0–0.1%
K_2O	3.5–4.5%	Sb_2O_3	0.2–0.5%
Al_2O_3	4–7.5%	CeO_2	0.01–0.03%

Light exposure (initiation), tempering/heat treatment and etching (development, structuring, fixation).

required for complete illumination will depend on the thickness of the glass. If the illumination is insufficient, the substrate is not illuminated right through.

The heat treatment brings the silver atoms together into clusters that act as the nuclei for the lithium metal silicate crystals. It has two critical thermal points: one hour at approximately 500 °C, during which the silver clusters group, and another hour at 600 °C. The lithium metal silicate crystals are generated at the higher temperature. The size of these crystals is between 1 and 10 μm, depending on the temperature profile employed and on the composition of the glass. The lithium metal silicate crystals are 10–60 times more soluble in dilute hydrofluoric acid as the glass would be if not illuminated. It is then possible to selectively dissolve the areas that have been exposed to light and become crystalline. Reinforcement with ultrasound is useful to remove remnants of glass from the edges of the crystallized areas. The following illustrations, Figure 2.15, show samples of structures.

When a reactor is being manufactured, both cavities and through-holes are required. In the case of photosensitive glass, it is difficult to make troughs. The necessary illumination goes right through the wafer and so there are crystallized areas going right through, which are then are dissolved out. Cavities and islands are thus impossible to create in this manner. The following illustrations, Figure 2.16, show the basic steps. For manufacturing islands, a wafer that has not been completely

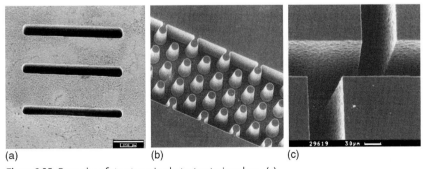

(a) (b) (c)

Figure 2.15 Examples of structures in photostructuring glass, (a) long hole, (b) columns and (c) interlocking channels. (With friendly permission of the Technical University of Ilmenau, Faculty of Mechanical Engineering, Department of Inorganic-Nonmetallic Materials).

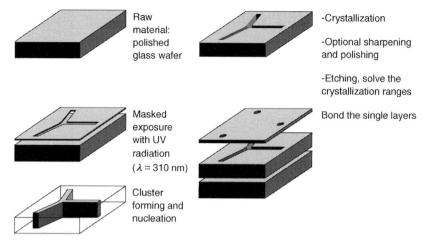

Figure 2.16 Basic steps for the photostructuring of glass.

etched through is necessary, or else the islands will drop out. It is possible to create them by interrupting the etching: once the necessary depth has been reached, the etching process is stopped. However, this is not desirable, as the wafer will now consist of crystallized areas and of glass, two components with different thermal and chemical properties. In Ref. [2] a description of all the other procedures using gray-scale masking and multiple exposure is given. Reactors are virtually monolithic, made completely out of the same material. It is thus only in exceptional cases that the manufacturing procedure includes interrupted etching.

The crystal phase and the surrounding glass have different coefficients of thermal expansion and different densities. The heat treatment and attendant critical transformation temperature cause distortion and some slight bulging of the wafer, which makes the subsequent joining procedures more difficult. Once it has been etched, the wafer bares delicate open structures and cannot be mechanically processed further. If the wafer is polished before it is etched, the bulging can be counteracted, so that even after etching it is more or less flat. Examples of structures are shown in Figure 2.15.

2.7
Chip Removing Processing

2.7.1
Drilling, Diamond Lapping, Ultrasonic Lapping

As described in more detail in Section 2.2, glass is so hard that it is mainly processed by machining using cutter mechanisms with indefinite geometry, such as ultrasound lapping (and, in older systems, ultrasound drilling). High-frequency electrical energy is converted by a piezo-ceramic sound converter into mechanical vibration at the same frequency. The amplitude of the longitudinal vibration of the sound converter is only 5 µm. For this reason, another unit is set between the shaping tool and the sound converter. This consists of an amplitude transformer and a sonotrode (also known as drill), which amplifies the value between 20 and 40 µm.

The four elements connected together are a vibratory system operating in resonance with each other. The system permits a process based on the hammering of the grains into the surface of the tool within the ultrasound range (19–22 kHz), which creates tiny cracks and eventually detaches small fragments. This process mainly takes place in the direction of movement of the ultrasound vibration. If possible, the use of rotating tools is advised, as they achieve better removal. The grains are contained in a liquid or paste, the lapping medium. This mixture is either poured or pressed continuously into the gap between shaping tool and working item. This gap should be approximately twice as wide as the average grain size. The choice of grain material usually falls on boron carbide because of its fracture capability and consequent self-sharpening. The boron carbide constitutes about 25–35% of the mass of the lapping medium and is usually in grain size 50–60 µm. As the tool is under the same strain as the worked item, it must be made of a material, which either is harder or does at least have adequate resistance to abrasion. For this reason, steels are usually used as they tend to be deformed elastically and plastically by the impact of the grains. The main area in which ultrasound lapping is used for manufacturing microreactors is in the drilling of external or internal connecting holes such as holes in glass lids used to seal fine structures made of silicon or glass.

2.7.2
Micropowder Blasting

Microabrasion using compressed air is a modification based on sandblasting, the micropowder blasting. This process enables all types of glass, ceramics and semiconductor materials, irrespective of their chemical composition and crystal structure, to be inexpensively processed down to the micrometer scale. The micropowder blasting is a masked procedure and works quasi-parallel on the whole substrate. A powder jet drives systematically over the substrate. Material is removed at the mask openings (see Figure 2.17).

The depth of structure achieved will depend on the processing time. The aspect ratio lies in the range 2–3, and the smallest structure that can be created is 50 µm. Masks are created with photolithography, which permits very fine tolerances of both shape and position and also allows as many lateral structures densely packed side by side as desired. The mechanical nature of the procedure means that the chemical composition of the material is unimportant. As the items are not exposed to heat, there is no distortion. One disadvantage is the angle of the abraded contours to the surface of the substrate. It will be 70–85°, depending on the depth sought. Deeper structures have steeper sides. Figure 2.18 shows different proceedings to the structuring. The depth of the structures is determined by the operating time. It is possible in one- and double-sided processing. By repeated masking structures can be

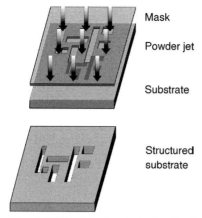

Mask

Powder jet

Substrate

Structured
substrate

Figure 2.17 Principle of masked microabrasives powder blasting.

realized, as shown in Figure 2.18c, with cavities and through-holes on both sides. The tolerances and surface quality for this procedure are similar to those of ultrasonic drilling/lapping.

Figure 2.19 gives examples of structures achieved by microabrasion. Cavities and through-holes are possible, using either a single-side or a double-side procedure.

(a)

(b)

(c)

(d)

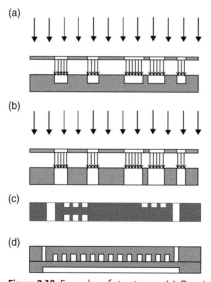

Figure 2.18 Examples of structures. (a) Quasi-parallel processing of the substrate by the mask. (b) The structure depth is determined by variation of the operating time, (c) double-sided adjusted processing, multiple masking and (d) adjusted bonding of many substrates.

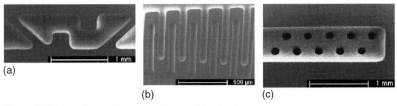

(a)

(b)

(c)

Figure 2.19 Manufactured structures by sandblasting for micromixers (a) Cavity for the setting up of a bar mixer. (b) Channels for a multi-lamination mixer. (c) Channel and drillings for a split and recombination mixer. (With friendly permission of the Little Things Factory GmbH, Ilmenau, Germany).

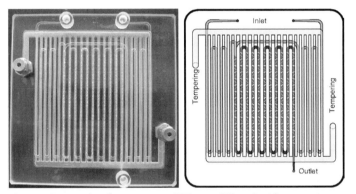

Figure 2.20 Examples of microreactors (150 mm × 150 mm), with two functional layers. Outside: tempering. Inside: preliminary heating, mixing, dwell channel (volumes approximately 3 ml), with friendly permission of the Little Things Factory GmbH, Ilmenau, Germany.

The roughness of the finished surface is $Rz \approx 2$–4 µm. Surfaces treated by lapping and microabrasion are similar. In contrast to what happens to photostructured glass, only the areas without a mask are processed; the covered parts of the wafer surface remain virtually intact. Figure 2.20 shows various examples of one-sided and double-sided structuring.

The ideal process can be selected for a particular application if one knows the nature of the various procedures. It is also possible to combine procedures.

2.7.3
Summary

The difficulties of processing glass have already been described. The following are the technologically significant methods.

2.8
Bonding Methods

Manufacturing multilayer systems requires suitable bonding methods. Attention must be paid not only to the mechanical stability of the bond but also to its resistance to chemicals and temperature changes. Bonding by means of modern glues is not impossible and is rarely used in the construction of microreactors. The prerequisite for tension-free bonding is similar coefficients of linear thermal expansion. The following bonding processes are important:

Anodic bonding of glass and silicon.
Silicon direct bonding (silicon fusion bonding).
Glass fusion bonding.

2.8.1
Anodic Bonding of Glass and Silicon

Metals or semiconductors can be joined to glass with a hermetic seal by anodic bonding. The process is used mainly to join silicon and borosilicate glass that contain a sufficiently high concentration of alkali, such as Borofloat 33 or Pyrex. It is a prerequisite that the surfaces are polished and clean and that the materials to be bonded have approximately the same coefficient of thermal expansion. They are brought into close contact with one another at a temperature of 400–500 °C under direct current at a voltage of 700–1000 V. The high temperature makes the glass conduct ions. The electrostatic forces generated by the electric current not only strengthen the contact but also cause a drift of sodium ions across the boundary from the glass anode to the silicon cathode. Unsaturated oxygen bonds are left behind. With these, silicon atoms form strong chemical SiO bridges that constitute the bond. This process is used to make a covering for structures formed in silicon.

2.8.2
Silicon Direct Bonding (Silicon Fusion Bonding)

Direct bonding is used as a procedure to join two or more wafers made of silicon. It is possible to bond the wafers in the oxidized state or to dip them first into a solution of $NH_4OH:H_2O_2:H_2O$ to make them hydrophilic; by this means, OH bonds are formed at the surface. Then one wafer is laid on the top of a second and they are pressed together. The wafers now adhere by virtue of van-der-Waals forces, as the surfaces are extremely smooth and even. The wafers are then heat treated in an oxidizing kiln, in which the atmosphere contains 5% O_2 and 95% N_2, for 60 min at 1050 °C. The results are firm Si–O bonds holding the wafers together.

If the wafers have been oxidized first no hydrophilization is required, as SiO_2 itself is hydrophilic.

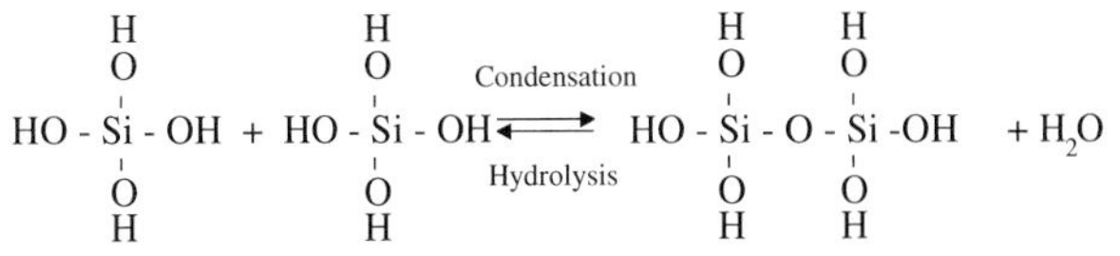

$$
\begin{array}{ccc}
\text{H} & \text{H} & \text{H} \quad \text{H} \\
\text{O} & \text{O} \qquad \text{Condensation} & \text{O} \quad \text{O} \\
| & | & | \quad\quad | \\
\text{HO - Si - OH} + \text{HO - Si - OH} \rightleftharpoons & \text{HO - Si - O - Si -OH} & + \text{H}_2\text{O} \\
| & | \qquad \text{Hydrolysis} & | \quad\quad | \\
\text{O} & \text{O} & \text{O} \quad \text{O} \\
\text{H} & \text{H} & \text{H} \quad \text{H}
\end{array}
$$

Figure 2.21 Condensation of the silica gel layer to siloxane bridges.

2.8.3
Glass Fusion Bonding

Thermal bonding of glass is similar to direct bonding of silicon. A distinction is made between two methods: with and without plastic deformation. For extremely high-precision components, the method without plastic deformation is employed. As in the case of silicon direct bonding, the hydrophilic wafers are first 'prebonded'. This brings the hydrophilic silica gel layer on each surface into contact with each other. At temperatures above 350 °C, this silica gel layer condenses and results in firm bridges across the surfaces that are in contact with one another (Figure 2.21). In practice, the wafers are given heat treatment for several hours between 400 and 450 °C. Because the temperature is lower than the transformation temperature, the wafers are still undistorted after the heat treatment. For all this to succeed, the surface quality and accuracy of shape are subject to very demanding specifications. The planarity must be better than 100 nm and the absolute roughness better than 1–2 nm. If this is not the case, and the wafers are not sufficiently even, or are too rough, as is the case with photostructurable glass after etching, the gap caused by the lack of planarity must be overcome by plastic deformation. This means that the wafers are bonded under slight pressure and at a bonding temperature exceeding the transformation temperature. The significant process parameters of temperature, pressure and bonding time will depend on the nature of the surfaces – their roughness or curvature.

The degree to which the whole component is deformed after bonding depends on the quality of the surface.

2.9
Establishing Fluid Contact

The connections for fluid contact have the task of ensuring ingress of the starting materials to the microreactor and egress of the products. They are normally minia-ture hoses made of PTFE or pipes made of chemically resistant stainless steel. Common diameters are 1/8 and 1/16 in fittings with 1/4 in UNF thread are employed for PTFE hoses and the steel tubing is fixed with jubilee clips. A variety of methods are used to fix compatible connection points on the microreactor. Here three of them will be described.

(1) *Casing*: the microreactors are embedded into a casing. Female threads are integrated into the casing and furnished with O-ring gasket to seal against

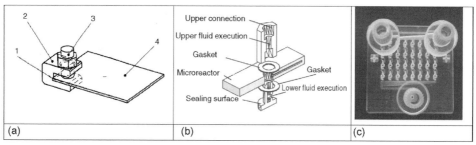

(a) (b) (c)

Figure 2.22 Possibilities for connecting fluidics to microreactors.
(a) 1: bedstop; 2: U form coupling body; 3: connection screw; and
4: microreactor [8]. (b) Self-supporting casing and hollow screws.
(c) Self-supporting casing with female threads.

leakage of the substances. Adapters for screw fittings or hose clips are screwed into the female thread using conventional seals.

(2) *Self-supporting casing*: Here the reactor has a self-supporting casing, that is the reactor is so solid that it is possible to fix the connections directly onto it. It does not need a separate casing. Sometimes partial frames are used, enclosing only the components at the ingress or egress points; for an exemplar U form coupling body, see in Figure 2.22a and Figure 2.23 where the tube seals directly on the component. Or, the connections are fixed with hollow screws and double muffs. On one side of the double muff there is the thread for the hollow screw and on the opposite side there is the 1/4 in UNF thread or a screw fitting, see Figure 2.22b.

(3) The right-sized female thread and the 1/4 in UNF are manufactured in glass, as a direct component of the self-supporting casing or in connector pieces that are

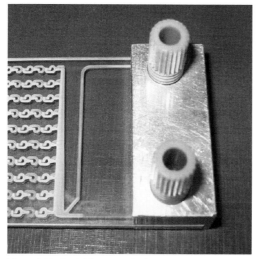

Figure 2.23 Partial frames with the microreactor.

bonded onto the reactor. This method is particularly suitable for thin micro-reactors, see Figure 2.22c.

2.10
Other Materials

Microreactors are sometimes made of ceramics besides being made of glass and/or silicon. In its unfired state, the ceramic material is 'green', meaning that the components, ceramic material, Al_2O_3 and glass, are made up of a film using a polymer binder. In the course of firing, the binder burns off and the green material sinters together. The reactor is constructed from individual layers as described above. The layer is structured in its green state – by stamping, imprinting or laser cutting. The layers are bonded together by a combined pressing and firing process. An LTC-type ceramic material is advantageous, as it will sinter already at 900 °C on account of its high glass content. Use of the ceramic material permits the additional inclusion of electronic components, which is known as hybrid assembly.

References

1 Bu, M., Melvin, T., Ensell, G.J., Wilkinson, J. S. and Evans, A.G.R. (2004) A new masking technology for deep glass etching and its microfluidic application, *Sensors and Actuators A*, **A115**, 476–482.

2 Beitrag zur Entwicklung von Herstellungstechnologien für komplexe Bauteile aus mikrostrukturierbarem Glas, Alf Harnisch, Dissertation, Ilmenau, Shaker Verlag, 1998.

3 Heuberger, A. (ed.) (1989) *Mikromechanik*, Springer Verlag, Berlin.

4 Mikromechanik: Einführung in die Technologie und Anwendungen, Stephanus Büttgenbach, Teubner Stuttgart, 1991.

5 Einführung in die Mikrosystemtechnik, F: Völklein, Th. Zetterer, Vieweg, Braunschweig, Wiesbaden, 2000.

6 Ri-Choudhury, P. (ed.) (2000) *MEMS and MOEMS Technologies and Applications*, SPIE Press, Washington.

7 Albrecht, A., Harnisch, A., Frank, Th. *et al.* Glass relief structuring system using fine linear scanner to irradiate photo-structurable glass, 12.10.1998 DE 19846751.

8 Wurziger, H. and Schwesinger, N. Anschlußkupplung für plättchenförmige Mikrokomponenten, 24.12.1998, DE 19860220.

9 Wikipedia (free online encyclopedia), 2007.

3
Properties and Use of Microreactors

David Barrow

3.1
Introduction

The distinctive fluid flow, thermal and chemical kinetic behavior, observed in microreactors, as well as their size and energy characteristics, contribute to their usefulness in diverse applications [1–3] including:

- highly exothermic reactions [4];
- screening for potential catalysts [5,6];
- precision particle manufacture [7];
- high-throughput materials synthesis [8];
- emulsification and microencapsulation [9];
- fuel cell construction [10];
- point-of-use, miniature and portable microplants [11].

These new application horizons are enabled by the following advantages: (i) size reduction through microfabrication; (ii) reduced diffusion distances; (iii) enhanced rates of thermal and mass transfer and consequent processing yields [12,13]; (iv) reduced reaction volumes; (v) controlled sealed systems avoiding contamination; (vi) use of solvents at elevated pressures and temperatures; (vii) reduced chemical consumption; and (viii) the facility for continuous synthesis [14]. Microreactor research and development has been particularly promoted for high-throughput synthesis in the pharmaceutical industry, where a large number of potential pharmaceutically beneficial compounds need to be generated, initially, in small quantities, as a component of the drug discovery process [15]. In this chapter, the key functional properties of microreactors will be reviewed in the context of use in diverse fields.

3.2
Physical Characteristics of Microreactors

3.2.1
Geometries

(a) *Size*: Microreactor systems incorporate structures for the directed transport or containment of gases or fluids, which have a dimensional property in at least one direction usually measured in micrometers, sometimes up to 1 mm. These structures may comprise microscale ducts (e.g. channels and slots) and pores, larger features (e.g. parallel plates) that cause fluid to flow in thin films and others that cause fluid to flow in microscale discontinuous multiphase flow (e.g. bubbles and emulsions). More specific details of these types of structure are given in Chapters 1 and 2. In addition, small containment structures such as microwells have been fabricated in an analogous format to traditional microtiter plates, rendering potential compatibility with existing robotic handling systems as used in many high-throughput screening laboratories. Extending the notion of a microreactor, an increasing number of studies are demonstrating how separated droplets may act as nanoscale-based reactors. For instance, the use of solvent droplets resulting from controlled segmented flow has been proposed as individual nanoliter reactors for organic synthesis [16,17]. Similarly, reverse micellar structures have been shown to provide reactors for the controlled systems of nanometer-scale particulates. Also, giant phospholipid liposomes ($\sim$10 μm diameter) have been utilized as miniature containers of reagents, which can be manipulated by various external mechanisms such as optical, electrical and mechanical displacement and fusion. Liposome-based microreactors, manipulated in this manner, hold the potential to enable highly controlled and multiplexed microreactions on a very small scale [18].

(b) *Architecture*: Geometries employed in microreactor design and fabrication may range from simple tubular structures, where perhaps two reagents are introduced to form a product to more sophisticated multicomponent circuits, where several functionalities may be performed, including reagent injection(s), mixing, incubation, quench addition, solvent exchange, crystallization, thermal management, extraction, encapsulation and phase separation.

(c) *Multiplicity*: Microreactors may comprise either single-element structures from which small quantities of reaction products may be obtained or a large number of parallel structures where output on an industrial scale can be realized. The engineering of such a 'numbering-up' solution for processes involving reactions with heat/mass transfer usually requires a distribution system from a common reactant source through many reaction microchannels to a common product outlet such that the same residence time is experienced in all the reaction microchannels. From an analytical comparison of bifurcating and consecutive source/outlet manifold structures, as resistance networks, design guidelines have been derived that consider manufacturing variations in microchannel geometry, microchannel aspect ratio and microchannel blockages incurred during function [19,20]. From this it has been shown that a distribution system of bifurcating ducts always produces flow equipartition as long as the length of the straight channel after each channel bend is sufficient for a symmetrical velocity profile to develop.

A key problem in the development of process chemistries is that a reaction scheme developed in a small bench-top flask may not scale up with the same output parameters when transferred to an industrial production reactor. Instead, this problem is potentially circumvented by arithmetically numbering up, in parallel operation, the multiplicity of the same microreactors to achieve a target output. However, this engineering challenge is not trivial, since many parallel reactors may be required to achieve significant volume outputs. Pioneering examples of those industrial processes that have successfully been achieved by using microreactor technology are described in Chapter 5. Those that have shown commercial success appear to represent mostly high-value, low-volume products that are particularly dangerous to manufacture and/or have a short shelf life.

3.2.2
Constructional Materials and Their Properties

Microfluidic devices, which may be suitable for chemical synthesis according to the processing conditions, have been fabricated from a range of materials [1], including glass [21], elastomers [22], silicon [23], quartz, fluoropolymers, metals, ceramics, employing the techniques of laser ablation, wet chemical etching, abrasive micro-machining, deep reactive ion etching, molding, embossing, casting and milling. Particular important aspects are described in more detail in Chapters 1 and 2.

Advanced microreactors for manufacturing-level chemicals production place de-manding requirements on their integrated functionality and durability. For instance, temporal stability of surface energy, surface chemistry and activity of incorporated catalysts and compatibility with sterilization protocol are just some of the several key considerations. These are further complicated by application- and process-specific requirements. For example, the excitation and control of reactions may require temporal and spatial modulation of applied energies such as UV, IR and microwave radiation. All these considerations place exacting specifications on the constructional materials in order that the required geometries may be fabricated in a cost-effective manner compatible with envisaged manufacturing-level scenarios. Where massively parallel microreactor systems are required for volume outputs, constructional materials must be appropriate to the economies and micromanufacturing processes of mass-fabricated parts. Equally, levels of specific functional integration must be equated with the overall system-level integration strategy and range from monolithic to hybrid solutions. Most preferably, and this is more of a long-term goal, reconfigurable and/or addressable component functions will allow for the creation of application-specific microreactor ensembles from a 'programmable' platform technology.

Although basic microreactors and arrays may be fabricated from either glass, polymers, metals or ceramics, advanced microreactors with multifunctional and reconfigurable capability will require construction from a set of diverse and integrated materials. For example, a focused microwave excitation delivered at multiple resonator nodes within a fluidic microreactor array will require constructional materials and associated machining processes suitable for both reaction chemistry and spatial distribution of microwave energy. For this, a materials set of glass, polymers and metals is required, each of which might be separately microstructured using one or more techniques of subtractive machining (etching and ablation), embossing, molding and casting. Industrial-scale processes in microreactors are often conducted under medium-to-high pressures and with the use and production of highly reactive chemicals. This may require the use of pressure-, solvent- and temperature-tolerant stainless steel, ceramic or glass with associated accessories such as gaskets and interconnects, sometimes fabricated from polytetrafluoroethylene (PTFE) and poly-etheretherketone. Although fluorous polymers might be an optimal choice for many applications including corrosive or other hazardous chemicals, their micromanufac-turing compatibility must be taken into account. For instance, PTFE does not lend itself to the mass fabrication technique of embossing, but can be microstructured by using reactive ion etching as used frequently at a wafer level with silicon. In contrast, perfluoroalkoxy, a thermoplastic variant that can be molded and is highly solvent resistant, has FDA approval for many applications and is sterilization compatible. In contrast to the requirements imposed by industrial application, experimental and laboratory chip-based devices for research purposes have also been fabricated from the same materials, but these may also include silicon, silicon–Pyrex and occasionally polymers such as polymethylmethacrylate.

Many chemical reactions performed in microreactors are conducted at room tempera-ture, but in others they require heating and/or cooling; therefore, thermal transfer to the microdevice is an important issue and impacts the selection of constructional materials. In this respect, cooling or heating units have been combined with microdevices to allow constant reaction temperatures or controlled temperature zones [24]. Along with the basic materials from which a microreactor is fabricated, there may be additional materials that are included as coatings or packings. For instance, in a glass–polymer composite continuous-flow microreactor, palladium particles have been loaded by ion exchange and reduced. This was used in a Heck reaction and demonstrated to be reusable for more than 20 times post-wash treatment [25]. Also, coating of the capillary channel of a microreactor with elemental palladium allowed palladium-catalyzed coupling reactions to be performed very efficiently, the metal coating also serving as recipient for microwaves allowing for a fast heating of the reaction solution [26].

3.3
Fluid Flow and Delivery Regimes

3.3.1
Fluid Flow

Monophasic fluid flow in capillary-scale ducts is characterized by a low Reynolds number, the flow in capillary-scale microreactors is generally laminar and transport phenomena are considered to be determined essentially by diffusion [27]. Without the use of special structures or active mechanisms, there is little turbulence-based mixing. Fick's law of diffusion states that

$$J = -D\Delta n, \tag{3.1}$$

where n is the particle density or concentration, D is the diffusion coefficient and Δ is the Laplace operator.

The diffusion time t is defined as the time taken by a molecule to travel distance x by diffusive processes:

$$t = \frac{x^2}{D}. \tag{3.2}$$

This means that for reactions limited by diffusion, reaction time is proportional to the square of the rate-limiting distance. Therefore, a reaction in a 10 cm diameter flask could take 1 000 000 times less if undertaken in a 100 μm diameter microreactor. Dramatically reduced reaction times have, arguably, been the most potent driving force behind research in microreactor technology.

While fluid flow is often continuous and laminar, other regimes, such as where immiscible fluids or phases are configured, provide contiguous 'trains' of fluid 'segments' or 'packets' (see also Chapter 4.3). The flow within these fluid segments may be configured to be such that there occurs an internal vortex that causes rapid mixing within segment contents (Figure 3.1) and counters the lack of mixing, normally the characteristic of microscale fluid flow [28–31]. Adjacent contiguous segments may enjoy a highly dynamic fluidic interface providing many opportunities for novel interfacial chemical and other reactions. This internal vortex and inter-packet dynamic interface may be readily switched to laminar flow (within packets) by simple modulation of the duct cross-sectional geometry, thereby changing the three-dimensional format of the individual fluid packets. Thus, dramatic alterations in mixing and mass transfer may be programmed within a given microreactor circuit configuration. The use of such solvent droplets resulting from controlled segmented flow has been proposed for individual nanoliter-scale reactors for organic synthesis [16,17]. Fluid flow segmentation may be generated for a wide range of immiscible fluid matrices. Fluid packets may (i) contain particulates, including solid support

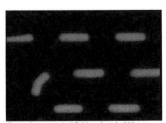

Dynamic segmented fluid packets in 200 micron width reactor duct photographed under UV illumination. One phase containing a fluorophore is visibile whilst the other is not seen.

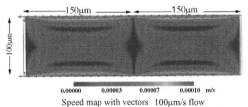

Computational simulation of velocity profile contiguous segmented fluid packets. An internal flow vortex causes rapid mixing within fluid packets and the dynamic interface beteen packets is illustrated

Figure 3.1 Internal vortex circulation in fluid segments.

beads, catalysts and separation media, (ii) be subject to sequential additional reagents through tributary ducts and channel injectors, (iii) be caused to split and/or coalesce and (iv) be provided with individual identity through the provision of addressable molecular photonic and other codes. Segmented fluid packets, as shown in Figure 3.1, may therefore be considered as 'test tubes on the move' that are transferred seamlessly from one functional high-throughput screening operation to another. The fluid packet format, for example, segmented by inert perfluorinated fluids can be combined with interpacket liquid–liquid or solid-phase extractions and microchannel contactor functions enabling many possibilities for compound trans- fer between the different solvent streams of hyphenated functional processes. Collectively, these tools pose a radically different opportunity for synthesis, assay and characterization procedures from the traditional high-throughput screening operations such as microtiter plate technology, storage and information handling. This is clearly a radically different platform paradigm with inherent opportunities and issues that require analysis through novel experimentation and modeling. For example, in a gas–liquid carbonylative coupling reaction, an annular flow regime was employed to generate a high interfacial surface area where a thin film of liquid was forced to the wall surfaces of a microreactor (5 m length, 75 µl capacity) by carbon monoxide gas flow through the center [32].

The laminar stationary flow of an incompressible viscous liquid through cylindrical tubes can be described by Poiseuille's law; this description was later extended to turbulent flow. Flowing patterns of two immiscible phases are more complex in microcapillaries; various patterns of liquid–liquid flow are described in more detail in Chapter 4.3, while liquid–gas flow and related applications are discussed in Chapter 4.4.

3.3.2
Fluid Delivery

(a) *Displacement*: Hydrodynamic pumping, using macro- or microscale peristaltic or positive displacement pumps, has been the main method of fluid delivery generally used in microreactor systems to date [33–35]. High pressures can be obtained and aggressive solvents can be used. However, peristaltic pumps suffer from fluid flow fluctuations at slow flow rates and syringe pumps require carefully engineered changeover or refill mechanisms to be able to be used in long-duration, continuous-flow synthesis schemes.

(b) *Electroosmotic flow (EOF)*: Fluid pumping in capillary-scale devices and systems may be readily enabled under certain conditions by electrokinetic flow, which has the advantage that low levels of hydrodynamic dispersion are observed [36–38]. A detailed theoretical description of chemical reactions in microreactors under electroosmotic and electrophoretic control has been given in the literature [39]. To enable EOF, electrodes are usually placed in reservoirs and voltages applied, most preferably under computer control, with the magnitude of the voltage being a function of several factors including reactor geometry. Electroosmotic flow pumping has been demonstrated in capillary-based flow reactors incorporating solid-supported reagents and catalysts [40,41]. Further, an array of parallel microreactors, packed with silica-supported sulfuric acid, was operated under EOF to produce several tetrahydropyranyl ethers, thus demonstrating arithmetic scale out of EOF-pumped microreactors [42]. However, EOF does place certain requirements on the microreactor design and surface properties of the constructional materials used. As an additional restriction, not every reaction can be performed in an electrical field as electrochemical side reactions can occur.

(c) *Centrifugal*: Centrifugal forces have for some time been harnessed for the controlled propulsion of reagents in spinning disk microreactors [43]. This mechanism has also been used to control the elution, mixing and incubation of reagents within enclosed reaction capillaries on rotating disk platforms [44]. This represents a very innovative approach to chemical synthesis since the technique makes use of both hardware and software systems already developed for a mass-produced commodity. Additionally, the use of centrifugal forces provides an elegant way in which these can be used in combination with hydrophobic, so-called burst, valves to control fluid flow and incubation regimes.

3.3.3
Mixing Mechanisms

Microreactors are usually characterized by geometries with a low Reynolds number. In such capillary-scale ducts, laminar flow is dominant and mixing relies essentially on diffusion unless special measures are taken, such as to cause turbulence or reduce diffusion time. Equally, laminar flow may be exploited such that laminar flow streams moving in parallel may contain reagents that are caused to interact by carefully controlling the flow rate and variations in the microreactor geometry. A range of passive and active techniques to induce mixing include (i) complex geometries within microfluidic manifolds to cause repeated fluid twisting and flattening [45], (ii) acoustic streaming [46], (iii) resonant diaphragms and (iv) acoustic cavitation microstreaming [47,48]. Passive techniques such as split and recombine suffer from the requirement that fluids must usually be in a state of flow, whereas active methodologies enable mixing where there is no flow, such as in microwell reactors and under temporary stopped-flow conditions in microchannel reactors. As a variant of this, a stopped-flow batch-mode technique has been employed to induce mixing on a centrifugal platform [49]. Not dissimilarly, pulsed flow in a microchannel has also been shown to be effective at causing accelerated mixing [50] and is dependent upon several factors including the Strouhal number, the Peclet number, phase difference, pulse-to-volume ratio and microchannel geometry. Microfabricated geometries within the microreactor design, which split and recombine fluids, have been shown to cause multilamination and thus reduced diffusion distances [51–53]. The chaotic advection may also be caused by channels, which either contain integral staggered, serial, asymmetric riblike structures [54] or are three-dimensionally twisted [55]. Active mechanisms for mixing based on energized ultrasonically induced transport have been demonstrated [56]. An interesting form of rapid micromixing may also be achieved in liquid–liquid multiphase flow microreactors where within serial contiguous fluid packets there exists an internal vortex flow that counters the laminar flow profile normally characteristic of low Reynolds number geometries.

3.4
Multifunctional Integration

Some argue that miniaturized tools for both chemical synthesis and analysis need to be integrated onto a single chip to gain the true benefits of miniaturization [57], not least because of the problems associated with subsystem interconnectivity, dead volumes and chip-to-world interfaces. Demonstrations toward such a goal include, for example, a hyphenated mixing reaction channel coupled to a capillary electrophoresis column [58].

Microdevices with other functionalities, as well as miniaturized reactors, extend the range of functional capabilities that may be achieved when a systems approach is considered [59]. Such microdevices may include mixers, separators, heat exchangers, heaters, coolers, photoreactors, analysis subsystems and devices for the application of pulsed electric fields [60]. Therefore, a wide range of processes including extractions (liquid–liquid, liquid–gas and solid-phase enhanced), crystallizations, purifications, conversions, phase changes, phase separations and identifications may be enabled. Thermal conditions may be more readily monitored throughout a microreaction system by employing a distributed reporter such as a thermochromic dye [61] that can provide less than 1 °C temperature variations, albeit over a limited dynamic temperature range.

Interconnects to and between such microdevices for laboratory-scale experimental apparatus have historically been problematic, since a mechanically sound, pressure-resistant and hermetic juncture with minimal dead volume is usually required. For robust, industry-ready chemicals production, equipment with stainless steel fittings is usually employed in microreactor systems. For laboratory-scale apparatus with more delicate chip-based microreactors, a range of microfabricated solutions have been explored [62,63] resulting in both miniaturized plug-and-play microconnectors [64] and macroscale interface housings. Although both adhesive [65] and mechanical [66] solutions have been developed, the mechanical solutions have been reported to possess at least an order of magnitude greater strength than the adhesive solutions. More widely, the issues surrounding the packaging and interconnectivity of microfluidics and associated devices have been comprehensively summarized by Velten *et al.* [67].

3.5
Uses of Microreactors

3.5.1
Overview

Microreactors offer a radical alternative platform for chemical synthesis, normally undertaken in macroscale flasks [68–70]. When reactions in microcapillary-scale reactors are compared with those in flask-scale batch reactors, they have been shown to offer yield, rate or selectivity advantages in a diversity of reactions schemes including carbonylative cross-coupling of arylhalides to secondary amides [32], oxidations [71], nitrations [4], fluorinations [72], hydrogenation [73] and many others described in detail in Chapters 4.1–4.5.

An important advantage of microreactor technology for organic synthesis and catalysis is that continuous-flow processing is enabled. This is often not possible with conventional macroscale reactors and batch production. For example, a multistep chemical synthesis of carbamates in a continuous-flow process has been demonstrated using three reaction steps and two separation steps in between the reaction steps. Using a serial cascade of three microreactors and two phase separators, to enable solvent switching and *in situ* generation and consumption of dangerous intermediates, the safe processing of high-energy chemistries and small-scale production of chemicals in a compact chip-based processing system was demonstrated [74,75]. One of the problems with continuous-flow microreactors is that of cross-contamination from (i) different reactions, where sequential reaction steps require solvent, reagent and other condition changes and (ii) parallel reactions, where similar types of reactions are performed using different combinations of reagents [76].

3.5.2
Unstable Intermediates Fast and Exothermic Reactions

Microreactors provide a safe means by which reactions, including multistage schemes, can be undertaken where, otherwise, products involving unstable intermediates may be formed. This is exemplified by Fortt who showed that for a serial diazonium salt formation and chlorination reaction performed in a microreactor under hydrodynamic pumping, significant yield enhancements (15–20%) could be observed and attributed them to enhanced heat and mass transfer [77]. This demonstrates the advantage of microreactor-based synthesis where diazonium salts are sensitive to electromagnetic radiation and static electricity, which in turn can lead to rapid decomposition. Microreactors facilitate the ability to achieve continuous-flow synthesis, which is often not possible with conventional macroscale reactors and batch production.

A key feature of microreactors is the comparatively large surface area they afford compared to conventional reactors. Surface-to-volume ratios of $20\,000\,\mathrm{m^2\,m^{-3}}$ may be possible for microreactors, whereas $1000\,\mathrm{m^2\,m^{-3}}$ may be more typical for a conventional reactor. The surface area may be further enhanced by (i) providing microfabricated pillared or ribbed structures within the reactor space, (ii) introducing packing materials and (iii) providing high specific area surfaces, such as those that can be obtained with porous silicon. In *catalytic* reactions in which competition exists between the rate of diffusion to the catalyst sites and the rate of the reaction, the mass transport resistance is usually eliminated in a microreactor [78]. Microreactors, therefore, provide excellent environments for catalytic reactions such as palladium catalysis as employed for Heck and Sonogashira couplings [79,80]. It has been highlighted that such catalysts usually form solid/liquid heterogeneous systems, rendering them difficult to employ in microscale channels. However, when room temperature ionic liquids are used to dissolve the catalyst, a liquid–liquid two-phase system can be successfully employed (as shown for a Sonogashira coupling), thus enabling fast catalyst screening [81]. Catalytic reactions performed in microreactors may be functionally extended by introducing external energy sources such as light. For instance, photocatalytic anatase titania films have been applied by slip casting onto the internal surface of planar glass microreactors and the dramatic improvement of photocatalytic function improved hugely by the addition of gaseous oxygen [82]. Planar chip-based microreactors for photocatalytic reactions offer the potential for improvements in the coupling (increased spatial homogeneity and reduced attenuation) of applied irradiation to reaction reagents and catalysts due to the short penetration path lengths that may be enabled by efficient design [83,84]. Such microreactors are fabricated ideally from Pyrex, amorphous fluoropolymers or quartz, most preferably the latter because it facilitates both employment of higher operational temperatures and lower light attenuation at lower ultraviolet wavelengths than Pyrex.

3.5.3
Precision Particle Manufacture

Multiphase flow in a tubular, chip-based or freestanding capillary microreactor may be controlled so that so-called segmented flow creates serial contiguous packets of immiscible fluids. The segmented flow condition may be created by several contrasting microreactor geometry configurations, including (i) simple T-junction, (ii) constriction junction, (iii) sheath flow junction and (iv) fluidic oscillator arrangements. The precision of fluid volume elution controlled by these means depends on several factors: particularly critical are temperature stability of microreactor and reagents used, a long-duration stable surface energy of the material from which the microreactor is fabricated and precision control of fluid flow rate(s). Indeed, segmented flow patterns are not always readily generated and gas–liquid flows produced; for example, from a T-junction, may result in annular flow instead [32]. Fluid packets created from segmented flow can exhibit a very narrow size distribution and may be converted to solid and semisolid microparticles of different morphology (e.g. spherical, discoid, fibrous and macroporous) by various means such as UV polymerization, freezing or chemical cross-linking.

3.5.4
Wider Industrial Context

3.5.4.1 Sustainability Agenda
The increasing motivation to develop desktop-scale, integrated, microreactor-based processing systems is led by several needs, including (i) a generic requirement for point-of-demand synthesis across several industrial sectors, (ii) individualized 'designer' products, (iii) point-of-use production of dangerous products, (iv) portable power plants and (v) universally, low carbon footprint production processes. Short 'shelf-life' products are good application candidates for production on demand in microplants. The sustainability agenda, in part, drives the need for process intensification where large-scale, expensive, energy-intensive equipment may be

replaced with others that are smaller, less costly, more efficient, multifunctional, can have a reduced environmental impact and provide improvements in safety and automation [11]. For example, the ecological advantages associated with the transfer of a chemical synthesis from a macroscale semicontinuous batch process to a continuous microscale setup were demonstrated for the two-step synthesis of *m*-anisaldehyde from *m*-bromoanisole [85]. This synthesis is highly exothermic and, ordinarily, can only be carried out with stringent safety precautions and a high cooling energy effort. Reaction temperatures of 223 and 193 K were used for the macroscale reaction chosen, whereas in the microreactor a continuous isothermal reaction was performed at 273 K. In the study, 11 pilot-scale microreactors were used in parallel to achieve comparable outputs from both the micro- and macroscale systems. A cradle-to-grave life cycle analysis demonstrated clear ecological benefits of employing arithmetically scaled-out microreactors to achieve comparable output rates to the macroscale reactor.

3.5.4.2 Point-of-Demand Synthesis
The progressive sophistication of tools for chemical analysis has led to the opportunity to create products that are designed for the specific genetic and phenotypic requirements of individuals, particularly for pharmaceuticals, nutriceuticals, foods, healthcare products and cosmetics. The manufacture of designer products at the point of demand could be exemplified by an in-store, plant-on-a-desk manufacturing setup at the point of sale; for instance, a cosmetic tuned to the individual requirements of a client's skincare status, personal (e.g. age, skin color) and other factors (e.g. prevailing UV strength). Products differentiated in this way will carry a significant price premium with cosmetics arguably representing the most likely application field for an early commercial investment return. Fast moving commodity goods are considered to represent an important, diverse, multisectorial industrial product platform that will drive the development of point-of-demand microplant systems. Equally, the sustainability agenda has led to the potential demand for a decentralized, distributed processing plant so that materials may be generated where and when they are required. This is particularly applicable where products have a short shelf-life, are exceptionally toxic, are costly (economically and environmentally) to dispose of if unused and are dangerous to transport. However, while such market applications might appear to represent profitable opportunities for commercial exploitation, other factors often prevent their appearance in the marketplace. Such factors include corporate resistance because of existing infrastructure investment and product brands, regulatory acceptance procedures and consumer acceptance. Instead, perhaps, there are completely new applications, for which no product exists, that could become the first successful, so-called killer, applications. For instance, microengineered components, including microreactors, are highly suited to the creation of lightweight, compact, integrated microscale power generators. The microscale dimensions enable low thermal and mass transport resistance rendering them highly efficient in fuel cell applications where highly endothermic and exothermic processes must be thermally coupled. Equally more exotic applications, such as microscale propellant generators for use in space exploration [86,87], can

lead to advances in microreactor technology owing to the exacting specifications for such applications.

References

1 Ehrfeld, W., Hessel, V. and Löwe, H. (2000) *Microreactors: New Technology for Modern Chemistry*, Wiley-VCH Verlag GmbH, Weinheim, Germany.

2 Fletcher, P.D.I., Haswell, S.J. and Zhang, X. (2002) *Lab on a Chip*, **2**, 102–112.

3 Jensen, K.F. (2001) *Chemical Engineering Science*, **56**, 293–303.

4 Halder, R., Lawal, A. and Damavarapu, R. (2007) *Catalysis Today*, **125**, 74–80.

5 Herweck, T., Hardt, S., Hessel, V., Löwe, H., Hofmann, C., Weise, F., Dietrich, T. and Freitag, A. (2001) Proceedings of the 5th International Conference on Microreaction Technology: IMRET 5 (eds M. Matlosz, W. Ehrfeld and J.P. Baselt), Springer-Verlag, Berlin, pp. 215–229.

6 Claus, P., Hönicke, D. and Zech, T. (2001) *Catalysis Today*, **67**, 319–339.

7 Sotowa, K.-I., Irie, K., Fukumori, T., Kusakabe, K. and Sugiyama, S. (2007) *Chemical Engineering and Technology*, **30**, 383–388.

8 Greenway, G.M., Haswell, S.J., Morgan, D.O., Skelton, V. and Styring, P. (2000) *Sensors and Actuators B: Chemical*, **63**, 153–158.

9 Nisisako, T., Torii, T. and Higuchi, T. (2002) *Lab on a Chip*, **2**, 24–26.

10 Mitchell, C.M., Kim, D.-P. and Kenis, P.J.A. (2006) *Journal of Catalysis*, **241**, 235–242.

11 Charpentier, J.-C. (2007) *Chemical Engineering Journal*, **134**, 84–92.

12 Mitchell, M.C., Spikmans, V. and de Mello, A.J. (2001) *Analyst*, **126**, 24–27.

13 Chambers, R.D. and Spink, R.C.H. (1999) *Chemical Communications*, 883–884.

14 Tokeshi, M., Minagawa, T., Uchiyama, K., Hibara, A., Sato, K., Hisamoto, H. and Kitamori, T. (2002) *Analytical Chemistry*, **74**, 1565–1568.

15 Watts, P. (2005) *Analytical and Bioanalytical Chemistry*, **382**, 865–867.

16 Gerdts, J., Sharoyan, D.E. and Ismagilov, R.F. (2004) *Journal of the American Chemical Society*, **126**, 6237–6331.

17 Holt, D.J., Payne, R.J., Hollfelder, F., Huck, W.T.S. and Bell, C. (2006) Proceedings of the 10th International Conference on Miniaturised Systems for Chemistry and Life Sciences (µTAS 2006), November 5–9, Tokyo, Japan, pp. 864–866.

18 Simone, K., Kishore, R., Helmerson, K. and Locascio, L. (2003) *Langmuir*, **19**, 8206–8210.

19 Amador, C., Gavriilidis, A. and Angeli, P. (2004) *Chemical Engineering Journal*, **101**, 379–390.

20 Amador, C., Wenn, D., Shaw, J., Gavriilidis, A. and Angeli, P. (2008) *Chemical Engineering Journal*, **135S**, S259–S269.

21 McCreedy, T. (2001) *Analytica Chimica Acta*, **427**, 39–43.

22 Unger, M.A., Chou, H.P., Thorsen, T., Scherer, A. and Quake, S.R. (2000) *Science*, **288**, 113–116.

23 Tiggelaar, R.M., van Male, P., Berenschot, J.W., Gardeniers, J.G.E., Oosterbroek, R.E., de Croon, M.H.J.M., Schouten, J.C., van den Berg, A. and Elwenspoek, M.C. (2005) *Sensors and Actuators A*, **119**, 196–205.

24 de Mello, A.J. (2003) *Nature*, **422**, 28–29.

25 Solodenko, W., Wen, H., Leue, S., Stuhlmann, F., Sourkouni-Argirusi, G., Jas, G., Schönfeld, H., Kunz, U. and Kirschning, A. (2004) *European Journal of Organic Chemistry*, 3601–3610.

26 Comer, E. and Organ, M.G. (2005) *Journal of the American Chemical Society*, **127**, 8160–8167.

27 Graveson, P., Branebjerg, J. and Jensen, O.S. (1993) *Journal of Micromechanics and Microengineering.*, **3**, 168–182.

28 Harries, N., Burns, J.R., Barrow, D.A. and Ramshaw, C.A. (2003) *Journal of Heat and Mass Transfer*, **46**, 3313–3322.

29 Dummann, G., Quittmann, U., Groschel, L., Agar, D.W., Worz, O. and Morgenschweis, K. (2003) *Catalysis Today*, **79-80**, 433–439.

30 Ahmed, B., Barrow, D. and Wirth, T. (2006) *Advanced Synthesis and Catalysis*, **348**, 1043–1048.

31 Burns, J. and Ramshaw, C. (2002) *Chemical Engineering Communications*, **189**, 1611–1628.

32 Miller, P.W., Long, N.J., de Mello, A.J., Vilar, R., Passchier, J. and Gee, A. (2006) *Chemical Communications*, 546–548.

33 Fernandez-Suarez, M., Wong, S.Y.F. and Warrington, B.H. (2002) *Lab on a Chip*, **2**, 170–174.

34 Kashid, M.N., Gerlach, I., Goetz, S., Franzke, J., Acker, J.F., Platte, F., Agar, D. W. and Turek, S. (2005) *Industrial & Engineering Chemistry Research*, **44**, 5003–5010.

35 Haswell, S.J., O'Sullivan, B. and Styring, P. (2001) *Lab on a Chip*, **1**, 164–166.

36 Fletcher, P.D.I., Haswell, S.J. and Zhang, X. (2001) *Lab on a Chip*, **1**, 115–121.

37 Fletcher, P.D.I., Haswell, S.J., Pombo-Villar, E., Warrington, B.H., Watts, P., Wong, S.Y.F. and Zhang, X.L. (2002) *Tetrahedon*, **58**, 4735–4757.

38 Kohlheyer, D., Besselink, G.A.J., Lammertink, R.J.H., Schlautmann, S., Unnikrishnan, S. and Schasfoort, R.B.M. (2005) *Microfluidics and Nanofluids*, **1**, 242–248.

39 Fletcher, P.D.I., Haswell, S.J. and Paunov, V.N. (1999) *Analyst*, **124**, 1273–1282.

40 Wiles, C., Watts, P. and Haswell, S.J. (2004) *Tetrahedon*, **60**, 8421–8429.

41 Wiles, C., Watts, P. and Haswell, S.J. (2005) *Tetrahedon*, **61**, 5209–5217.

42 Wiles, C. (2006) Proceedings of the 10th International Conference on Miniaturised Systems for Chemistry and Life Sciences (μTAS 2006), November 5–9, Tokyo, Japan, 861–863.

43 Boodhoo, K.V.K. and Jachuck, R.J.J. (2000) *Green Chemistry*, **2**, 235–244.

44 Ducrée, J., Haeberle, S., Lutz, S., Pausch, S., von Stetten, F. and Zengerle, R. (2007) *Journal of Micromechanics and Microengineering*, **17**, S103–S115.

45 Lee, Y.-K., Deval, J., Tabeling, P. and Ho, C.-M. (2001) Proceedings of the 14th IEEE Workshop on MEMS, Interlaken, Switzerland, pp. 483–486.

46 Sritharan, K., Strobl, C.J., Schneider, A., Wixworth, A. and Guttenberg, Z. (2006) *Applied Physics Letters*, **88**, 054102.

47 Liu, R.H., Lenigk, R., Druyor-Sanchez, R.L., Yang, J.N. and Grodzinski, P. (2003) *Analytical Chemistry*, **75**, 1911–1917.

48 Liu, R.H., Robin, H., Lenigk, R. and Grodzinski, P. (2006) *Journal of Microlithography, Microfabrication, and Microsystems*, **2**, 178–184.

49 Grunmann, M., Geipel, A., Riegger, L., Zengerle, R. and Ducrée, J. (2005) *Lab on a Chip*, **5**, 560–565.

50 Glasgow, I., Lieber, S. and Aubry, N. (2004) *Analytical Chemistry*, **76**, 4825–4832.

51 Branebjerg, J., Graveson, P., Krog, J.P. and Nielsen, C.R. (1996) Proceedings of the EEEE MEMS Workshop, San Diego, CA, pp. 441–446.

52 Hinsmann, P., Frank, J., Svasek, P., Harasek, M. and Lendl, B. (2001) *Lab on a Chip*, **1**, 16–21.

53 Schönfeld, F., Hessel, V. and Hofmann, C. (2004) *Lab on a Chip*, **4**, 65–69.

54 Strook, A.D., Dertinger, S.K.W., Ajdari, A., Mezi, I., Stone, H.A. and Whitesides, G.M. (2002) *Science*, **295**, 647–651.

55 Che-Hsin, L., Chien-Hsiung, T., Chih-Wen, P. and Lung-Ming, F. (2006) *Biomedical Microdevices*, **9**, 1572–8781.

56 Lee, N.Y., Yamada, M. and Seki, M. (2005) *Analytical and Bioanalytical Chemistry*, **383**, 776–782.

57 Belder, D. (2006) *Analytical and Bioanalytical Chemistry*, **385**, 416–418.

58 Belder, D., Ludwig, M., Wang, L.-W. and Reetz, M.T. (2006) *Angewandte Chemie-International Edition*, **45**, 2463–2466.

59 Gravesen, P., Branebjerg, J. and Jensen, C. (1993) *Journal of Micromechanics and Microengineering*, **3**, 168–182.

60 Fox, M., Esveld, E., Luttge, R. and Boom, R. (2005) *Lab on a Chip*, **9**, 943–948.

61 Iles, A., Fortt, R. and de Mello, A.J. (2005) *Lab on a Chip*, **5**, 540–544.

62 González, C., Collins, S.D. and Smith, R.L. (1998) *Sensors and Actuators B*, **49**, 40–45.

63 Meng, E., Wu, S. and Tai, Y.-C. (2000) Micro Total Analysis Systems 2000: Proceedings of the μTAS 2000 Symposium (eds A. van den Berg, W. Olthius and P. Bergveld), Kluwer Academic Publishers, Dordrecht, pp. 41–44.

64 Gray, B.L., Jaeggi, D., Mourlas, N.J., van Drieënhuizen, B.P., Williams, K.R., Maluf, N.I. and Kovacs, G.T.A. (1999) *Sensors and Actuators A*, **77**, 57–65.

65 Morrissey, A., Kelly, G. and Alderman, J. (1998) *Sensors and Actuators A*, **68**, 404–409.

66 Nittis, V., Fortt, R., Legge, C.H. and de Mello, A.J. (2001) *Lab on a Chip*, **1**, 148–152.

67 Velten, T., Ruf, H.H., Barrow, D., Aspragathos, N., Lazarou, P., Jung, E., Malek, C.K., Richter, M., Kruckow, J. and Wackerle, M. (2005) *IEEE Transactions on Advanced Packaging*, **28**, 533–546.

68 Hessel, V. and Löwe, H. (2005) *Chemical Engineering & Technology*, **28**, 267–284.

69 Watts, P. and Haswell, S.J. (2005) *Chemical Engineering & Technology*, **28**, 290–301.

70 Brivio, M., Verboom, W. and Reinhoudt, D.N. (2006) *Lab on a Chip*, **6**, 329–344.

71 de Mas, N., Jackman, R.J., Schmidt, M.A. and Jensen, K.F. (2001) Proceedings of the 5th International Conference on Microreaction Technology (eds M. Matloz, W. Ehrfeld and J.P. Baselt), p. 60.

72 Chambers, R.D., Fox, M.A., Sandford, G., Trmcic, J. and Goeta, A. (2007) *Journal of Fluorine Chemistry*, **128**, 29–33.

73 Kobayashi, J., Mori, Y., Okamoto, K., Akiyama, R., Ueno, M., Kitamori, T. and Kobayashi, S. (2004) *Science*, **304**, 1305–1308.

74 Sahoo, H.R., Kralj, J.G. and Jensen, K.F. (2006) Proceedings of the 10th International Conference on Miniaturised Systems for Chemistry and Life Sciences (μTAS 2006), November 5–9, Tokyo, Japan, pp. 1029–1031.

75 Sahoo, H.R., Kralj, J.G. and Jensen, K.F. (2007) *Angewandte Chemie-International Edition*, **46**, 5704–5708.

76 Sui, G. and Tseng, H.-R. (2006) Proceedings of the 10th International Conference on Miniaturised Systems for Chemistry and Life Sciences (μTAS 2006), November 5–9, Tokyo, Japan, pp. 1023–1024.

77 Fortt, R., Wootton, C.R. and de Mello, A.J. (2002) Proceedings of the Micro Total Analysis Systems, vol. 2 (eds Y. Baba, S. Shoji and A. van den Berg), Kluwer Academic Publishers, The Netherlands, pp. 850–852.

78 McGovern, S., Gadre, H., Pai, C.S., Mansfield, W., Pau, S. and Besser, R.S. (2006) Proceedings of the 2006 Spring Meeting, American Institute of Chemical Engineers.

79 Fukuyama, T., Shinmen, N., Nishitani, S., Sato, M. and Ryu, I. (2002) *Organic Letters*, **4**, 1691–1694.

80 Liu, S., Fukuyama, T. and Ryu, I. (2004) *Organic Process Research & Development*, **8**, 477–481.

81 Nieuwland, P.J., Koch, K., van Hest, J.C.M. and Rutjes, F.P.J.T. (2006) Proceedings of the 10th International Conference on Miniaturised Systems for Chemistry and Life Sciences (μTAS 2006), November 5–9, Tokyo, Japan, 870–872.

82 Iles, H., Lindstrom, R. and Wooton, R. (2006) Proceedings of the 10th International Conference on Miniaturised Systems for Chemistry and life Sciences (μTAS 2006), November 5–9, Tokyo, Japan, pp. 873–875.

83 Matsushita, Y., Kumada, S., Wakabayashi, K., Sakeda, K. and Ichimura, T. (2006) *Chemistry Letters*, **35**, 410–411.

84 Ichimura, T., Matsushita, Y., Sakeda, K. and Suzuki, T. (2006) Photoreactions, in *Microchemical Engineering in Practice* (ed. T. R. Dietrich), Blackwell Publishing.

85 Kralisch, D. and Kreisela, G. (2007) *Chemical Engineering Science*, **62**, 1094–1100.

86 Holladay, J.D., Brooks, K.P., Wegeng, R., Hua, J., Sanders, J. and Baird, S. (2007) *Catalysis Today*, **120**, 35–44.

87 Besser, R.S., Ouyang, X. and Surangalikar, H. (2003) *Chemical Engineering Science*, **58**, 19–26.

4
Organic Chemistry in Microreactors

4.1
Homogeneous Reactions

Takahide Fukuyama, Md Taifur Rahman, Ilhyong Ryu

4.1.1
Acid-Promoted Reactions

The nitration reaction is one of the fundamental reactions in organic synthesis and is industrially important. Standard batch processes frequently employ a combination of sulfuric and nitric acids, which often become violent and hazardous when scaled up. Because nitration reactions are highly exothermic, they are temperature sensitive. This led chemists to examine the feasibility of running nitration reactions in a continuous-flow manner using microreactors, which would obviate the safety issue with excellent temperature control in a tiny reaction space. Taghavi-Moghadam and coworkers reported that continuous-flow nitration using microreactors represents a safe and controllable method for the nitration of aromatic compounds [1]. For example, the nitration of pyrazole-5-carboxylic acid was carried out using microchannels 100 µm in width (CYTOS Lab System, CPC GmbH), which allowed strict temperature control at 90 °C. This process resulted in the formation of the desired nitration product in good yield (Scheme 4.1). In the batch system, the nitration of pyrazole-5-carboxylic acids suffers from decarboxylation by heat evolution; however, this was not the case for this microflow system as the reaction temperature could be kept strictly below the decarboxylation limit of 100 °C.

Ducry and Roberge reported controlled nitration of phenol in a glass microreactor with a channel width of 500 µm and an internal volume of 2.0 ml [2]. Nitration was most efficient and controlled under nearly solvent-free conditions at 20 °C without the addition of sulfuric acid or acetic acid (Scheme 4.2). Under these concentrated conditions, autocatalysis spontaneously started in the mixing zone, allowing safe control of the reaction. Undesirable polymer formation, which is significant in batch reactions, was effectively suppressed by a factor of 10.

Microreactors in Organic Synthesis and Catalysis. Edited by Thomas Wirth

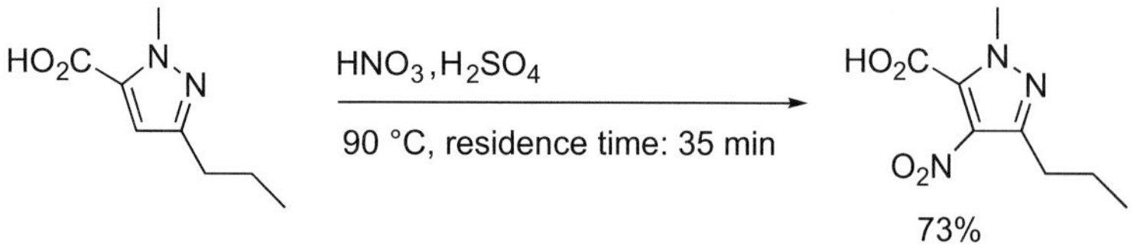

Scheme 4.1

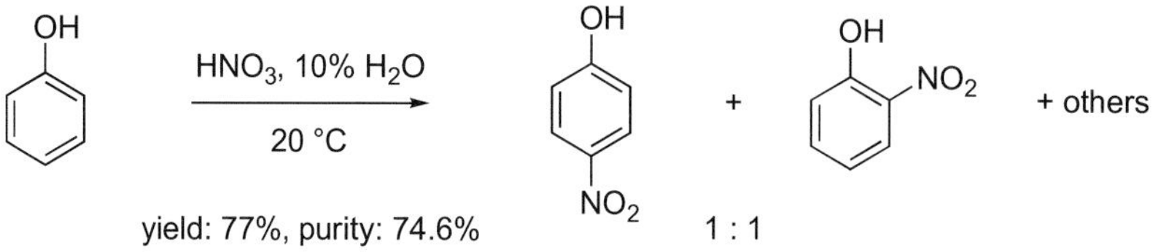

yield: 77%, purity: 74.6%   1 : 1

Scheme 4.2

Synthetic organic chemists have now recognized the potential of microreactors to increase reaction efficiency in organic synthesis. Fukase and coworkers reported that *p*-toluenesulfonic acid-catalyzed dehydration of allylic alcohols proceeds effectively in a microreactor system consisting of an IMM micromixer (channel width = 40 μm) and an additional residence time unit (diameter = 1 mm and length = 1 m) [3]. Thus, dehydration of an allylic alcohol, which was prepared by two steps from farnesol, produced the corresponding conjugate diene at 80% yield (Scheme 4.3). The corresponding batch reaction gave lower yields, especially when the reaction was conducted on a larger scale. A multikilogram synthesis of pristane, an adjuvant for monoclonal production, was attained using a continuous method for dehydration followed by reduction. A parallel system was created using 10 micromixers (Comet X-01), and the continuous-flow reactions occurred over 3–4 days. Hydrogenation of the product, using 10% Pd/C, provided about 5 kg of the desired pristane.

For synthetic organic chemists, the tedious process of optimizing reaction conditions is generally inevitable. Methods for optimization frequently require large quantities of valuable starting materials. In this regard, microreactor-based optimization is advantageous, especially for glycosylation that notoriously lacks reliable general conditions. Seeberger, Jensen and coworkers used microreactors to optimize glycosyla-

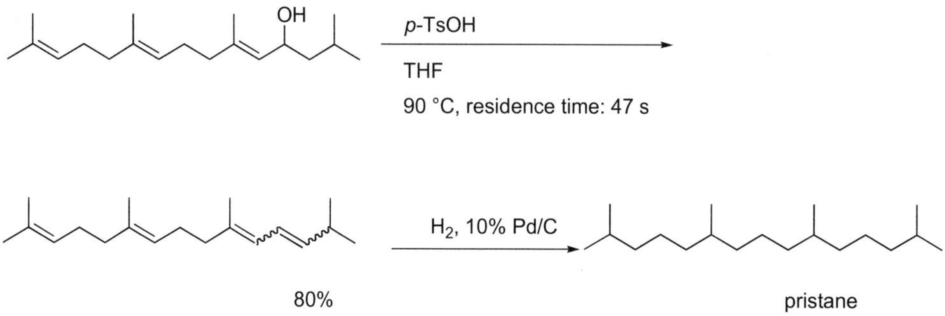

80%   pristane

Scheme 4.3

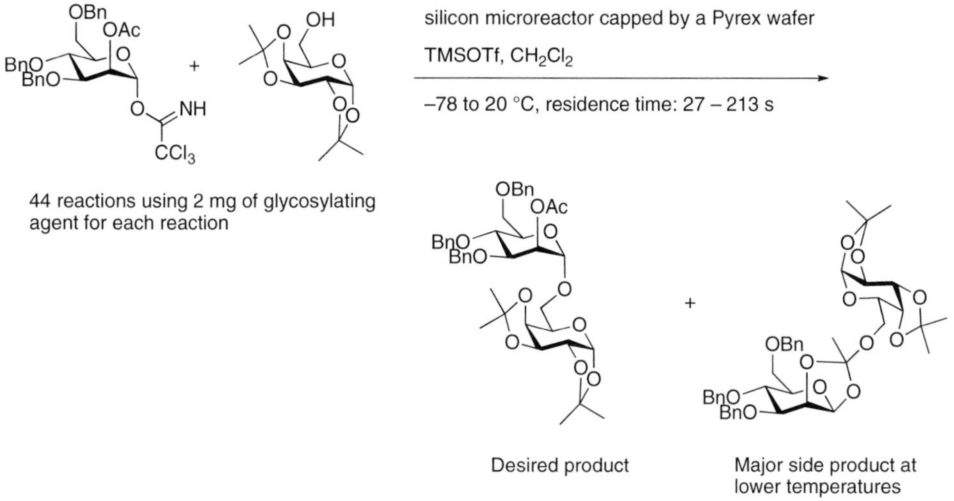

44 reactions using 2 mg of glycosylating agent for each reaction

silicon microreactor capped by a Pyrex wafer

TMSOTf, CH$_2$Cl$_2$

−78 to 20 °C, residence time: 27 – 213 s

Desired product

Major side product at lower temperatures

Scheme 4.4

tion reactions [4]. With a single preparation of reagents, 44 reactions were completed at varying temperatures and reaction times, requiring just over 2 mg of glycosylating agent for each reaction (Scheme 4.4).

Asymmetric cyanosilylation of ketones and aldehydes is important because the cyanohydrin product can be easily converted into optically active aminoalcohols by reduction. Moberg, Haswell and coworkers reported on a microflow version of the catalytic cyanosilylation of aldehydes using Pybox [5]/lanthanoid triflates as the catalyst for chiral induction. A T-shaped borosilicate microreactor with channel dimensions of 100 μm × 50 μm was used in this study [6]. Electroosmotic flow (EOF) was employed to pump an acetonitrile solution of phenyl-Pybox, LnCl$_3$ and benzaldehyde (reservoir A) and an acetonitrile solution of TMSCN (reservoir B). LuCl$_3$-catalyzed microflow reactions gave similar enantioselectivity to that observed in analogous batch reactions. However, lower enantioselectivity was observed for the YbCl$_3$-catalyzed microflow reactions than that observed for the batch reaction (Scheme 4.5). It is possible that the oxophilic Yb binds to the silicon oxide surface of the channels.

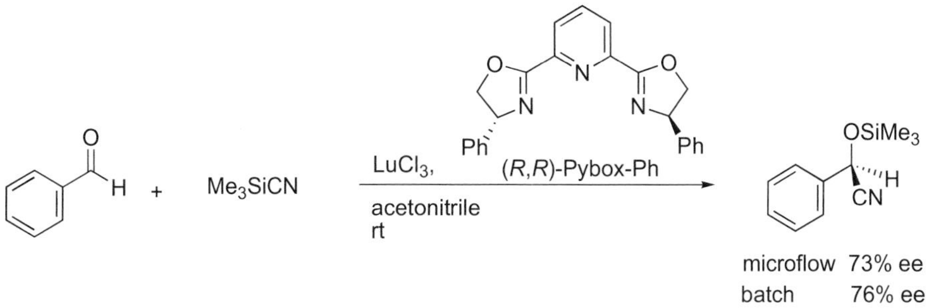

(R,R)-Pybox-Ph

LuCl$_3$,

acetonitrile
rt

microflow 73% ee
batch   76% ee

Scheme 4.5

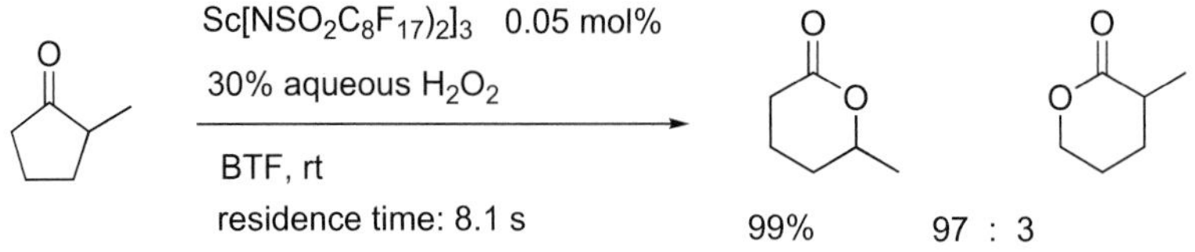

Sc[NSO$_2$C$_8$F$_{17}$)$_2$]$_3$ 0.05 mol%

30% aqueous H$_2$O$_2$

BTF, rt

residence time: 8.1 s

99%   97 : 3

Scheme 4.6

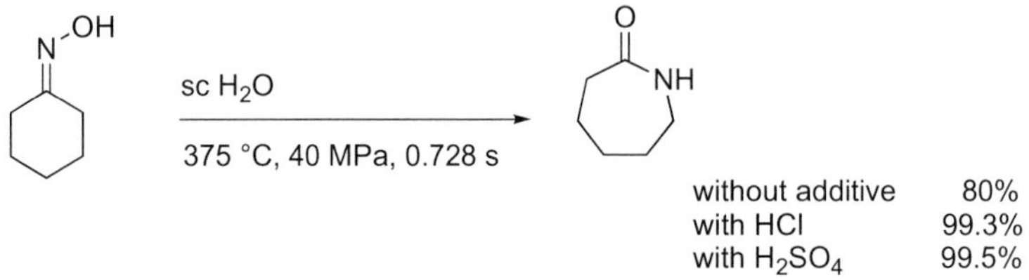

sc H$_2$O

375 °C, 40 MPa, 0.728 s

without additive	80%
with HCl	99.3%
with H$_2$SO$_4$	99.5%

Scheme 4.7

Mikami and coworkers examined the Bayer–Villiger reaction of cyclic ketones with aqueous hydrogen peroxide using a borosilicate microreactor that was 3 cm in length, 30 μm in depth and 30 μm in width [7]. A BTF (benzotrifluoride) solution of a fluorous lanthanide catalyst, Sc[N(SO$_2$C$_8$F$_{17}$)$_2$]$_3$, was used (Scheme 4.6). Extremely low flow rates (25–200 nl/min), which the authors termed 'nanoflow', were well suited for the short length of the channel (3 cm). The residence time under optimal conditions was only 8.1 s. In several cases, regioisomeric ratios observed using a microreactor were superior to those observed obtained using a batch reactor. Effective precomplexation in microchannels of the scandium catalyst with hydrogen peroxide to generate Sc peroxide species seems to account for better selectivity.

The Beckman rearrangement of cyclohexanone oxime into ε-caprolactam is an industrially important process because ring-opening polymerization of ε-caprolactam produces nylon 6. Ikushima and coworkers invented a new continuous-microflow process for an effective Beckman rearrangement of cyclohexanone oxime using supercritical water [8]. With the use of a flow system consisting of a Hastelloy C-276 alloy, the reaction proceeded well without any additives, whereas the addition of a small amount of hydrochloric or sulfuric acid gave higher yields with reaction time less than 1 s (Scheme 4.7).

4.1.2
Base-Promoted Reactions

The combination of fluoride ions and enol silyl ethers provides a useful method for the generation of enolate anions [9]. Watts, Haswell and coworkers applied a borosilicate glass microreactor, having channel dimensions of 100 μm × 50 μm and equipped with an electrosmotic flow (EOF) pumping system, to the C-acylation of enolate anions, which leads to 1,3-diketones [10,11]. A THF solution of tetrabutyl-ammonium fluoride (TBAF) was placed in reservoir A, a THF solution of benzoyl

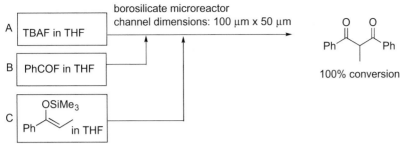

Scheme 4.8

fluoride in reservoir B and enol silyl ether of propiophenone was placed in reservoir C. The desired 1,3-diketone was formed at 100% conversion (Scheme 4.8). As for enol silyl ether of acetophenone, benzoyl cyanide was used to obtain the corresponding 1,3-diketone because the use of benzoyl fluoride resulted in the selective formation of the corresponding vinyl acetate, the O-acylation product. A similar EOF-based microreactor was tested for use in Michael addition of β-diketones and diethyl malonate to ethyl propiolate and methyl vinyl ketone. Ethyldiisopropylamine in ethanol was used as the solvent, and a stopped-flow technique was employed to increase conversion efficiency [11]. More recently, Löwe and coworkers reported a detailed study on the addition of secondary amines to ethyl acrylate and acrylonitrile using a continuous microreaction process based on an IMM micromixer and a tabular reactor [12]. In the best case, space–time yields (g/ml h) for the microflow system were much higher than those for the batch system by a factor of approximately 650.

A borosilicate glass microreactor (152 μm (width), 51 μm (depth) and 2.3 cm (length)) that was connected to a T-shaped PEEK (poly(ether–ether–ketone)) unit (MicroTee, Upchurch Scientific) was used for the formation of sodium enolate of

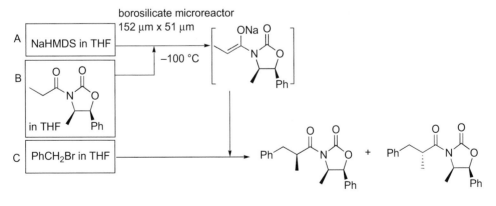

91 : 9
85 : 15 (batch reactor)

Scheme 4.9

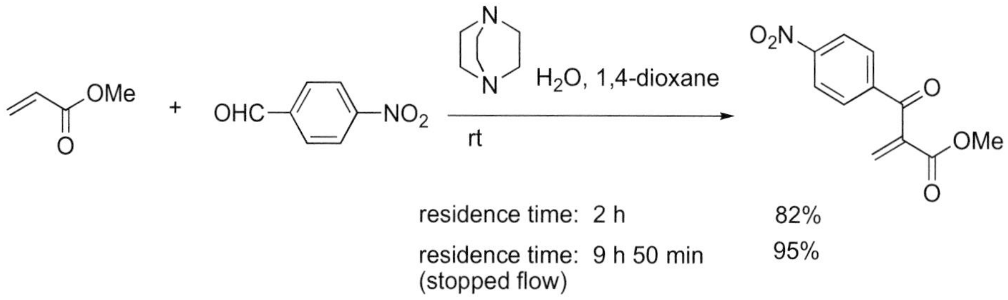
OMe
+
OHC
NO2
N
N
H2O, 1,4-dioxane
rt
O2N
O
OMe
O
residence time: 2 h
residence time: 9 h 50 min
(stopped flow)
82%
95%

Scheme 4.10

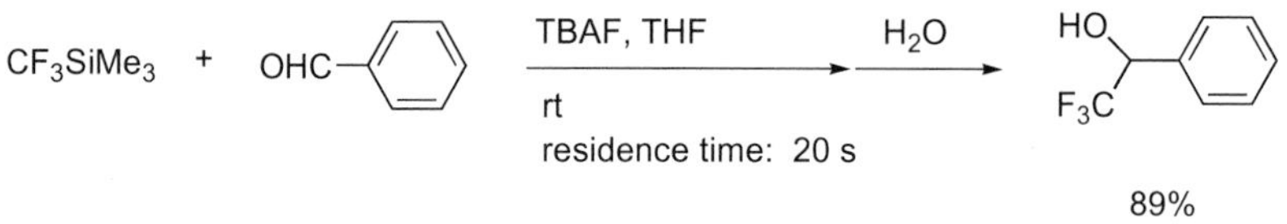
CF3SiMe3
+
OHC
TBAF, THF
rt
residence time: 20 s
H2O
HO
F3C
89%

Scheme 4.11

N-propionyloxazolidinone and the subsequent diastereoselective alkylation with benzyl bromide at −100 °C. The observed diastereomeric ratio of 91 : 9 was superior to that of 85 : 15 observed in a batch reactor (Scheme 4.9) [11,13].

The DABCO-promoted Baylis–Hillman reaction was run continuously using the CYTOS College System (CPC Systems GmbH) with yields comparable to those of the batch reaction and with a significant reduction in reaction time (Scheme 4.10) [14]. Coupled with the stopped-flow technique, an almost complete conversion was achieved.

Kitazume and coworkers used microreactors with microchannels 100 μm wide and 40 μm deep for the synthesis of a series of organofluorine compounds [15,16]. The silylation of 4,4,4-trifluorobutan-2-one and the Mukaiyama-type aldol reaction of the resulting enol silyl ether with acetals gave good yields of the desired products [16]. They also described nitro-aldol reactions of 2,2-difluoro-1-ethoxyethanol and Michael additions of nitroalkanes to ethyl 4,4,4-trifluorocrotonate and ethyl 4,4-difluorocrotonate [15,16]. Reactions were carried out at room temperature, and the yields were generally comparable to those obtained in batch reactions. The following example demonstrates trifluoromethylation of benzaldehyde using trifluoromethyl(trimethyl)silane in the presence of TBAF (Scheme 4.11) [15].

4.1.3 Condensation Reactions

Haswell and coworkers carried out the Wittig reaction of 2-nitrobenzyl triphenylphosphonium bromide with methyl 4-formylbenzoate in a microflow system [17,18]. They used a borosilicate glass microreactor with T-shaped channels (width = 200 μm and depth = 100 μm), and the reagents were added via EOF by applying a constant

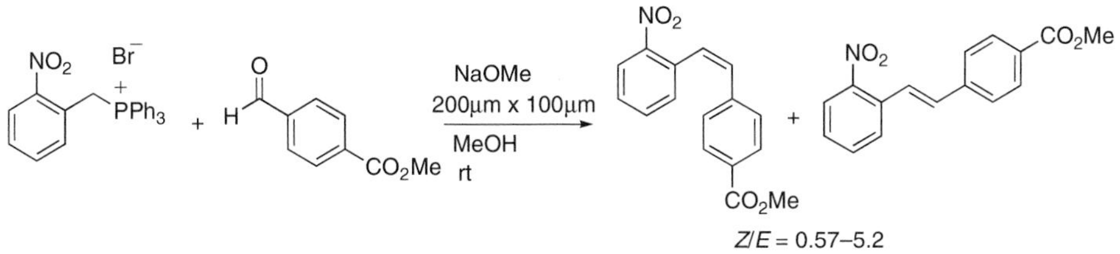

Scheme 4.12

and controlled voltage. The Z/E ratio varied between 0.57 and 5.2 in the microflow system (Scheme 4.12), whereas the batch reaction using varying reactant ratios (phosphonium bromide : aldehyde $= 2:1–1:10$) gave Z/E ratios in the range of 2.8–3.0 [18]. Haswell *et al.* also reported enamine [19] and ester synthesis [20] using a similar microreaction system.

Applications of microreactors to the synthesis of *N*-heterocyclic compounds were also investigated. Garcia-Egido and coworkers reported on the Hantzsch synthesis of 2-aminothiazoles using a microflow system (width $= 300\,\mu$m and depth $= 115\,\mu$m) under EOF-driven flow using *N*-methyl-2-pyrrolidinone (NMP) as the solvent [21]. The reaction temperature was kept at 70 °C using a Peltier heater. Fanetizole, a pharmacological agent with an activity for the treatment of rheumatoid arthritis, was prepared (Scheme 4.13).

Condensation of 1,3-diketones with hydrazines or hydroxyamines was conducted in a microflow system to give pyrazoles and isoxazoles in good yields [22]. High-throughput synthesis of pyrrole by the Paal–Knorr condensation of ethanolamine and acetonylacetone was achieved using the CPC CYTOS Lab System [23]. The running of the system for 165 min resulted in 714 g of the pyrrole (Scheme 4.14).

Fernandez-Suarez and coworkers investigated a domino reaction using a glass microreactor with a channel 74 μm in width [24]. The reaction of citronellal with

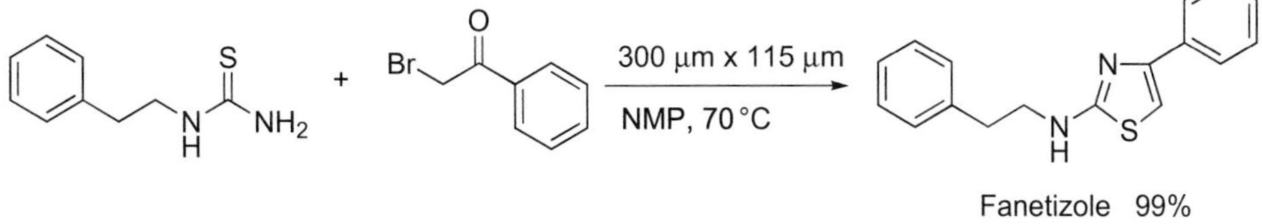

Fanetizole 99%

Scheme 4.13

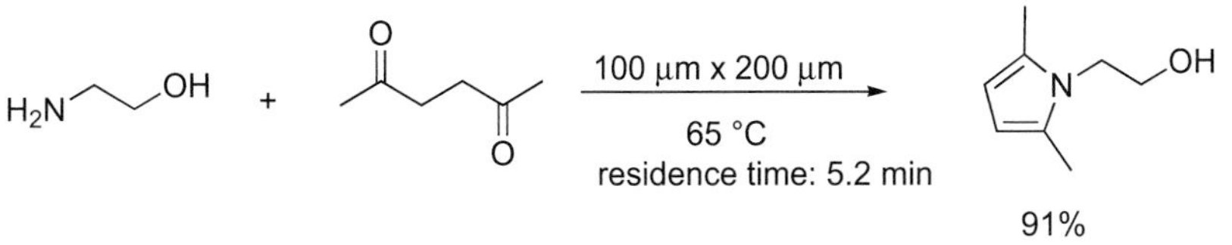

91%

Scheme 4.14

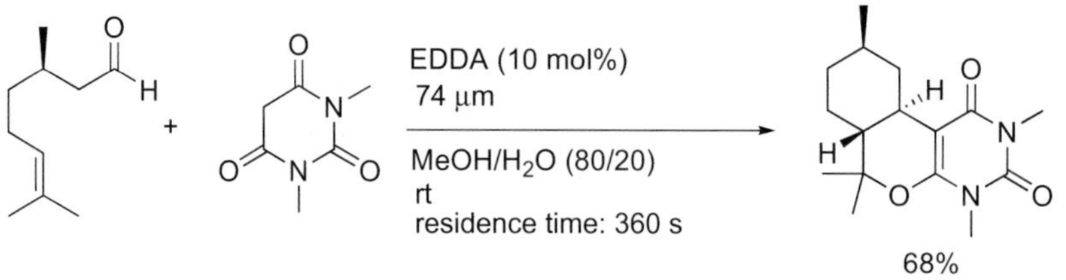

Scheme 4.15

1,3-dimethylbarbituric acid, in the presence of ethylenediamine acetate (EDDA) and with a residence time of 360 s, gave the tricyclic compound at 68% yield (Scheme 4.15). Knoevenagel condensation and intramolecular Diels–Alder reaction took place subsequently, and the synthesis of three different cycloadducts in a parallel way was also demonstrated.

Microwave irradiation has been proven useful in accelerating chemical reactions. A unique approach to multicomponent reactions – the combination of microwave irradiation and microreactors – was developed by Organ and Bremner [25]. The three-component coupling reaction of amino pyrazole with an aldehyde and diketone in a glass capillary tube microflow system (1180 µm i.d.) under microwave irradiation (170 W) proceeded smoothly to give the desired quinolinone in high yield (Scheme 4.16). Without microwave irradiation, the reaction efficiency was very low.

Microreaction technology has also been applied to peptide synthesis. Haswell and coworkers demonstrated that, using a borosilicate glass microreactor, the desired

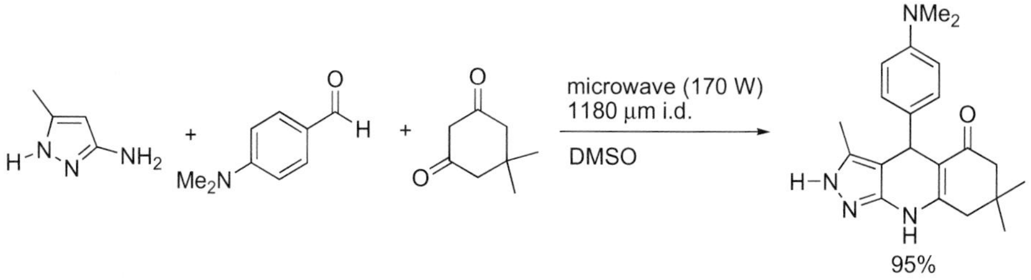

Scheme 4.16

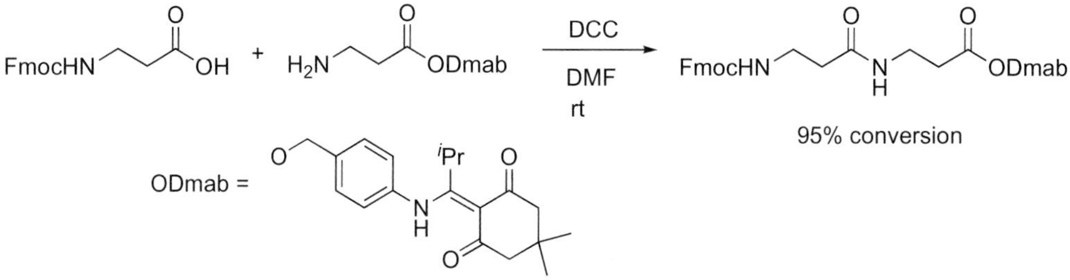

Scheme 4.17

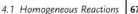

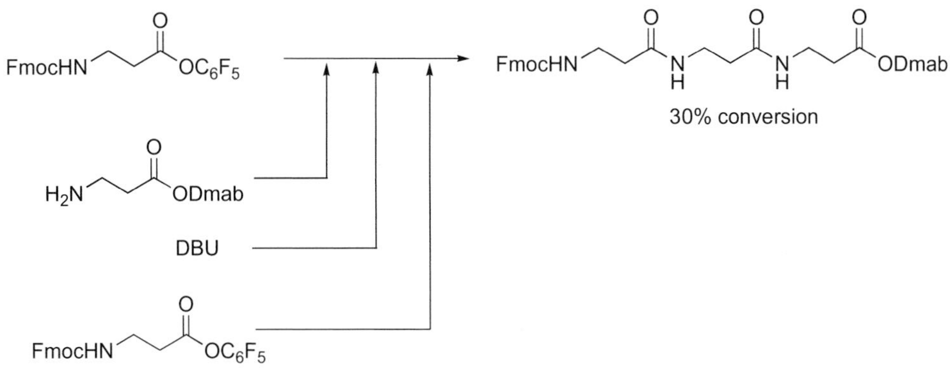

Scheme 4.18

dipetide was obtained by the reaction of *N*-Fmoc-β-alanine with β-alanine Dmab ester in the presence of DCC (1,3-cyclohexylcarbodiimide) (Scheme 4.17) [26,27]. It was also demonstrated that the Fmoc group could be removed by DBU (1,8-diazabicyclo [5.4.0]undec-7-ene), using a microreactor, to yield the free amine. A multistep synthesis of a tripeptide was also carried out using a microflow system, though the overall conversion was rather modest (Scheme 4.18) [27]. Using a microreactor, the degree of racemization for the α-peptide synthesis was slightly less than that observed in bulk reactions [28].

Seeberger and coworkers prepared synthetically useful amounts of β-peptides (0.2–0.6 mmol) by using a microreactor (reactor volume = 78.3 µl). The reaction of acid fluoride and the TFA salt of amino acid benzyl ester in the presence of *N*-methylmorpholine (NMM) at 90 °C (3 min residence time) gave the dipeptide in 92% yield (Scheme 4.19). A fluorous tag method was used for an efficient synthesis of tetrapeptides. Amino acid esters having fluorous tags were used to facilitate purification by fluorous solid-phase extraction (FSPE) (Scheme 4.20).

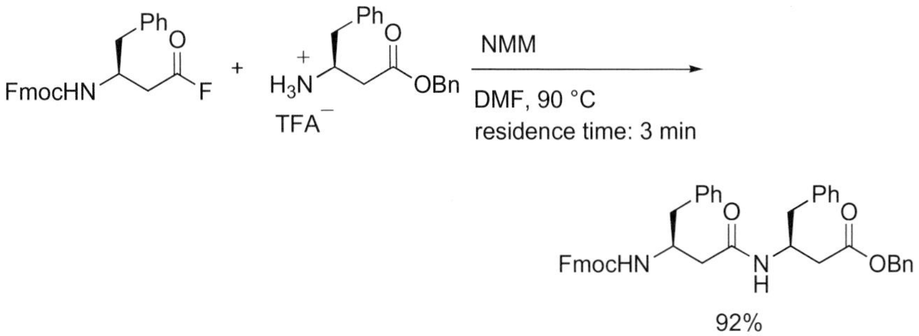

Scheme 4.19

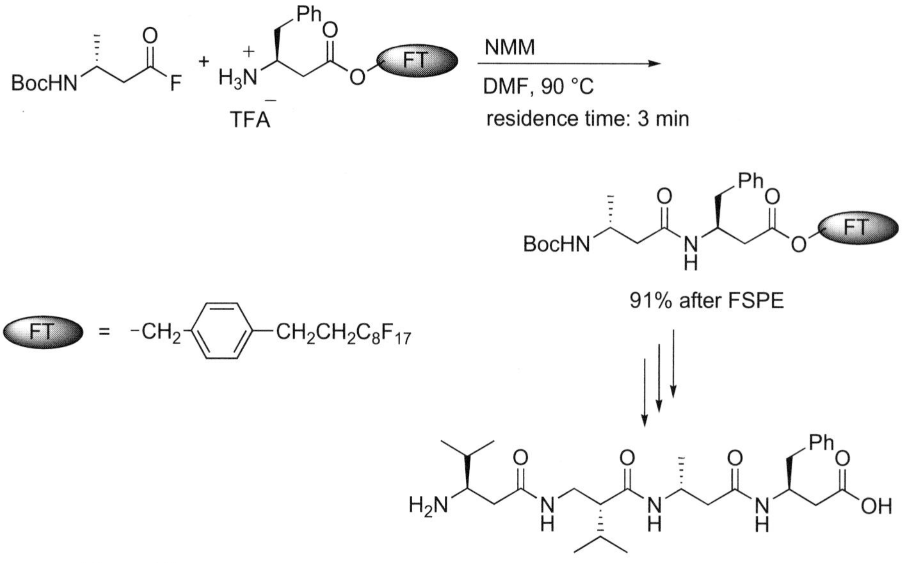

Scheme 4.20

4.1.4
Metal-Catalyzed Reactions

Although heterogeneous catalysis capitalizes on the high volume-to-surface ratio ensured by microchannels, the potential of microreaction technology in homogeneous catalysis is now the focus of many researchers.

The Sonogashira coupling reaction of aryl or vinyl halides with terminal alkynes is typically carried out in the presence of a Pd catalyst and a Cu cocatalyst. In 2002, Ryu and coworkers reported that the Sonogashira coupling reaction using ionic liquid [bmim]PF$_6$ (1-butyl-3-methylimidazolium hexafluorophosphate) proceeded smoothly without a Cu cocatalyst. The Cu-free Sonogashira coupling reaction using an IMM interdigitated microreactor with channels 40 μm channel wide was also examined [30]. When a solution of [bmim]PF$_6$ containing a Pd N-heterocyclic carbene complex and a mixture of iodobenzene, phenylacetylene and dibutylamine was mixed using a IMM micromixer at 110 °C with 10 min residence time, the coupling product was obtained in 93% yield (Scheme 4.21). After successive biphasic treatment with

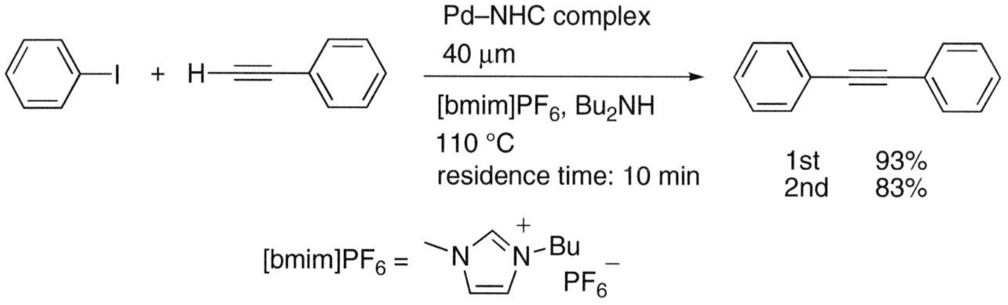

Scheme 4.21

hexane and water, the ionic liquid solution containing Pd catalyst could be reused leading to 83% yield of diphenylacetylene in the second reaction. A microflow system requiring lower CO pressures than that required in a batch system was used for the carbonylative Sonogashira coupling reaction yielding acetylenic ketones [31].

The catalyst recycling system was further developed by combining microreaction and microextraction using the Mizoroki–Heck reaction as a model reaction in which the integration of all the basic steps, that is: reaction, separation of the product from the ionic liquid phase and recycling of the ionic liquid containing the Pd catalyst, could be realized in a completely continuous fashion [32]. A 'bench-top' continuous production system was constructed using a continuous-microflow reactor, a CPC CYTOS Lab System and an originally developed workup protocol based on a dual microextraction system using T-shaped micromixers (300 μm i.d.). Low-viscosity ionic liquid, [bmim]NTf$_2$, ensured a smooth flow of the catalyst solution. After extraction using double T-shaped mixers, the ionic liquid containing the Pd catalyst was continuously recycled using a pump. After running the system for 11.5 h, during which time 144.8 g (0.71 mol) of iodobenzene was consumed, 115.3 g of *trans*-butyl cinnamate was obtained (80% yield, 10 g/h) (Scheme 4.22). The ionic liquid with the Pd catalyst was recycled about five times during this catalytic reaction.

Lee, Valiyaveettil and coworkers reported the use of a glass capillary microreactor (400 μm i.d.) for the Suzuki–Miyaura coupling reaction, catalyzed by Pd nanoparticles [33]. Organ and Comer reported the microwave-assisted Suzuki–Miyaura coupling reaction in a microflow system [34]. The continuous-flow design consisted of a stainless steel holding/mixing chamber with three inlet ports connected to a simple glass capillary tube (1150 μm i.d.) located in the irradiation chamber. The

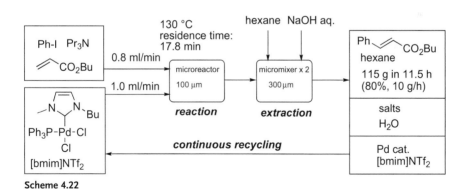

Scheme 4.22

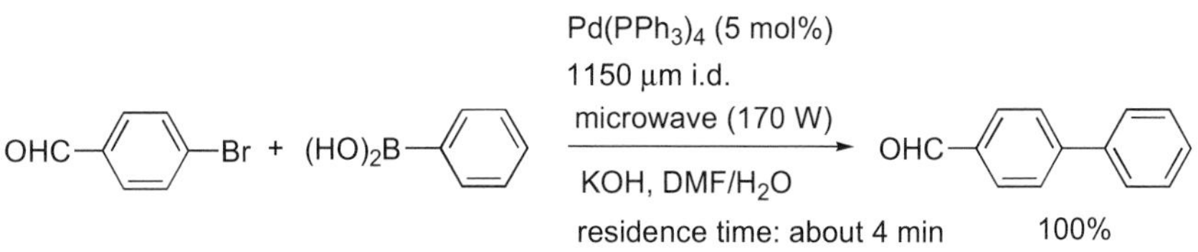

Scheme 4.23

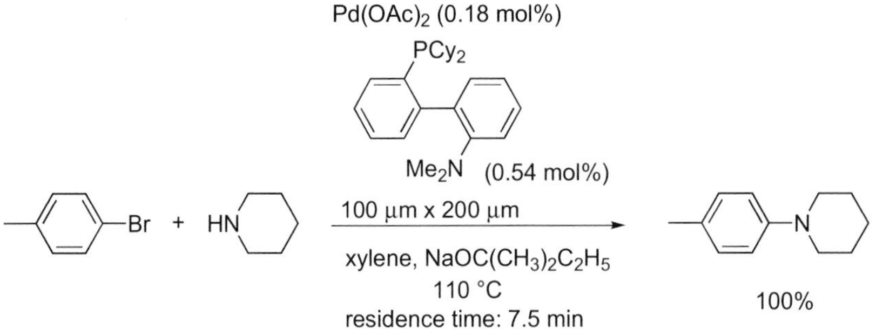

Pd(OAc)$_2$ (0.18 mol%)

PCy$_2$

Me$_2$N   (0.54 mol%)

100 µm x 200 µm

xylene, NaOC(CH$_3$)$_2$C$_2$H$_5$
110 °C
residence time: 7.5 min

100%

Scheme 4.24

reaction of *p*-bromobenzaldehyde with phenyl boronic acid using Pd(PPh$_3$)$_4$ as the catalyst and KOH as the base gave quantitative yields of the coupling product with about 4 min residence time (Scheme 4.23). When the reaction was carried out using Pd(OAc)$_2$, the Pd catalyst decomposed and coated the capillary wall with a thin metal film, which also catalyzed the coupling reaction. This microwave-assisted microflow system has also been applied to ruthenium-catalyzed ring-closing metathesis.

Palladium-catalyzed amination of aromatic halides, developed independently by Buchwald and Hartwig, is a useful tool for the construction of C–N bonds. Caravieilhes and coworkers reported using a microreactor for aromatic amination [35]. With the use of CPC CYTOS Lab System, the reaction of *p*-bromotoluene with piperidine gave the desired coupling product in quantitative yields with a residence time of 7.5 min (Scheme 4.24).

Suresh, Lee and coworkers demonstrated oxidation of cyclohexene catalyzed by Mn or Cu complexes using H$_2$O$_2$ in aqueous phase in a microreactor (width = 200 µm and depth = 50 µm) [36]. Water-soluble ionic liquid [bmim]BF$_4$ was added (0.5%, v/v) to improve the solubility of cyclohexene in the reaction buffer. With the use of a reduced Schiff base–Cu complex, 2-hydroxycyclohexanone was obtained as the major product with 25 min residence time, whereas the bulk scale reaction gave 2-cyclohexenol as the major reaction product.

4.1.5
Photochemical Reactions

Photochemical organic transformations are considered to be attractive synthetic routes. However, when scaling up photochemical reactions, a number of factors ought to be considered, such as scalability of the light sources, heat and mass transfer and safety concerns. Photomicroreactors have the potential to overcome the difficulties associated with conventional photochemical reactors. The most intriguing features of a photomicroreactor are as follows (i) very short optical path lengths that allow extensive and homogeneous illumination of the reaction mixture, (ii) avoidance/minimization of undesirable side reactions or follow-up reactions by adjusting the residence time and (iii) an effective use of light energy saving energy.

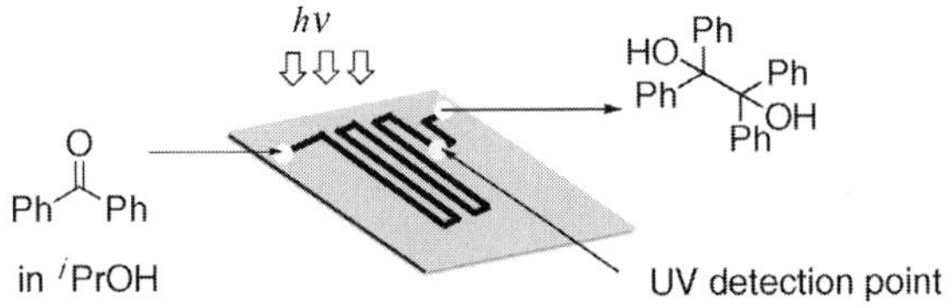

Scheme 4.25

In their pioneering work, Jensen *et al.* demonstrated that photochemical transformation can be carried out in a microfabricated reactor [37]. The photomicroreactor had a single serpentine-shaped microchannel (having a width of 500 μm and a depth of 250 or 500 μm, and etched on a silicon chip) covered by a transparent window (Pyrex or quartz) (Scheme 4.25). A miniature UV light source and an online UV analysis probe were integrated to the device. Jensen *et al.* studied the radical photopinacolization of benzophenone in isopropanol. Substantial conversion of benzophenone was observed for a 0.5 M benzophenone solution in this microflow system. Such a high concentration of benzophenone would present a challenge in macroscale reactors. This microreaction device provided an opportunity for fast process optimization by online analysis of the reaction mixture.

Jähnisch *et al.* used an IMM falling-film microreactor for photochlorination of toluene-2,4-diisocyanate [38] (see also Chapter 4.4.3.3, page 161). As a result of efficient mass transfer and photon penetration, chlorine radicals were well distributed throughout the entire film volume, improving selectivity (side chain versus aromatic ring chlorination by radical versus electrophilic mechanism) and space–time-based yields of 1-chloromethyl-2,4-diisocyanatobenzene compared to those obtained using a conventional batch reactor.

Fukuyama, Ryu and coworkers reported intermolecular [2 + 2]-type cycloaddition of various cyclohexenone derivatives and alkenes using a microreactor made entirely of glass, which was supplied by Mikroglas (Scheme 4.26) [39]. The device was equipped with a heat exchanger channel system through which water flowed to maintain isothermal reaction conditions. The remarkable photochemical efficiency of this device was manifested in rapid cycloaddition of vinyl acetate to cyclohex-2-enone. With this device, the desired product was obtained in 88% yield after 2 h, whereas the same reaction carried out in a Pyrex flask was very sluggish (only 8%

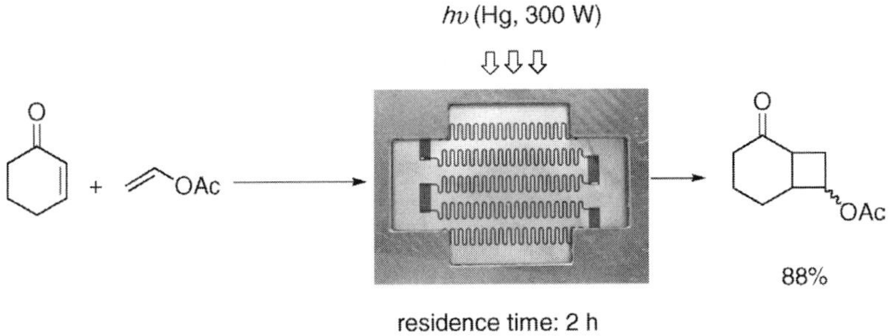

Scheme 4.26

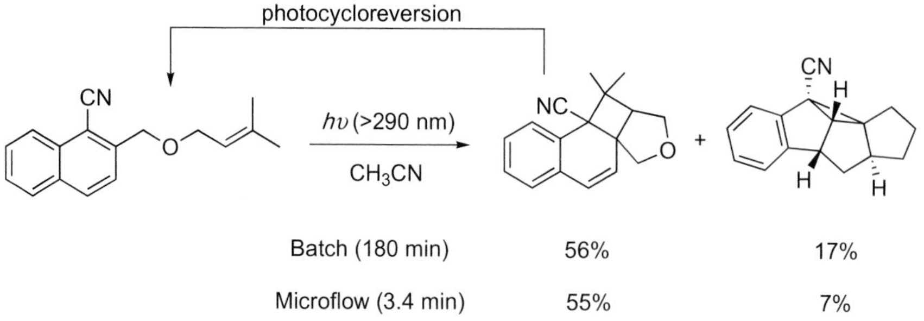

Scheme 4.27

yield after 2 h). The remarkable difference in performance was attributed to the short optical path length (500 μm) of the microreactor.

Mizuno et al. demonstrated an intramolecular version of [2 + 2] photocycloaddition using a microreactor made of PDMS [poly(dimethoxysilane)] (channel dimensions: 300 μm wide, 50 μm deep and 45 or 202 mm long) [40]. Because one of the products photochemically reverts to the starting material, while the other does not, a much shorter residence time, that is, 3.4 min (batch reaction time = 3 h), inside the microchannel reduces the possibility of the reverse reaction. The difference in residence times explains the slight difference in regioselectivity between the microflow and batch systems (Scheme 4.27).

The application of photochemically generated singlet oxygen is a powerful synthetic method in organic chemistry. However, the explosive nature of oxygen-rich organic solvents as well as of the products (e.g. endoperoxides) may pose serious hazards. Considering that microflow systems require only small volumes of oxygenated organic solvent, the hazard potential can be substantially reduced. These factors prompted a number of research groups to use microreactors in singlet oxygen chemistry.

De Mello et al. reported the use of microreactor having serpentine microchannels etched on a glass chip for the addition of singlet oxygen to α-terpine, yielding ascaridole (Scheme 4.28) [41]. A very shallow effective optical path length (50 μm) allowed the use of a low-intensity light source (20 W tungsten lamp) as well as a high concentration of sensitizer (5 mM, Rose Bengal). Both features were advantageous in the context of reduced sample heating and efficient photosensitization.

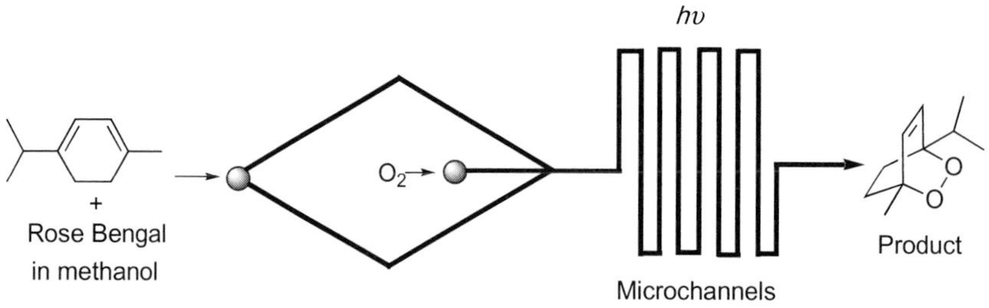

Scheme 4.28

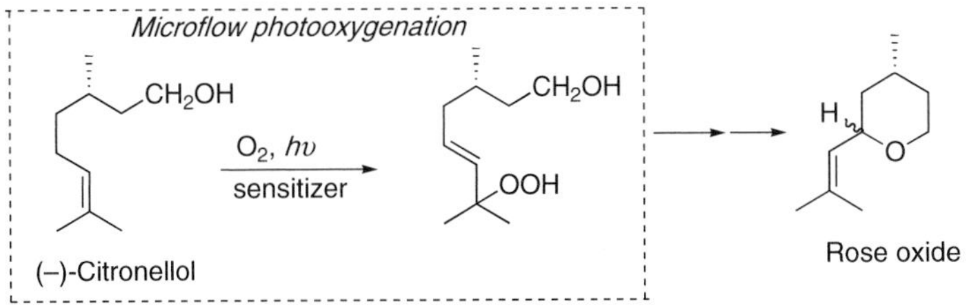

Scheme 4.29

Jähnisch and Dingerdissen demonstrated that a falling-film microreactor (IMM, reaction plate with 32 parallel microchannels, 66 mm long, 600 μm wide and 300 μm deep) is also viable for the generation of singlet oxygen [42]. They reported a [4 + 2]-type cycloaddition of singlet oxygen to cyclopentadiene. The explosive endoperoxide intermediate was immediately converted to a pharmaceutically important compound, 2-cyclopenten-1,4-diol, by reduction with thiourea.

Meyer et al. described photosensitized oxidation of citronellol in an HT-residence glass microreactor having meandering, semicircular channels 1 mm wide [43]. They found that (S)-$(-)$-$β$-citronellol, in the presence of Ru(^{t}bpy)$_3$Cl$_2$ photosensitizer, was oxidized by singlet oxygen via an ene reaction to give (E)-7-7-hydroperoxy-3,7-dimethyloct-5-en-2-ol along with by-products (Scheme 4.29). The obtained hydroperoxide was a precursor to $(-)$-rose oxide, an elemental substance in the perfume industry. In their system, an LED array was used as the light source. Their system showed a greater photonic efficiency (4.8×10^{-2}) than the Schlenk reactor (2.2×10^{-2}).

Photocatalytic reduction using a TiO$_2$-coated microchannel device was reported by Ichimura et al. [44]. By using a quartz microreactor (microchannel, 500 μm wide, 100 μm deep and 40 mm long) and a 365-nm UV-LED light source, benzaldehyde was reduced to benzyl alcohol (yield of 11%) and p-nitrotoluene to p-toluidine (yield of 46%) after 1 min in the presence of ethanol (Scheme 4.30).

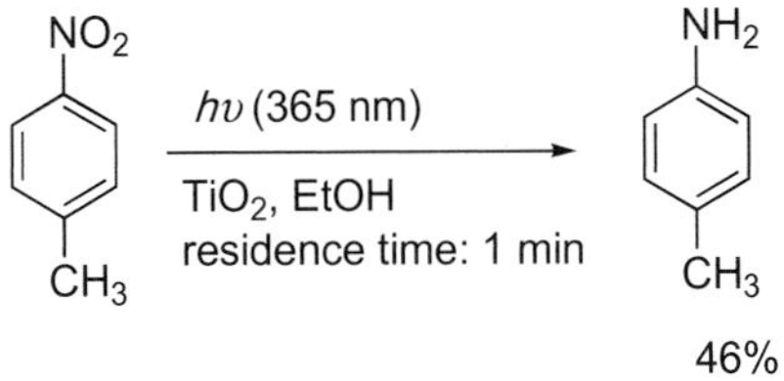

Scheme 4.30

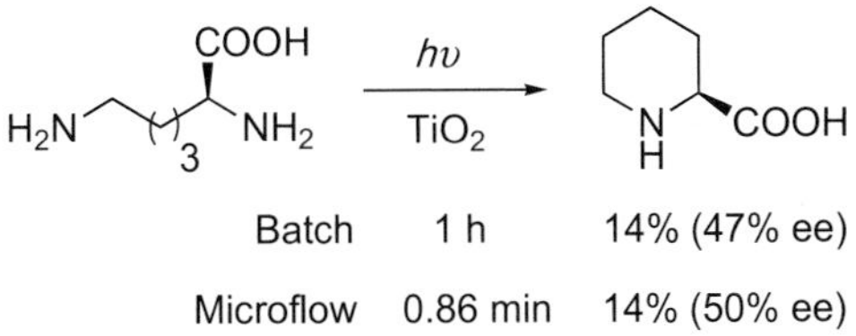

Scheme 4.31

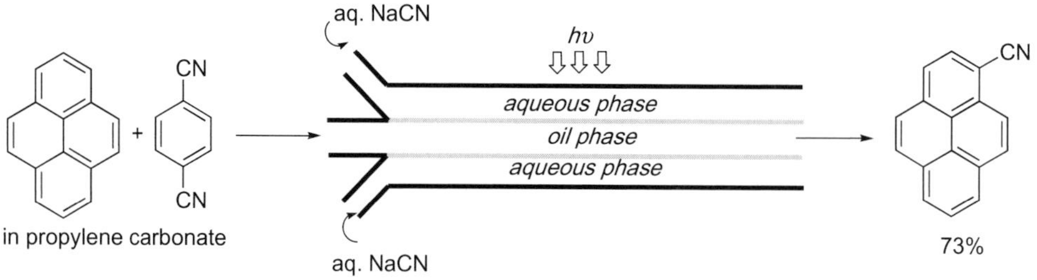

Scheme 4.32

Kitamori and coworkers reported the use of a TiO_2-modified microchannel chip reactor (TMC, Pyrex glass chip, having branched channels 770 μm wide and 3.5 μm deep) for photocatalytic redox-combined synthesis of L-pipecolinic acid from L-lysine (Scheme 4.31) [45]. Although both batch and microflow systems gave comparable yield and enantiomeric excess of the product, the conversion rate was significantly higher for the microflow reactor than for the batch system.

Kitamura and coworkers demonstrated the photocyanation of pyrene by manipulating stable organic/aqueous (oil/water) laminar flow inside the microchannel (polystyrol microchannel chip with a channel, 100 μm wide, 20 μm deep and 350 mm long) [46]. This two-layer oil/water system gave only 28% of the desired cyanated pyrene in 210 s under irradiation by a 300 W high-pressure Hg lamp (Scheme 4.32). However, the yield was improved (73%) by using a water/oil/water three-layer flow system. The 2.5-fold increase in yield was attributed to the greater surface area-to-volume ratio in the three-layer system.

Ryu and coworkers reported that the Barton reaction of a steroidal substrate to a key intermediate in the synthesis of an endothelin receptor antagonist was successfully carried out in a continuous-microflow system [47]. A Pyrex-covered stainless steel microreactor of Dainippon Screen Mfg. Co., Ltd. having a serpentine single channel (1000 μm wide, 107 μm deep and 2.2 m long) was used (Scheme 4.33). A highly energy-efficient black light (15 W) was used as the light source. By serially connecting two microreactors with larger holding volumes and continuously running the system, a multigram-scale production was realized. The increasing of the number of microreactors was the first approach to scale up photochemical reactions.

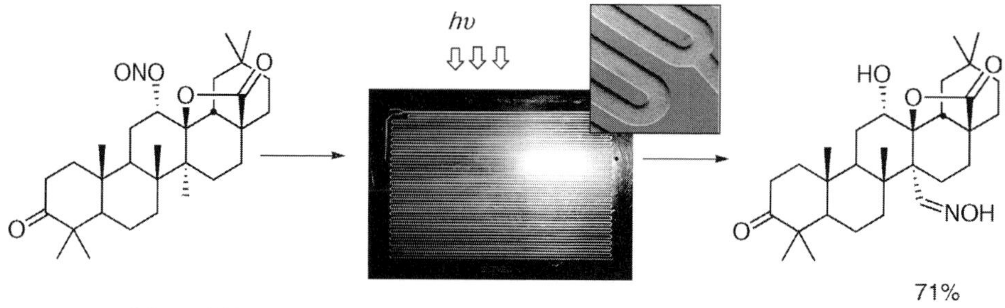

Scheme 4.33

4.1.6
Electrochemical Reactions

Electrolysis offers an alternative route for organic synthesis via the formation of anion and cation radical intermediates. However, traditional electrolytic methods suffer from a number of limitations such as heterogeneity of the electric field, thermal loss due to heating and obligatory use of supporting electrolytes. These factors either hamper electrosynthetic efficiency or make the separation process cumbersome. The combination of electrosynthesis and microreaction technology effectively overcomes these difficulties.

Löwe and Ehrfeld devised a microelectrochemical cell for methoxylation of 4-methoxytoluene, albeit in the presence of KF as the supporting electrolyte [48]. The device consisted of a 75-μm thick polyimide foil containing microstructured slits (250 μm wide and 45 mm long) sandwiched between the working and counter electrodes (Scheme 4.34). The reaction took place inside the microchannels that were in contact with both electrodes. To maintain a constant reaction temperature, a heat exchanger block was also mounted above the working electrode. This micro-reactor-based electrolysis afforded 98% product selectivity, which was higher than that for common industrial processes (about 85%).

Paddon *et al.* reported that a simple thin-layer flow cell having closely spaced electrodes (50 μm) does not require added electrolytes (Scheme 4.35) [49]. Such close proximity of the electrodes allows the two diffusion layers of the electrodes to overlap or 'couple'. Hence, ions generated at one electrode may act as the charge carrier, obviating the need to add electrolytes. A 'self-supported' two-electron/two-proton reduction of tetra(ethoxycarbonyl)ethylene to tetra(ethoxycarbonyl)ethane in a micro-flow system was demonstrated as a model reaction.

Haswell *et al.* demonstrated self-supported microflow cathodic coupling of activated olefins with benzyl bromide (Scheme 4.36) [50]. A DMF solution of the

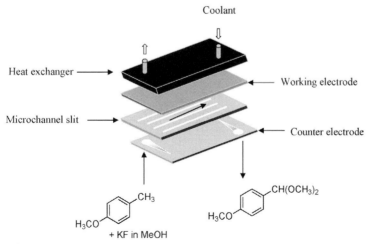

Scheme 4.34

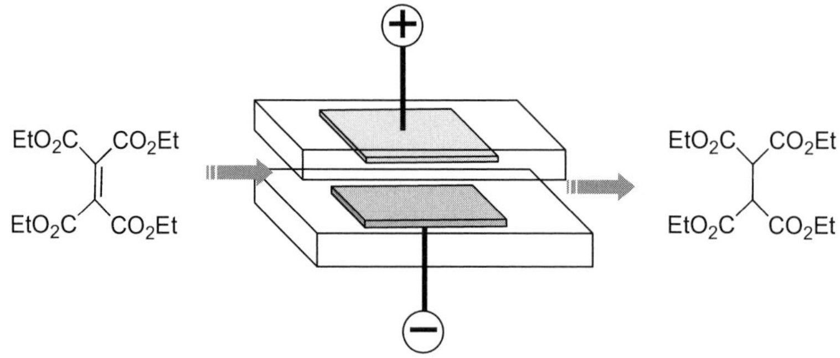

Scheme 4.35

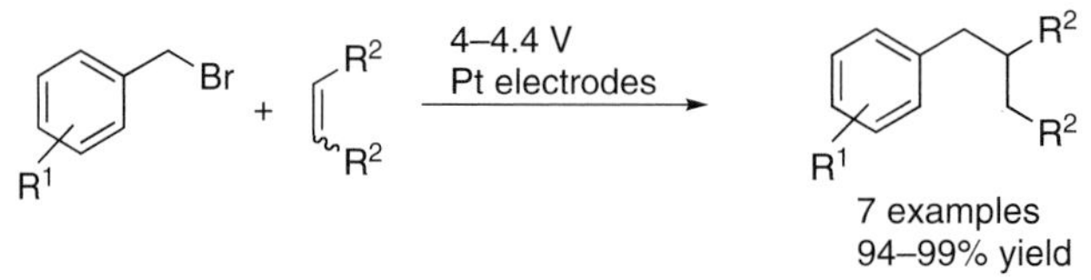

7 examples
94–99% yield

Scheme 4.36

reactants was allowed to flow through two platinum electrodes with an interelectrode gap of either 160 or 320 μm. A delicate balance of interelectrode gap, current intensity and flow rate was crucial for the minimization of unwanted side products and optimization of yields. Using a similar type of thin-layer flow cell (Pt electrodes, interelectrode gap of 320 μm), Haswell and coworkers performed highly efficient and self-supported reductive homodimerization of 4-nitrobenzyl bromide (Scheme 4.37) [51]. Another report from this group described a simple process for the self-supported synthesis of phenyl-2-propanone and its derivatives on the basis of a one-step electrochemical acylation reaction [52]. Microflow electroreductive coupling of benzyl bromide derivatives and acetic anhydride in a microgap cell (Pt electrodes, interelectrode gap of 160 μm and 7.2 μl cell volume) gave the products in only 43 s residence time (Scheme 4.38). However, a fairly large excess of acetic anhydride was required to suppress the side reactions. They scaled up the process by connecting four identical cells in parallel, increasing the throughput by fourfold.

Girault *et al.* developed a ceramic electrochemical microreactor (CEM) in which an array of platinum interdigitated band electrodes (gap between electrodes: 500 μm) was screen printed on the ceramic surface (Scheme 4.39) [53]. A methanolic solution

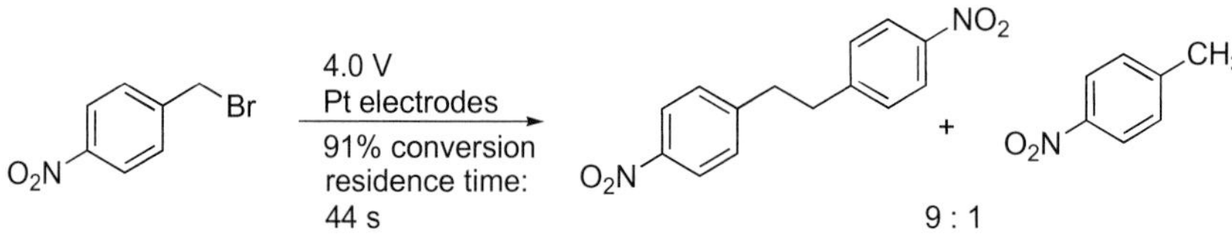

4.0 V
Pt electrodes
91% conversion
residence time:
44 s

9 : 1

Scheme 4.37

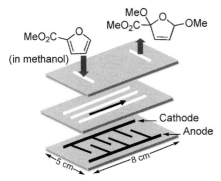

Scheme 4.38

Scheme 4.39

of methyl-2-furoate flowed through a network of microchannels, which were in contact with the electrode array. The CEM was directly connected to a mass spectrometer for online monitoring of the methoxylation process.

Küpper *et al.* carried out a methoxylation reaction of 4-methoxytoluene in an electrochemical microreactor in which a glass carbon anode and a stainless steel cathode were separated by a microchannel foil 25 μm thick [54]. The chemical resistance of the microchannel foils was very important because of the evolution of hydrogen and oxygen gases and the strong pH shifts during electrolysis. PEEK was found to be the most robust material. They also observed that selectivity of the oxidation of 4-methoxytoluene in acidified methanolic solution (pH 1, sulfuric acid) was influenced by the current density and flow rate.

Yoshida *et al.* developed a novel self-supported electrochemical microflow reactor consisting of porous carbon fiber anodes and cathodes, which were separated by a hydrophobic PTEF membrane 75 μm thick (Scheme 4.40) [55]. Owing to the porous nature of the electrodes, the effective electrode surface area was greater than that of solid electrode plates. After feeding the substrate solution (at a flow rate of 2 ml/h) into the anodic chamber, protons and acetals were formed under the applied potential and were immediately transferred to the cathodic chamber through the porous membrane. At the cathode, protons combine with electrons and cause evolution of hydrogen gas. The diffusion of protons from the anode to the cathode through the porous membrane maintains the required conductivity of the cell, thereby rendering the process self-supporting.

PTFE
membrane
anode
cathode
CH_3
OMe
2 x MeOH
OMe
H^+
product
$2 H_2$
4 x e
75 μm
4 x e

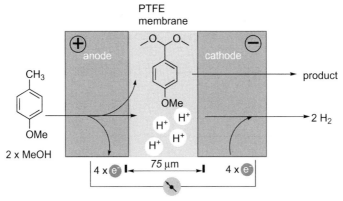

Scheme 4.40

Fuchigami, Marken and coworkers also reported self-supported anodic methoxylation and acetoxylation of several aromatic compounds using a simple thin-layer flow cell reactor (interelectrode gap of 80 μm) (Scheme 4.41) [56]. The current efficiency (CE) of this process was 10% at best because of oxidation of methanol at flow rates lower than 0.03 ml/min. Even though CE increased at a faster flow rate (0.5 ml/min), the yield decreased sharply. The importance of selecting an appropriate choice of electrode material also was noted.

Yoshida *et al.* reported that generation and online detection of highly reactive carbocations from carbamates were accomplished by integrating an electrochemical microreactor with an FTIR spectrometer [57]. They also demonstrated that both the carbocations and nucleophiles could be generated using the paired electrochemical flow system to give the coupling products in reasonable yields (Scheme 4.42) [58].

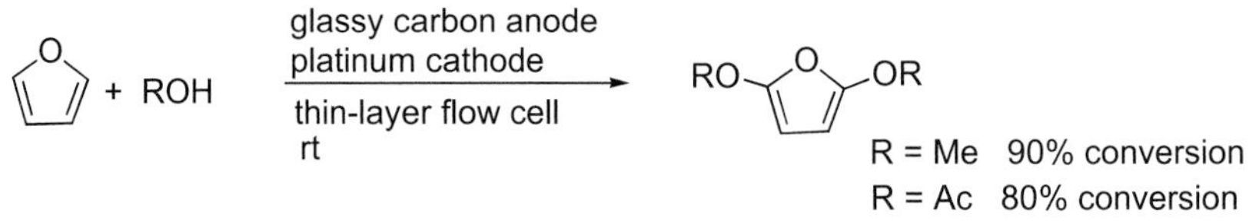

+ ROH
glassy carbon anode
platinum cathode
thin-layer flow cell
rt
RO
OR
R = Me 90% conversion
R = Ac 80% conversion

Scheme 4.41

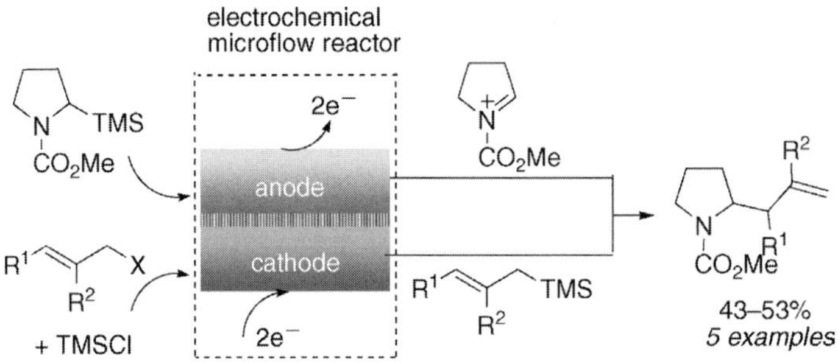

electrochemical
microflow reactor
TMS
CO_2Me
$2e^-$
anode
cathode
$2e^-$
R^1
X
R^2
+ TMSCl
CO_2Me
R^1
TMS
R^2
R^2
R^1
CO_2Me
43–53%
5 examples

Scheme 4.42

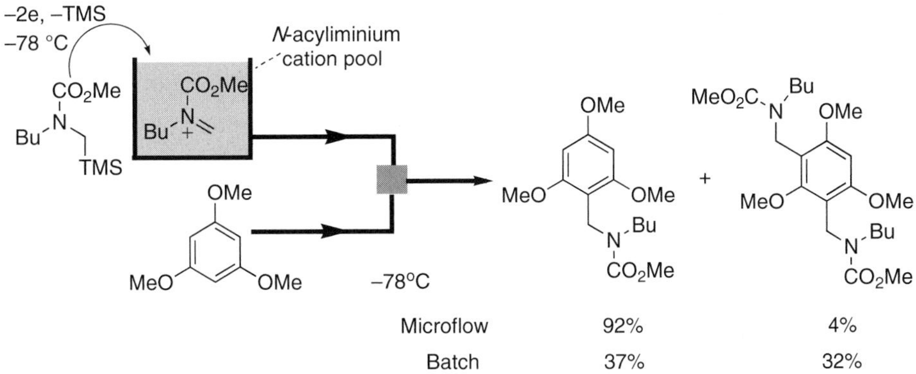

Scheme 4.43

Friedel–Crafts alkylation of aromatic compounds with electrochemically generated carbocations is typically faster than mixing. Inefficient mixing by conventional mechanical stirring methods causes local deviations in reagent concentrations, resulting in poor selectivity. Yoshida and coworkers demonstrated that when N-acyliminium cations (electrochemically generated and accumulated as a cation pool) were allowed to mix with aromatic compounds in a multilamination micromixer, a remarkable selectivity for monoalkylation over dialkylation was observed (Scheme 4.43) [59]. Similarly, this micromixing protocol was extended to [4 + 2] cycloaddition of N-acyliminium ion and dienophiles, yielding the cycloadducts in higher amounts than the batch methods [60]. They also showed that micromixing of electrochemically generated I^+ with electron-rich benzene derivatives gave monoiodinated products in a highly selective manner (Scheme 4.44) [61].

Yoshida and coworkers also developed a microreaction system for cation pool-initiated polymerization [62]. Significant control of the molecular weight distribution (M_w/M_n) was achieved when N-acyliminium ion-initiated polymerization of butyl vinyl ether was carried out in a microflow system (an IMM micromixer and a microtube reactor). Initiator and monomer were mixed using a micromixer, which was connected to a microtube reactor for the propagation step. The polymerization reaction was quenched by an amine in a second micromixer. The tighter molecular weight distribution $(M_w/M_n = 1.14)$ in the microflow system compared with that of the batch system $(M_w/M_n > 2)$ was attributed to the very rapid mixing and precise control of the polymerization temperature in the microflow system.

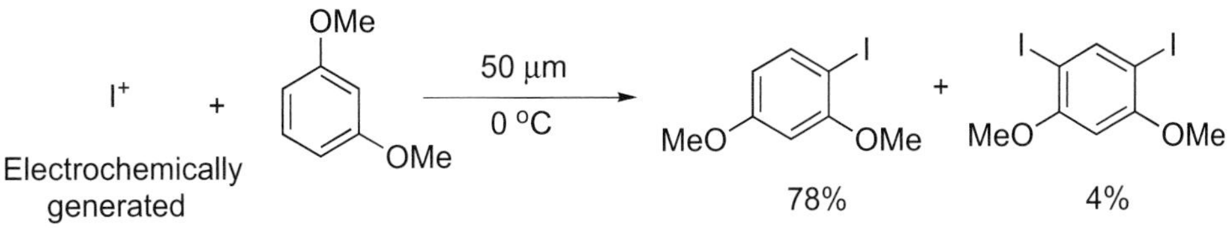

Scheme 4.44

4.1.7
Miscellaneous

4.1.7.1 **Swern Oxidation**
Oxidation of primary and secondary alcohols to aldehydes and ketones with DMSO, oxalyl chloride and a base is known as Swern oxidation. When using oxalyl chloride as the dehydration agent, the reaction must be kept colder than $-60\,°C$ to avoid side reactions such as Pummerer rearrangement. In contrast, when trifluoroacetic anhydride is used instead of oxalyl chloride, the reaction can be warmed to $-30\,°C$ without side reactions. Microflow systems offer a smarter approach for controlling the reaction temperature and allow precise control of the residence time. Yoshida and coworkers reported a microflow Swern oxidation that worked well even at room temperature [63]. By the sequential mixing of DMSO, alcohols and reagents, with the aid of a serially connected network of micromixers (IMM) and microtube reactors, carbonyl products were obtained in good to excellent yields and formation of undesired side products was minimized as well (Scheme 4.45). The reaction was efficient even at $20\,°C$, provided that the residence time was extremely short (2–5 s). An extremely short reaction time ensures immediate transfer of the highly unstable intermediates for the subsequent reaction before their decomposition. In addition, the process withstood extended run times (run time $= 3\,h$) allowing for potential scale-up.

4.1.7.2 **Grignard Exchange Reaction**
Grignard exchange reaction is used for preparing Grignard reagents, which are difficult to prepare from by reacting organic halides and metallic magnesium. Yoshida and coworkers aimed to perform the rapid and highly exothermic Grignard exchange reaction of ethylmagnesium bromide and bromopentafluorobenzene in a microflow system (Scheme 4.46) [64]. Different types of micromixers (T-shaped mixer, IMM multilamination mixer and Toray Hi-mixer) were examined. A 'shell and tube' microheat exchanger, consisting of 55 microtubes (internal diameter $= 490\,\mu m$ and length $= 200\,mm$) bundled together and placed in a shell (internal diameter 16.7 mm and length $= 200\,mm$), was employed for the residence time unit. By

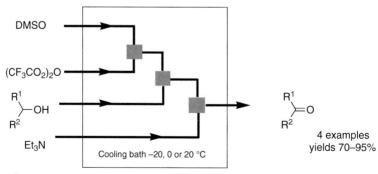

Scheme 4.45

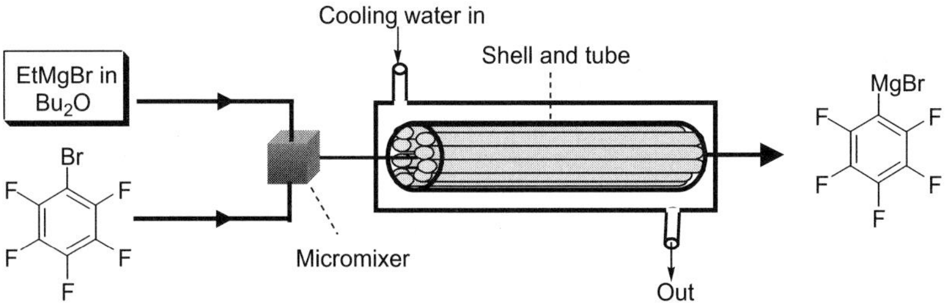

Scheme 4.46

combining a Toray Hi-mixer with the shell and tube heat exchanger, multikilograms of the product were obtained after 24 h of continuous operation.

4.1.7.3 Lithium–Halogen Exchange Reaction

Halogen–lithium exchange reactions are common procedures in organic synthesis. However, large-scale use demands safety measures and strict temperature control. Because efficient control over exothermic reactions was accomplished by means of microreaction technology, Schmalz *et al.* investigated the microflow generation of aryllithium compounds from bromoaromatics using *n*-BuLi and subsequent quenching with fenchone in THF at −17 °C (Scheme 4.47) [65]. This microflow protocol had a high yield (e.g. when X = N, yield was 93%) and required a brief reaction time, for example, 5 min.

4.1.7.4 Phenyl Boronic Acid Synthesis

Typical industrial process for the synthesis of phenyl boronic acid from phenylmagnesium bromide and boronic acid trimethoxy ester requires strict temperature control (−25 to −55 °C) to minimize the formation of side products. Recently, Hessel and coworkers reported that a micromixer (width 40 μm and depth 300 μm)/tubular reactor system gave the phenyl boronic acid at high yield (>80%) even at higher temperatures (22 or 50 °C) with minimum amounts of the side products (Scheme 4.48) [66]. They also achieved a pilot-scale production by employing a caterpillar minimixer (width range 600–1700 μm and depth range 1200–2400 μm).

Microreaction technology has already shown a great deal of promises for homogeneous reactions, be thermal, photochemical or electrochemical. Efficient mixing, precise control of reaction temperature and residence time enable one to manipulate the selectivity issue, tame an ultrafast reaction or even conduct a highly exothermic

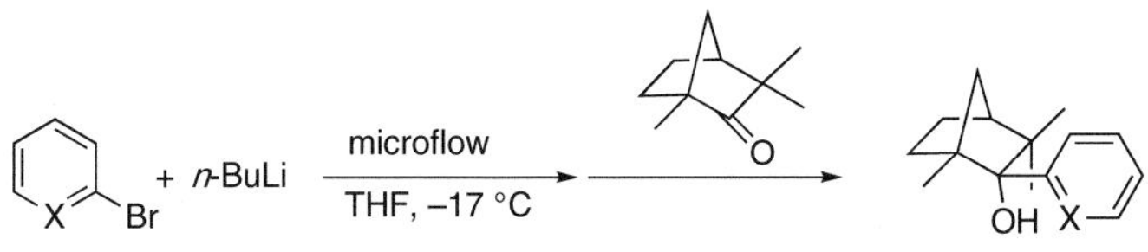

X = N, CH, C-OMe

Scheme 4.47

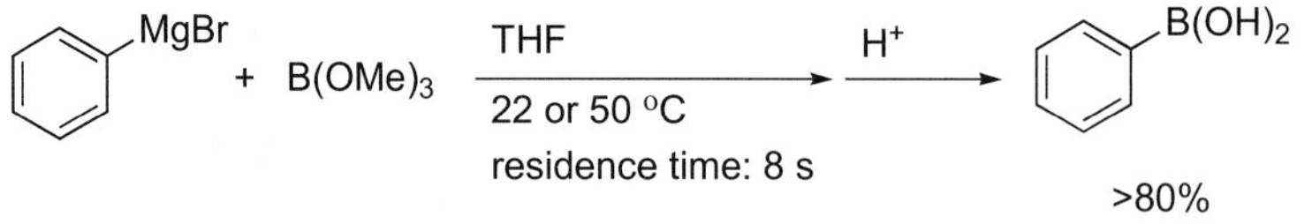

Scheme 4.48

reaction at room temperature. To conclude, doors for many fascinating properties of microreaction technology are now open for organic chemists.

References

1 Panke, G., Schwalbe, T., Stirner, W., Taghavi-Moghadam, S. and Wille, G. (2003) *Synthesis*, 2827.

2 Ducry, L. and Roberge, D.M. (2005) *Angewandte Chemie-International Edition*, **44**, 7972.

3 Tanaka, K., Motomatsu, S., Koyama, K., Tanaka, S.I. and Fukase, K. (2007) *Organic Letters*, **9**, 299.

4 Ratner, D.M., Murphy, E.R., Jhunjhunwala, M., Snyder, D.A., Jensen, K.F. and Seeberger, P.H. (2005) *Chemical Communications*, 578.

5 Nishiyama, H., Kondo, M., Nakamura, T. and Itoh, K. (1991) *Organometallics*, **10**, 500.

6 Jönsson, C., Lundgren, S., Haswell, S.J. and Moberg, C. (2004) *Tetrahedron*, **60**, 10515.

7 Mikami, K., Yamanaka, M., Islam, MdN., Tonoi, T., Itoh, Y., Shinoda, M. and Kudo, K. (2006) *Journal of Fluorine Chemistry*, **127**, 592.

8 Ikushima, Y., Hatakeda, K., Sato, M., Sato, O. and Arai, M. (2002) *Chemical Communications*, 2208.

9 Kuwajima, I. and Nakamura, E. (1985) *Accounts of Chemical Research*, **18**, 181.

10 Wiles, C., Watts, P., Haswell, S.J. and Pombo-Villar, E. (2002) *Chemical Communications*, 1034.

11 Wiles, C., Watts, P., Haswell, S.J. and Pombo-Villar, E. (2005) *Tetrahedron*, **61**, 10757.

12 Löwe, H., Hessel, V., Löb, P. and Hubbard, S. (2006) *Organic Process Research & Development*, **10**, 1144.

13 Wiles, C., Watts, P., Haswell, S.J. and Pombo-Villar, E. (2004) *Lab on a Chip*, **4**, 171.

14 Acke, D.R.J. and Stevens, C.V. (2006) *Organic Process Research & Development*, **10**, 417.

15 Miyake, N. and Kitazume, T. (2003) *Journal of Fluorine Chemistry*, **122**, 243.

16 Kawai, K., Ebata, T. and Kitazume, T. (2005) *Journal of Fluorine Chemistry*, **126**, 956.

17 Skelton, V., Greenway, G.M., Haswell, S.J., Styring, P., Morgan, D.O., Warrington, B. and Wong, S.Y.F. (2001) *Analyst*, **126**, 7.

18 Skelton, V., Greenway, G.M., Haswell, S.J., Styring, P., Morgan, D.O., Warrington, B.H. and Wong, S.Y.F. (2001) *Analyst*, **126**, 11.

19 Sands, M., Haswell, S.J., Kelly, S.M., Skelton, V., Morgan, D.O., Styring, P., Warrington, B. (2001) *Lab on a Chip*, **1**, 64.

20 Wiles, C., Watts, P., Haswell, S.J. and Pombo-Villar, E. (2003) *Tetrahedron*, **59**, 10173.

21 Garcia-Egido, E., Wong, S.Y.F. and Warrington, B.H. (2002) *Lab on a Chip*, **2**, 31.

22 Wiles, C., Watts, P. and Haswell, S.J. (2004) *Organic Process Research & Development*, **8**, 28.

23 Schwalbe, T., Autze, V., Hohmann, M. and Stirner, W. (2004) *Organic Process Research & Development*, **8**, 440.

24 Fernandez-Suarez, M., Wong, S.Y.F. and Warrington, B.H. (2002) *Lab on a Chip*, **2**, 170.

25 Bremner, W.S. and Organ, M.G. (2007) *Journal of Combinatorial Chemistry*, **9**, 14.

26 Watts, P., Wiles, C., Haswell, S.J., Pombo-Viller, E. and Styring, P. (2001) *Chemical Communications*, 990.

27 Watts, P., Wiles, C., Haswell, S.J. and Pombo-Villar, E. (2002) *Tetrahedron*, **58**, 5427.

28 Watts, P., Wiles, C., Haswell, S.J. and Pombo-Villar, E. (2002) *Lab on a Chip*, **2**, 141.

29 Flögel, O., Codée, J.D.C., Seebach, D. and Seeberger, P.H. (2006) *Angewandte Chemie-International Edition*, **45**, 7000.

30 Fukuyama, T., Shinmen, M., Nishitani, S., Sato, M. and Ryu, I. (2002) *Organic Letters*, **4**, 1691.

31 Rahman, M.T., Fukuyama, T., Kamata, N., Sato, M. and Ryu, I. (2006) *Chemical Communications*, 2236.

32 Liu, S., Fukuyama, T., Sato, M. and Ryu, I. (2004) *Organic Process Research & Development*, **8**, 477.

33 Basheer, C., Hussain, F.S.J., Lee, H.K. and Valiyaveettil, S. (2004) *Tetrahedron Letters*, **45**, 7297.

34 Comer, E. and Organ, M.G. (2005) *Journal of the American Chemical Society*, **127**, 8160.

35 Mauger, C., Buisine, O., Caravieilhes, S. and Mignani, G. (2005) *Journal of Organometallic Chemistry*, **690**, 3627.

36 Basheer, C., Vetrichelvan, M., Suresh, V. and Lee, H.K. (2006) *Tetrahedron Letters*, **47**, 957.

37 Lu, H., Schmidt, M.A. and Jensen, K.F. (2001) *Lab on a Chip*, **1**, 22.

38 Ehrich, H., Linke, D., Morgenschweis, K., Baerns, M. and Jähnisch, K. (2002) *Chimia*, **56**, 647.

39 Fukuyama, T., Hino, Y., Kamata, N. and Ryu, I. (2004) *Chemistry Letters*, **33**, 1430.

40 (a) Maeda, H., Mukae, H. and Mizuno, K. (2005) *Chemistry Letters*, **34**, 66. (b) Mukae, H., Maeda, H. and Mizuno, K. (2006) *Angewandte Chemie-International Edition*, **45**, 6558.

41 Wootton, R.C.R., Fortt, R. and de Mello, A.J. (2002) *Organic Process Research & Development*, **60**, 187.

42 Jähnisch, K. and Dingerdissen, U. (2005) *Chemical Engineering & Technology*, **28**, 426.

43 Meyer, S., Tietze, D., Rau, S., Schäfer, B. and Kreisel, G. (2007) *Journal of Photochemistry and Photobiology A: Chemistry*, **186**, 248.

44 Matsushita, Y., Kumada, S., Wakabayashi, K., Sakeda, K., Ichimura, T. (2006) *Chemistry Letters*, **35**, 410.

45 Takei, G., Kitamori, T. and Kim, H.-B. (2005) *Catalysis Communications*, **6**, 357.

46 Ueno, K., Kitagawa, F. and Kitamura, N. (2002) *Lab on a Chip*, **2**, 231.

47 Sugimoto, A., Sumino, Y., Takagi, M., Fukuyama, T. and Ryu, I. (2006) *Tetrahedron Letters*, **47**, 6197.

48 Löwe, H. and Ehrfeld, W. (1999) *Electrochimica Acta*, **44**, 3679.

49 Paddon, C.A., Pritchard, G.J., Thiemann, T. and Marken, F. (2002) *Electrochemical Communications*, **4**, 825.

50 He, P., Watts, P., Marken, F. and Haswell, S.J. (2006) *Angewandte Chemie-International Edition*, **45**, 4146.

51 He, P., Watts, P., Marken, F. and Haswell, S.J. (2005) *Electrochemical Communications*, **7**, 918.

52 He, P., Watts, P., Marken, F. and Haswell, S.J. (2007) *Green Chemistry*, **9**, 20.

53 Mengeaud, V., Bagel, O., Ferrigno, R., Girault, H.G. and Haider, A. (2002) *Lab on a Chip*, **2**, 39.

54 Küpper, M., Hessel, V., Löwe, H., Stark, W., Kinkel, J., Michel, M. and Schmidt-Traub, H. (2003) *Electrochimica Acta*, **48**, 2889.

55 Horcajada, R., Okajima, M., Suga, S. and Yoshida, J. (2005) *Chemical Communications*, 1303.

56 Horii, D., Atobe, M., Fuchigami, T. and Marken, F. (2006) *Journal of the Electrochemical Society*, **153**, D143.

57 Suga, S., Okajima, M., Fujiwara, K. and Yoshida, J. (2001) *Journal of the American Chemical Society*, **123**, 7941.

58 Suga, S., Okajima, M., Fujiwara, K. and Yoshida, J. (2005) *QSAR & Combinatorial Science*, **24**, 728.

59 (a) Suga, S., Nagaki, A. and Yoshida, J. (2003) *Chemical Communications*, 1303. (b) Nagaki, A., Togai, M., Suga, S., Akoi, N.,

Mae, K. and Yoshida, J. (2005) *Journal of the American Chemical Society*, **127**, 11666.

60 Suga, S., Nagaki, A., Tsutsui, Y., Yoshida, J. (2003) *Organic Letters*, **5**, 945.

61 Midorikawa, K., Suga, S. and Yoshida, J. (2006) *Chemical Communications*, 3794.

62 Nagaki, A., Kawamura, K., Suga, S., Ando, T., Sawamoto, M. and Yoshida, J. (2004) *Journal of the American Chemical Society*, **126**, 14702.

63 Kawaguchi, T., Miyata, H., Ataka, K., Mae, K. and Yoshida, J. (2005) *Angewandte Chemie-International Edition*, **44**, 2413.

64 Wakami, H. and Yoshida, J. (2005) *Organic Process Research & Development*, **9**, 787.

65 ElSheikh, S. and Schmalz, H.G. (2004) *Current Opinion in Drug Discovery & Development*, **7**, 882.

66 Hessel, V., Hofmann, C., Löwe, H., Meudt, A., Scheres, S., Schönfeld, F. and Werner, B. (2004) *Organic Process Research & Development*, **8**, 511.

4.2
Heterogeneous Reactions

Ian R. Baxendale, John J. Hayward, Steve Lanners, Steven V. Ley, Christopher D. Smith

4.2.1
Introduction

Modern organic synthesis is undergoing a period of rapid change because of the many demands it is facing today. The need for greater speed in the discovery process requires high-throughput methods and inevitably more advanced automation. However, safety factors, sustainable procedures and costs are all key issues that must be considered. This is especially true if we wish to retain all the required levels of synthetic flexibility necessary to assemble the vast range of products and materials that are needed and at the same time be able to stimulate further innovation and invention.

Until now much of the literature has concentrated on equipment and esoteric single-step applications rather than appreciating the need for more general platforms that deliver multistep syntheses leading to pure products for biological evaluation or the determination of some other fundamental property. For this reason, knowledge of the science of synthesis – which underpins the chemical reactivity and the product outcome – is of fundamental importance. This chapter therefore emphasizes new drivers and tools designed to deliver a quality product from a flow process through the application of immobilized reagents to effect chemical transformations, scavenge impurities from reaction streams, 'catch and release' products and bring about phase switching purification. These reagents can take the form of surface coatings, the use of polymer and silica supports in packed reactor columns or as composite lithographs or similar porous architectures. A special emphasis will be placed on the applications and the quality of the product because these are ultimately the goals of the users rather than the devices themselves, which are simply the servants of synthesis.

The very term microreactor has been grossly misused or even abused because it has been used to encompass everything from a trickle column reactor, T-piece mixer, a simple syringe pump-driven device and circulating systems, to elaborate pumping and separation equipment. More often than not there is an ineffective and inadequate description of the reagent and substrate mixing and little enough to justify the terms 'micro' or 'reactor'. The enormous range of volumes, quantities and purities of products produced in these flow-derived arrangements does not therefore naturally fit with a single definition. From our perspective, a microreactor operates across a range from sub-milligrams to grams without significant alteration of the reactor's engineering. However, any further scaling often requires larger pumps, columns and plumbing and is better suited to the term 'mesoreactor' where quantities ranging from multigrams to kilograms are produced. Anything larger involving continuous-flow reactors for the bulk production of chemicals is the domain of the process chemist and the engineer and often requires dedicated plants.

Although the first reported use of a heterogeneous reagent under flow conditions was in 1932 [67], in the sporadic communications since few have defined the field – because of either their one-step nature or their lack of general acceptance in the multivariant research laboratory. Nevertheless, the situation has changed dramatically in recent years, owing to commercial availability of many immobilized reagents [68–72] and their use in multistep processes [73–80]. This, together with one-pass scavenging techniques and the beginnings of truly workable, modular, off-the-shelf flow research equipment, has begun a step change in the attitudes to high-throughput chemistry concepts [81–86]. This chapter not only delineates some of these principles but also tries to give a wider perspective of the future of organic synthesis, relegating many of the routine tasks and repetitive scale-up events to flow-based machinery. Furthermore, it also provides a new environment for discovery by creating a fresh set of variable synthetic parameters that simultaneously release skilled operator time for in-depth synthesis planning. The future of flow chemistry also hangs on the ability to adapt rapidly to scale, thereby allowing for reproducible production of appropriate quantities and qualities of material in short time frames, from high-throughput screens to pre-process early toxicology testing. Of particular interest will be the concept of 'make and screen' that removes traditional bottlenecks in the discovery process. Indeed, linking the synthesis component to rapid biological evaluation or property determination will become crucial. For this concept to be realized, the delivery of clean materials is essential, and immobilization techniques – in terms of both synthetic reagents and particularly heterogeneous purification agents – will have a central role to play.

4.2.2
Concepts in Flow Mode Synthesis

There are many possible arrangements used for the application of immobilized reagents in flow chemistry; several of the common setups are illustrated in Scheme 4.49. (a) The first and the simplest system involves a linear series of solid-supported reagents within column reactors; by using large columns and running the system for longer time, the reaction can be easily scaled up. As with all these setups, in-line monitoring of the reaction at any stage is possible; this is typically achieved using UV, either in-line or as part of an HPLC system, enabling immediate evaluation and optimization of the reaction conditions. (b) A recirculating setup can be used if the transformation requires a longer reaction time than is practical by simple flow rate changes; the solution is then continuously recirculated

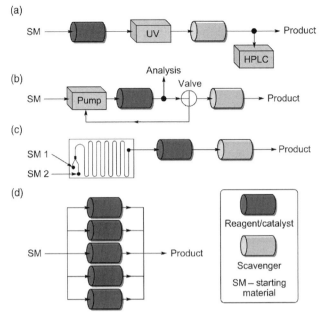

Scheme 4.49 Different concepts for flow processing.

through the reagent column. Moreover, surface-to-substrate interaction should be higher based on an appreciable recirculating flow rate in comparison to a virtually static flow velocity. When the reaction has reached completion, the flow stream can then be directed into the next in-line reagent column. (c) Two or more reagent streams can be combined within a chip or coil microreactor, followed by solid-supported reagent columns providing further functionalization and/or acting as a scavenger to ensure clean products. (d) A parallel setup of reactors [87] can be used to scale out the reaction if a larger amount of compound is desired and can be a significant improvement over the use of a single, large reagent column.

The systems described in Scheme 4.49 are for simple single-step transformations, typically with in-line purification; however, the real opportunities presented by flow techniques will be multistep transformations occurring both in series and in parallel. Initial work in this field has already resulted in the total synthesis of natural products, such as grossamide [88] and (±)-oxomaritidine [89], and the drug candidate BMS-275291 [90], which are discussed later.

4.2.3
Methods of Conducting Flow Chemistry

4.2.3.1 On-Bead Synthesis
The dehydration of diethylcarbinol in 1932 was the first example of synthetic chemistry attempted in flow mode using a heterogeneous reagent [67]; although no experimental details were given, it was reported that the use of acidic silica gel at high temperature promoted the dehydration of the compound to various alkenes.

However, it was not until Merrifield developed a polystyrene matrix for peptide synthesis that a broad range of functionalized solid supports became available [91], leading to solid phase organic synthesis (SPOS) and eventually to polymer-assisted solution synthesis (PASS). Both of these techniques are heavily utilized by industry, especially for rapid library synthesis.

Despite the breakthrough associated with Merrifield's approach, there are several limitations such as the discontinuous nature of the reaction, the need for large excesses of reagent and the mechanical instability of the polymer matrix. An early solution to the restrictions imposed by Merrifield's polystyrene supported batch process was the use of commercially available benzyl alcohol-functionalized silica (used for HPLC columns). This was initially derivatized with the first member of the peptide chain to be propagated. The synthesis of a tetrapeptide in flow was completed in half the time required for the equivalent batch mode assembly and required significantly smaller excesses of the solution-phase reagent [92].

A flow synthesis of peptides using a macroporous polymer resin soon followed additionally incorporating an in-line UV detector, which constantly monitored the progress of the reaction by analyzing the output of the column; the reaction was recirculated until it reached completion. However, backpressures of 500–1000 psi were encountered, making these systems only suitable for metal-encased columns [93].

Although current commercial polymer architectures are well defined and characterized, the early polymer supports suffered from numerous disadvantages, such as partial solubility, mechanical weakness and broad range of particle sizes. One approach to overcome these difficulties was to polymerize a methylacrylamide resin within porous Kieselguhr particles, providing an inorganic matrix to support the resin; the polydimethylacrylamide–Kieselguhr system was then functionalized and used to synthesize an octapeptide [94]. Reaction monitoring has always been the most complex task of on-bead synthesis, and although qualitative monitoring of the reaction is possible using IR spectroscopy, to accurately monitor a reaction under these conditions a sample must be cleaved from the polymer beads; inevitably this results in reduced yields and is a time consuming procedure. This difficulty was resolved when UV detection was used to monitor consumption of the excess reagents used for on-bead peptide synthesis. Once the UV level had stabilized the reaction was judged to have reached completion and the next step in the synthesis was performed [95,96].

4.2.3.2 Solution-Phase Synthesis

Many of the early pioneers of flow chemistry opted for jacketed gravity-fed columns [82], similar to those used for flash chromatography filled with solid reagents. One example is the use of a macroporous peracid–functionalized polymer **1** to oxidize sulfides, including several penicillins (Scheme 4.50); for example, a solution of penicillin G (**2**) was passed through the reagent column and the resultant sulfoxide **3** isolated in 87% yield with a residence time of 10 min at 40 °C, compared with 85% yield after 2 h at 20 °C in the batch system. Importantly, the resin could be regenerated to 99% of its previous efficiency, though after several cycles the polymer began to degrade irreversibly [97].

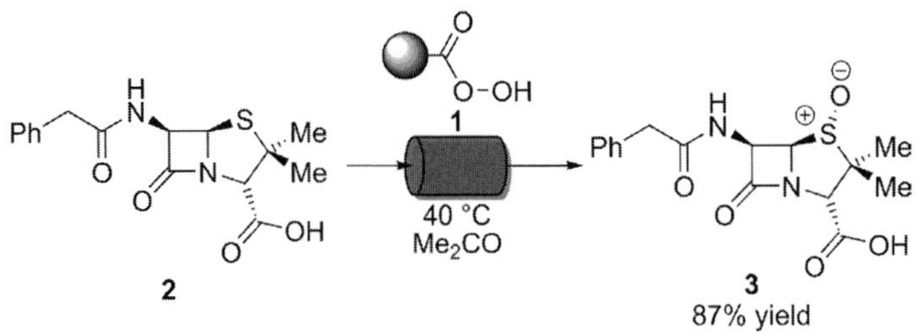

Scheme 4.50 Flow oxidation of penicillin G.

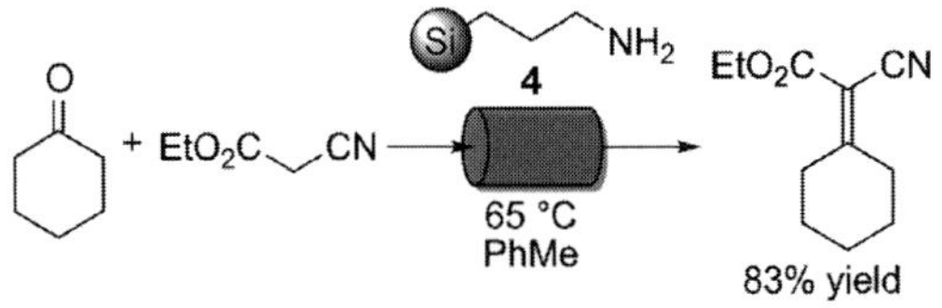

Scheme 4.51 The Knoevenagel condensation in flow.

Venturello provided an early example of a Knoevenagel condensation using the aminopropyl-functionalized silica gel 4 (Scheme 4.51); the yields varied from 98% for the best substrates to 41% for the more difficult. A 10 mmol sample of each reagent was added to a gravity-fed column containing the functionalized silica gel, providing a catalytic system that could be reused six times with no apparent loss of catalytic ability; this simple setup could be used to generate gram amounts of compound. For a more difficult substrate, the column was heated to 65 °C, which improved the conversion from 86 to 98% and the yield from 66 to 83% [98,99].

The first enantioselective process within the flow domain was the reduction of valerophenone (5) by borane in the presence of the polymer-supported amino alcohol catalyst 6 (Scheme 4.52) as reported by Itsuno [100]. Solutions of 5 and borane were mixed into the bottom of the column using long syringe needles, and the product 7 flowed from the top. Following washing with THF and water, acidic workup and bulb-to-bulb distillation, 1.8 g of 7 was isolated in 84% yield and in 83–91% ee, depending on the fraction analyzed.

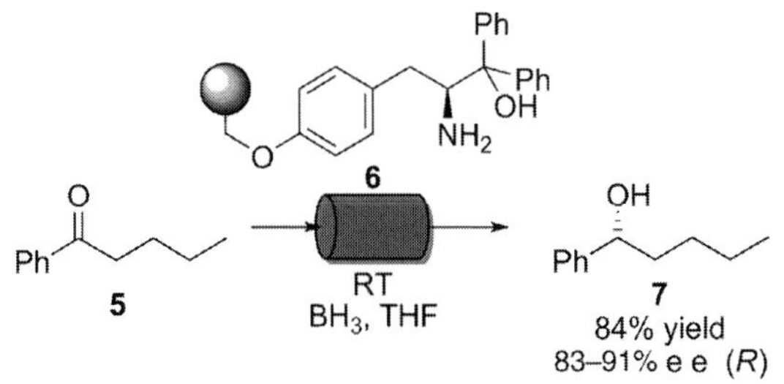

Scheme 4.52 Itsuno's enantioselective reduction of valerophenone.

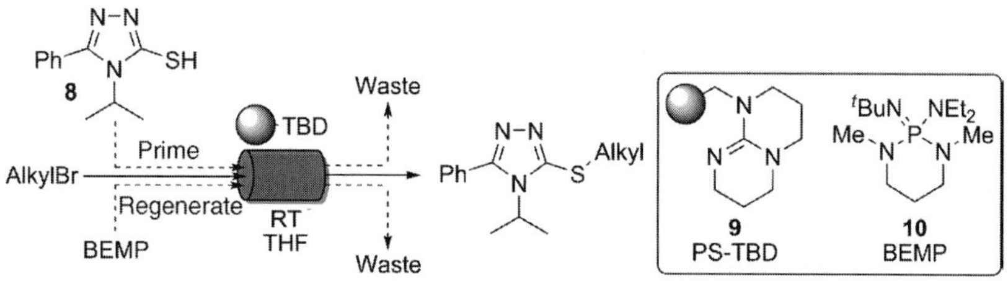

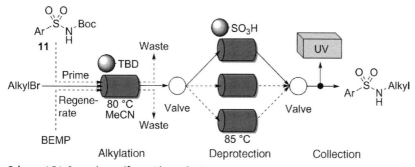

Scheme 4.53 Thioether synthesis.

4.2.3.3 Library Synthesis in Flow

The great versatility of flow chemistry lies in its ability to perform reactions reproducibly, either one continuous reaction to produce gram quantities of material or many sequential reactions on smaller scale. By using suitable front- and back-end liquid handling and detection systems, sequential plugs of reagent followed by appropriate solvent washes can be used to synthesize compound libraries. For example, a series of thiol-functionalized heterocycles (**8**) was deprotonated using a column of PS-TBD (**9**), forming a stable ion pair. A second flow stream then carried a substoichiometric plug of alkylating agent through the column, reacting with the thiolate and washing the newly formed thioether from the resin to be collected by a UV-directed liquid handler (Scheme 4.53). The remaining thiolate was removed by directing a solution of allyl bromide (as an inexpensive alkylating agent) through the column to waste. The column was then regenerated by flowing a solution of BEMP (**10**) through the system followed by a wash step; in this way, the column could be regenerated and reused over 30 times. A collection of 44 compounds (>75% yield and 95% purity) was synthesized in an automated fashion overnight, with only a single manual transfer of the vials to the drying station required for compound isolation [101].

In an analogous approach, a second library of sulfonamides was prepared. N-Boc protected sulfonamides (**11**) were ionized by a column of PS-TBD heated to 80 °C, and the alkylation was performed as before; the exit stream passed onto a column of PS-SO$_3$H heated to 85 °C, which deprotected the Boc group and afforded the secondary sulfonamide products (Scheme 4.54). This method was applied to the synthesis of two collections of compounds, a 24-membered array and the second comprising a 48-compound set, in both high purities and yields. No cross-contamination was

Scheme 4.54 Secondary sulfonamide synthesis.

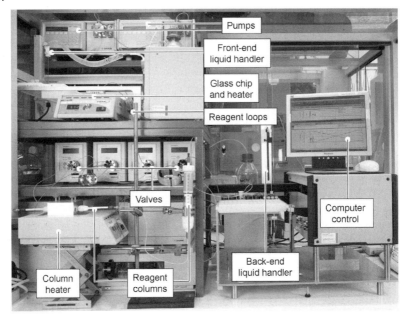

Figure 4.1 Bespoke modular flow reactor.

observed, though collection using the UV-directed liquid handler resulted in reduced yields (but retained high purities) for one series of the sulfonamide monomers because of difficulties in identifying an appropriate UV chromophore [102].

4.2.3.4 Heterocycle Synthesis

Our group has investigated the formation of 4,5-disubstituted oxazoles (**12**) in flow using the apparatus shown in Figure 4.1. An isocyanide and acid chloride were mixed on a glass chip, typically heated to 60 °C, forming a reactive adduct. The stream then flowed through a column of PS-BEMP, closing the ring and forming the oxazole (Scheme 4.55). The excess acid chloride used in the reaction was scavenged by a column of Quadrapure-BZA (QP-BZA, **13**, a macroporous benzyl amine resin). A library of 36 compounds was generated, with yields of 83–98% and in high purities; no further purification or workup was required. It should be remarked upon that when the reaction was run in batch mode the yields were poor, typically around 50%, demonstrating the

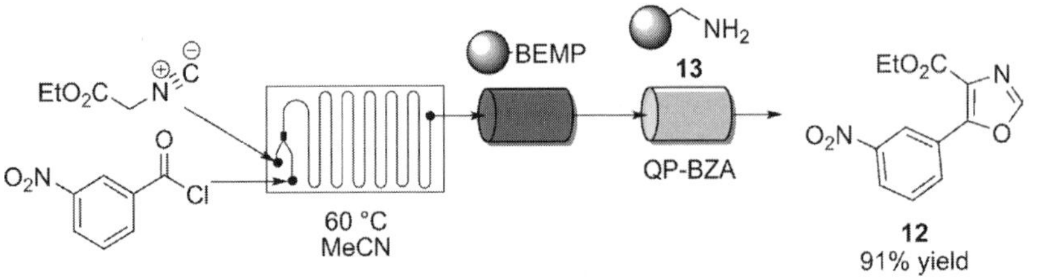

Scheme 4.55 Flow synthesis of 4,5-disubstituted oxazoles.

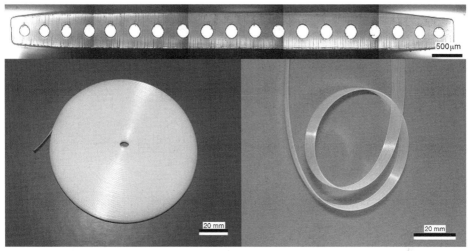

Figure 4.2 MFD providing longer residence times.

different properties of flow and batch chemistry. Scaled synthesis generating 10–20 g of compound could be achieved by simply using larger columns of supported reagents and allowing the system to run for longer periods (~12 h), clearly illustrating the versatility of the instrumentation and the scalability of this technology [103].

Lower yields and purities were obtained when the formation of the reactive adduct did not go to completion before encountering the PS-BEMP and several examples required residence times that could not be easily provided by using a microfluidic glass chip. To overcome this limitation, longer residence times were obtained by using a microcapillary flow disc (MFD) reactor (Figure 4.2). This inexpensive and disposable alternative to glass reactor chips has been made from a variety of thermoplastic materials that can be extruded up to 40 m in length, thereby providing a much larger reactor volume and hence much longer residence times. Furthermore, the residence time within the MFD was long enough that certain examples did not require heating; in a number of examples, this resulted in improved purities [87].

This 4,5-disubstituted oxazole synthesis was further expanded by using isothiocyanates and carbon disulfide as electrophiles, providing a bifurcated route to the preparation of thiazoles and imidazoles in a similar modular flow reactor. When aryl isothiocyanates were used, typically ~50% yield of the desired thiazole **13** was obtained (Scheme 4.56). When an electrophile (an α-bromo ketone in the example shown) was subsequently flowed through the PS-BEMP column, the remainder of the material was eluted as the imidazole **14**, providing combined yields of 76% to quantitative. An unidentified open chain species was sequestered onto the resin, and only after ring closure by alkylation was the secondary product eluted. Alkyl isothiocyanates and carbon disulfide furnished only the corresponding thiazoles, again in good yields (72–97%) and in excellent purities [104].

Pyrazoles have found common usage as pharmaceutical and agrochemical agents; consequently, we have developed an effective method for the synthesis of 5-amino-4-

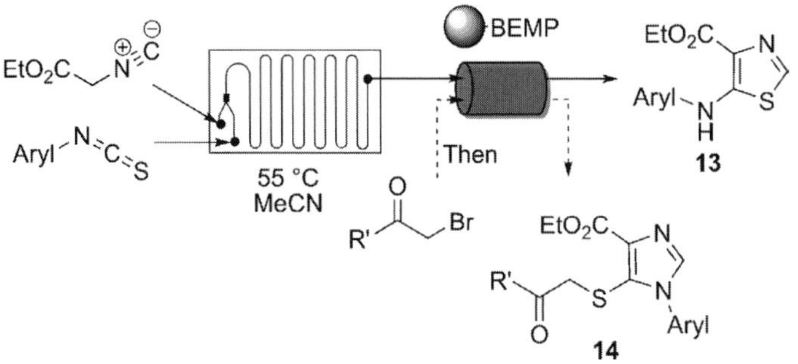

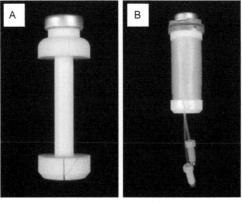

Scheme 4.56 Synthesis of thiazoles and immidazoles in flow.

cyanopyrazoles (**15**) under continuous microwave conditions. The microreactor consists of a Teflon spigot (Figure 4.3a) around which was wrapped 1/16 in. fluoropolymer tubing (11.5 m), which could then be placed within a microwave cavity. A methanol solution containing an organic hydrazine and ethoxymethylene malononitrile was then flowed through the reactor and heated to 100–120 °C. The resultant solution was then directed through a column of QP-BZA, removing the starting electrophile or any remaining uncyclized intermediates, followed by a column of activated carbon to remove colored or polymeric impurities (Scheme 4.57). This sequence provided the desired products in 62–95% yield and >95% purity, enabling them to be used in further transformations without any additional purification [105].

Triazole formation by the copper(I)-mediated coupling of organic azides with terminal acetylenes is of much current interest, with applications in many different areas of chemistry from cell biology to materials science [106]. The typical batch reaction involves crystallization of the product from the reaction solution, frequently trapping large amounts of the copper catalyst, and thus requiring further processing before the pure compound can be isolated. Our group has used a series of

Figure 4.3 Flow microwave coil. (a) Teflon spigot. (b) Wrapped coil system.

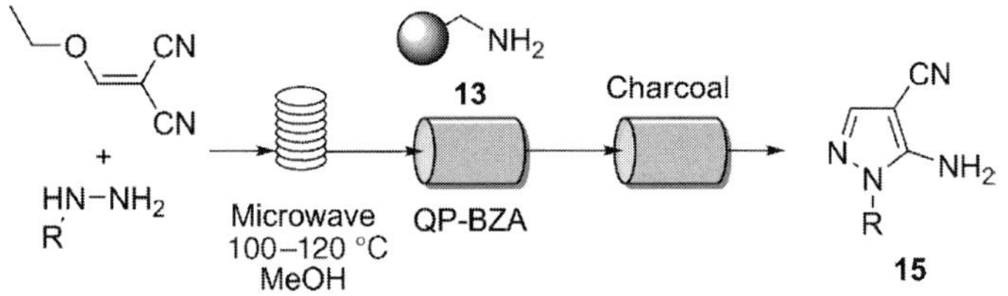

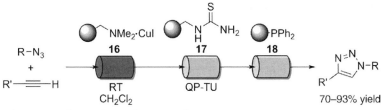

Scheme 4.57 Synthesis of pyrazoles under flow microwave conditions.

three columns of reagent to allow formation of the 1,2,3-triazoles in a single pass (Scheme 4.58). The cycloaddition was catalyzed using Amberlyst 21 preloaded with copper(I) iodide (**16**) with the flow stream being directed through a subsequent cartridge of Quadrapure-TU (QP-TU, **17**) [107] to remove any copper(I) residues that inevitably leach from the preceding column. Finally, PS-PPh$_2$ (**18**) was used to remove the excess azide that was used to drive the reaction to completion. This system afforded the desired compounds in high purity; no chromatography was required and no homo-coupling was observed [108].

4.2.4
Introduction to Monoliths

Despite the advances in materials since Merrifield's early investigations, microporous polymers are still mechanically weak. Their volume and properties – such as apparent loading – are altered when the solvent is changed, and, furthermore, packed columns have very large void areas, as an inevitable result of the particulate nature of the packing. Even when perfectly organized ~30% of the volume consists of interstitial voids and, as a result, a significant portion of the reactor is underused [109].

When smaller porous particles are used, the result is a highly efficient packed column with smaller interspatial voids; however, this leads to lower permeability of the column, and high backpressures are often generated. To overcome these problems, polymeric monoliths have been developed; these are a continuous phase of porous material that can be used without generating the high backpressures observed with fine particles. These systems, popularized by Fréchet and Svec, have found application both in separation devices [110] and, more recently, as flow reactors [111–113].

Scheme 4.58 Flow synthesis of 1,2,3-triazoles.

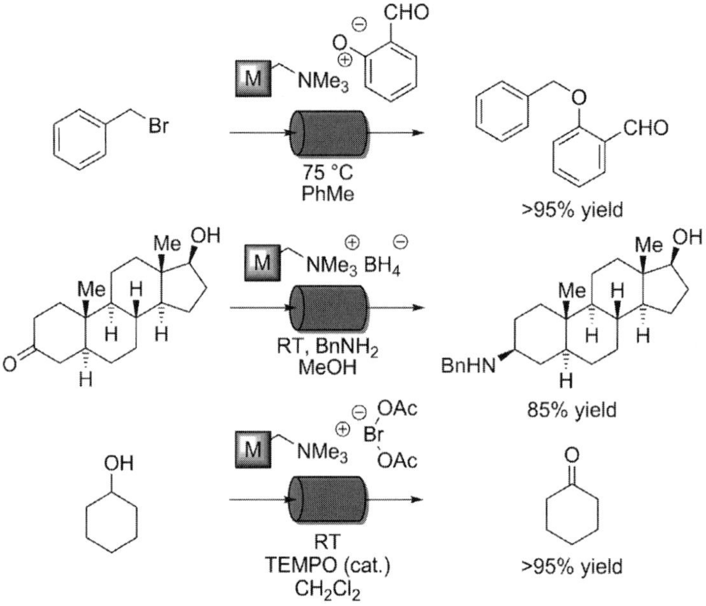

Scheme 4.59 Monolith reactors employed as stoichiometric reagents.

Although monoliths have many attractive features, their synthesis can prove troublesome. It is often a matter of trial and error to find the correct combination of monomer, cross-linking agent, poragen, initiator and polymerization temperature to obtain workable results. This optimization process must be repeated each time until ideal conditions are found [114,115]. However, this process allows for the formation of a monolith of virtually any shape desired, normally some form of secondary containment, such as a column.

A noteworthy example of the monolith concept is the PASSflow [116] system, which has been used in a variety of synthetic applications. This differs from the Fréchet monoliths as the polymerization step occurs within a porous glass structure, resulting in a strong and stable rod. The incorporation of derivatized styrene monomers allows for the introduction of functionality to the monolith; alternatively, unfunctionalized polystyrene can be sulfonylated with an appropriate reagent. This system has been used as an immobilized reagent system for simple transformations including reduction, oxidation, deprotection of silyl groups, substitution and reductive amination (Scheme 4.59). In most cases, the reactions did not proceed to completion in a single pass and thus the reaction mixture was recirculated through the monolith until the reaction was completed.

A similar monolith was also used in the Horner–Wadsworth–Emmons (HWE) olefination (Scheme 4.60). Initial reactions using the hydroxide-functionalized column **19** resulted in partial hydrolysis of the ester groups, but it was found that thorough drying of the column (under vacuum with P_2O_5) and the use of anhydrous THF prevented this

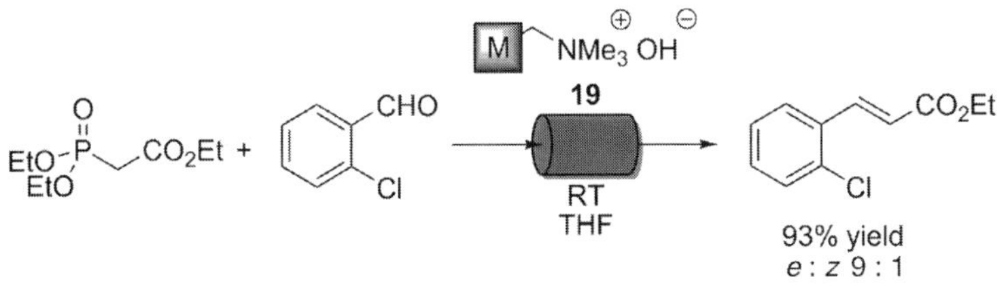

Scheme 4.60 Monolith reactor used in the HWE olefination.

side reaction. In addition to acting as the base for this process, the monolith also acted to scavenge the diethyl phosphate by-product onto the column, resulting in a workup-free reaction that produces compounds in both high yields and purities [117].

4.2.5
Transition Metal Chemistry Under Flow Conditions

Transition metal catalysis is of great importance in modern synthetic organic chemistry, possessing enormous versatility and broad ranging applications [118]. However, the usefulness of transition metals for the construction of pharmaceutically active molecules is limited by a number of factors. Firstly, catalyst residues can be difficult to remove from the product, as regulatory authorities set acceptable levels below 10 ppm for metal contamination in final goods. Secondly, precious metal catalysts are expensive, and methods for their efficient recovery and regeneration are crucial. Many investigators have attempted to address these issues through the use of solid-supported reagents, and examples such as perovskites [119–121], monoliths [122], functionalized polystyrenes [69], polyurea matrices [123–125] and dendrimers [126] are known and have been well reported. Additionally, the use of scavenging resins to improve the quality of the product has also gained popularity in recent years. The natural evolution of these methods has been to apply them sequentially in a flow process, allowing for the development of new, high-purity transformations.

4.2.5.1 Reduction
The biggest advance in laboratory hydrogenation in recent years has been the development of the commercially available H-Cube from ThalesNano Technologies (Figure 4.4). This system consists of an immobilized metal catalyst cartridge (such as Pd, Rh and Raney Ni) and uses electrolysis of water to generate hydrogen gas *in situ*. The hydrogen in the system can be pressurized up to 100 bar and pumped with the substrate through the catalyst cartridge, which can be heated safely to 100 °C. This combination of pressure and temperature ensures that reactions can be driven to completion in only one pass and eliminates the need for high-pressure cylinders of flammable gas. In comparison to traditional batch hydrogenation methods optimization is rapid, as the small plugs of substrate can be quickly collected and analyzed using standard methods. Once this process has been optimized, the scale of the reaction can be increased to multigram levels with no further modification required. The system has been used in a variety of applications [127,128] including imine bond

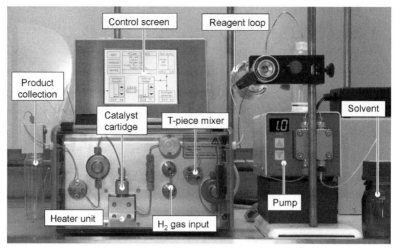

Figure 4.4 The H-Cube system.

reductions [129,89] and benzyl/Cbz deprotections (Scheme 4.61) [130,131]; a description of the best practices for the use of the H-Cube has also been published [132]. The use of a metal-scavenging column post-emergence from the H-Cube ensures that any metals that have leached from the catalyst cartridge are effectively removed. The addition of front- and back-end liquid handling robots [128] allows for rapid library synthesis after initial optimization of conditions.

The PASSflow system described previously has also been functionalized with nanoparticular metal species (Pd, Ni and Pt) [116,133,122]. These nanoclusters are formed by first flowing a metal chloride salt through an anion-exchange monolith

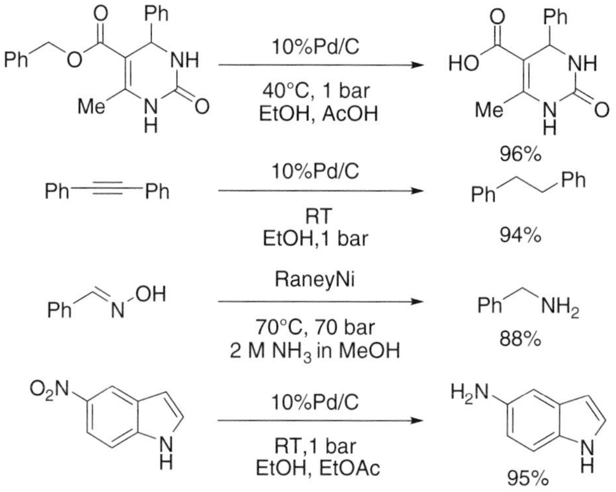

Scheme 4.61 Examples of H-Cube use.

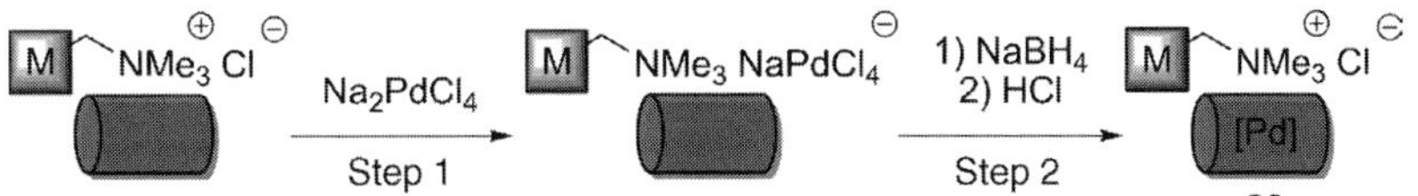

Scheme 4.62 Formation of nanoparticular palladium monoliths.

(Step 1, Scheme 4.62) displacing the chloride ion; subsequently, the metal is then reduced (typically using sodium borohydride) to produce small, highly dispersed metal particles throughout the monolith (**20**). These nanoparticular systems have been used for reduction and de-benzylation reactions [134], using 1,4-cyclohexadiene or lithium (or ammonium) formate under hydrogen transfer conditions, thereby avoiding the use of flammable gases in flow devices.

An alternative approach has been used to immobilize palladium metal onto the glass walls of a chip-like device [135]. A hydrogen gas stream was metered through a mass-flow controller and mixed with the substrate solution in the microreactor. At low flow rates, inefficient plug flow (alternating bubbles of gas and liquid) was observed, and at higher hydrogen flow rates 'pipe flow' was observed where the gas stream is forced through the center of the microchannel and the liquid phase is concentrated along the walls, resulting in better mixing. Consequently efficient debenzylation and reduction of alkenes and alkynes was possible. In a follow-up paper, supercritical carbon dioxide (scCO$_2$) was used as the substrate solvent, allowing substantially more material to be processed per hour; this is believed to be a consequence of the increased solubility of hydrogen gas in scCO$_2$ [136]. For larger amounts of material to be processed, the system was changed from a chip microreactor to a capillary tube (200 μm i.d., 40 cm length) coated with palladium; this setup could be more readily scaled out by using a bundle of identical capillaries. Unfortunately, the THF and gas phases were not uniformly distributed to each line, resulting in poor conversions. This problem was overcome by increasing the amount of palladium in each of the capillaries, driving the reaction to completion and providing 124 mg of product in 17 min (2.8 mmol/h), in quantitative yield with apparently no significant palladium leaching (Scheme 4.63) [137].

There are a number of different approaches to performing enantioselective reductions of ketones within the flow domain, using either a borane-derived hydride transfer agent such as that described previously or modified transition metal hydrogenations; an example of the latter involved a column of Pt/Al$_2$O$_3$ modified with O-methyl cinchonidine (**21**) to induce chirality in the product (Scheme 4.64). Continuous monitoring showed that a 30 min induction period was required before the optimal reaction rate and ee could be obtained. This was ascribed to the need for

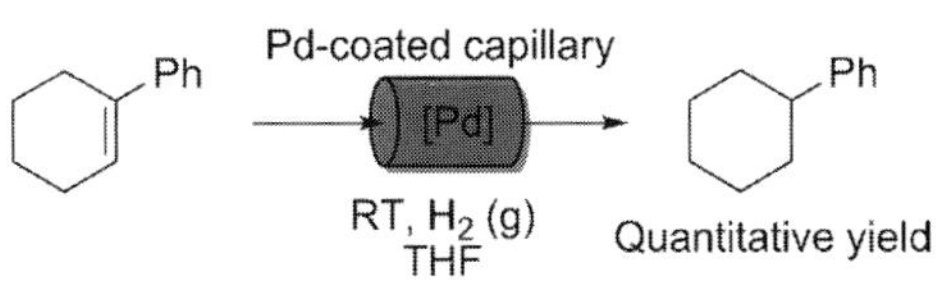

Scheme 4.63 Reduction using palladium-coated capillaries.

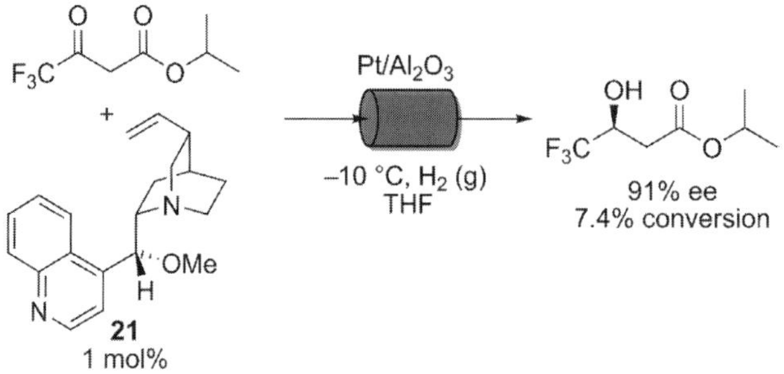

Scheme 4.64 Enantioselective hydrogenation in flow.

the establishment of a chirally modified surface during this time. A considerable amount of time and effort was dedicated to the optimization of the reaction conditions, including variation of temperature, hydrogen pressure, trifluoroacetic acid (TFA) concentration, chiral modifier and flow rate, though flow rates were purposely kept high to ensure conversion was less than 10%. The highest ee of 91% was found when the reaction was run at $-10\,°C$ and no TFA was used as additive [138].

A different approach that has been used is a ruthenium-catalyzed Meerwein–Ponndorf–Verley-type reduction of ketones using the silica-supported amino alcohol ligand **22** (Scheme 4.65). It was found necessary to cap the remaining free silica hydroxyl sites to alkylsilane derivatives to prevent catalyst deactivation. Initial studies found that slower flow rates resulted in lower ee because of equilibration back to the starting materials – after optimization, the best conditions were found to be 1400 µl/h providing a 95% conversion and 90% ee. The stability of the catalyst was investigated over time, during which a constant formation of 175 µmol/h was obtained; only after a period of 7 days was some decrease in activity observed. The extended lifetime of the

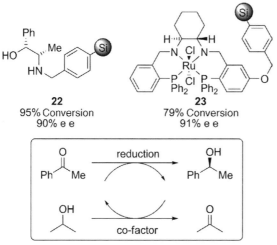

Scheme 4.65 Acetophenone reduction under hydrogen transfer conditions.

immobilized catalyst versus that of the homogeneous catalyst was ascribed to site isolation preventing clustering, which leads to a deterioration in catalyst activity [139].

This same reaction was also applied in a membrane reactor by using dendritic catalyst 23 (Scheme 4.65). A membrane reactor uses the size exclusion principle and is an alternative approach to phase transfer catalysis, whereby the catalyst is homogeneous – thus enabling faster reaction kinetics – but is easily separable from the starting material and products. The system had the potential to make 578 g/l/day, corresponding to 79% conversion and possessing an ee of 91%. When compared side-by-side to a similar enzymatic system, the chemenzyme was five times more productive and did not require an additional cofactor reducing agent; however, the enzyme could achieve an impressive 99% ee [140].

4.2.5.2 Oxidation

Oxidations in flow have been achieved through the use of several different strategies such as the previously described PASSflow TEMPO oxidation and a number of metal-mediated reactions. An oxidation of benzyl alcohol has been performed using a polymer-incarcerated ruthenium catalyst 24 (PI-Ru, Scheme 4.66a) by flowing a solution of the alcohol and N-methyl morpholine oxide (NMO, the stoichiometric oxidant) through a mixed bed of PI-Ru and $MgSO_4$, providing the benzaldehyde product. The reaction requires the absolute exclusion of moisture and air from the system to prevent catalyst degradation. This proved to be possible in flow by using $MgSO_4$ to remove any water present, and the closed flow system necessarily excludes oxygen. The system was run for 8 h and still functioned at 94% conversion and 92% yield; rather surprisingly, no ruthenium leaching was reported in these reactions. This catalyst system was also used for other oxidations, including the oxidation of sulfides to sulfoxides or sulfones and the conversion of an acetylene to a 1,2-diketone; however, no attempt was made to perform these reactions within the flow domain [141].

Silica-supported Jones reagent has also been used for the oxidation of benzyl alcohol (Scheme 4.66b). When this reagent was used in batch mode, the reaction

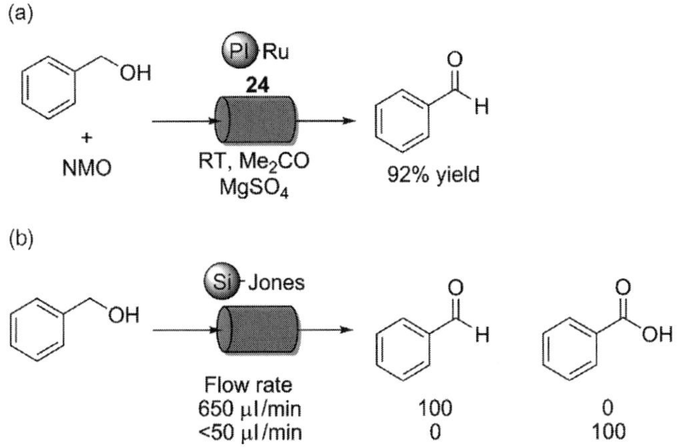

(a)

(b)

Scheme 4.66 Oxidations.

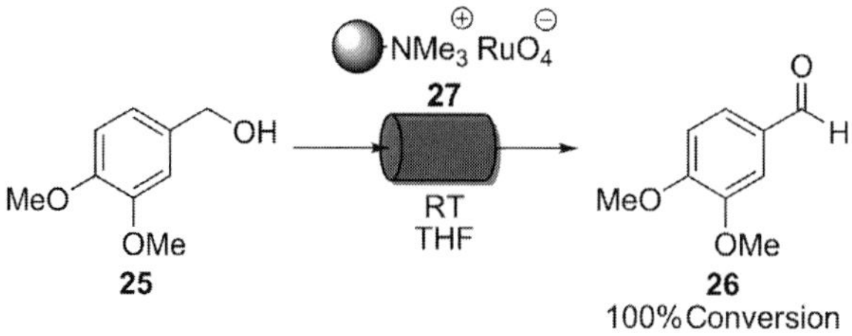

Scheme 4.67 Oxidations in flow.

resulted in a mixture of starting material, benzaldehyde and benzoic acid; however, when used in flow mode it was possible to isolate a single product depending on the flow rate through the reagent column. At 650 μl/min (10 s residence time) only benzaldehyde was isolated. Conversely, when the flow rate was less than 50 μl/min (>126 s residence time) only benzoic acid was obtained. These conditions were then applied to a series of 15 primary alcohols (selectively forming either the aldehyde or the acid as desired) and 15 secondary alcohols forming the respective ketones. This clearly demonstrates the possibilities inherent to flow chemistry for chemoselective reactions simply by varying the flow rate. In all cases other than primary aliphatic alcohols leaching of chromium was found to be less than 69 ppm. When primary aliphatic alcohols were used substantial chromium leaching was observed, which was attributed to the polarity of the substrates, making this system only suitable for aromatic alcohols [142].

The seven-step flow synthesis of (±)-oxomaritidine included the oxidation of benzylic alcohol **25** to aldehyde **26** (Scheme 4.67) using polymer-supported tetra-*N*-propylammonium perruthenate (TPAP) **27**. Although this reaction is stoichiometric in ruthenium, the Ru(VII) species can be readily regenerated by flowing a solution of NMO through the spent reagent cartridge [89].

4.2.5.3 Cross-Coupling Reactions

The PASSflow monolith using nanoparticular palladium (**20**) has been used in a variety of coupling reactions [134]. A Sonogashira reaction was attempted (Scheme 4.68a), furnishing **28** in 81% yield accompanied by approximately 9% of the homocoupling impurity, which complicates isolation of the pure product. A number of Heck reactions were attempted, and the monolith proved to be a good catalyst (requiring 0.5 mol% palladium), though recirculating conditions were required. The reusability of the column was investigated using the reaction shown in Scheme 4.68b. Each run was stopped after 3 h and gave the following isolated yields of **29**: 78% (first run), 92% (second run), 90% (third run), 90% (fourth run), 88% (fifth run), 84% (sixth run) and 71% (seventh run).

In addition, Suzuki couplings were also investigated. For these examples the anion-exchange moiety of the palladium monolith **20** was converted to the hydroxide form **30** by washing with a sodium hydroxide solution. A solution of boronic acid and aryl halide was recirculated through the column furnishing the desired coupled

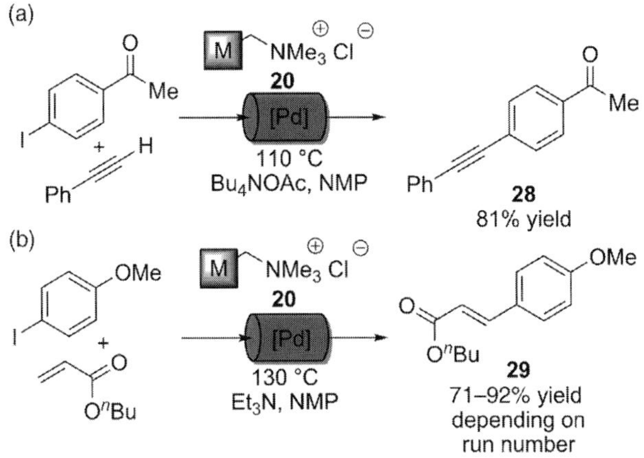

(a)

(b)

Scheme 4.68 Heck and Sonogashira reactions applied in flow.

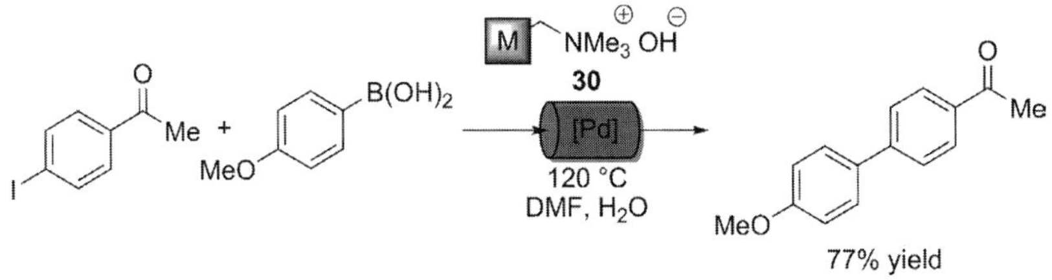

Scheme 4.69 Flow Suzuki couplings.

products in 67–99% yield after chromatography (Scheme 4.69). The column could be regenerated by washing with sodium hydroxide ready for the next reaction [143].

A system for the Heck reaction between methyl acrylate and iodobenzene has been developed comprising an imidazolium-functionalized polystyrene monolith, initially for use in batch. This system could be reused six times before any reduction in yield was observed. Accordingly, a continuous-flow reactor system was developed using DMF at 200 °C with a residence time of 3–4 min to achieve full conversion. This system was characterized by very low palladium loadings (0.02 mol%), and ICP-MS of the solution aliquots showed leaching of less than 1 ppm. Attempts at using EtOH as solvent to provide a more environmentally and procedurally benign protocol resulted in maximum yields of 85% [144].

Our group has also performed Heck reactions in flow using monoliths similar to those developed by Fréchet and Svec; the nanoparticular nature of the palladium species was observed by TEM and HRTEM. The system demonstrated several technological advances: first, a Vapourtec R-4 flow reactor heater [145] was used for both the monolith polymerization step and the subsequent cross-coupling reactions, providing accurate and safe heating; second, the resulting outflow from the reactor was directed through a second column containing QP-TU, to scavenge any leached

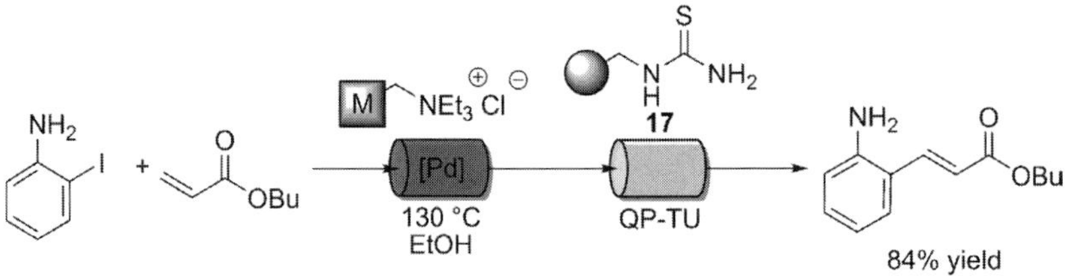

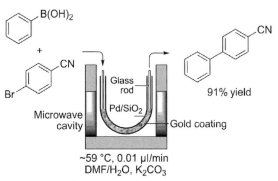

Scheme 4.70 Flow Heck reactions.

metal to <5 ppm (Scheme 4.70). Additionally, a change of solvent from DMF to EtOH was made possible using a 100 psi backpressure regulator, despite a column temperature of 130 °C; thereby greatly facilitating the isolation of the desired product. This system was sufficiently active to ensure complete conversion of a variety of aryl iodides in only one pass, with residence times of approximately 25 min; the columns were reused 25 times without loss of activity [146].

A Pd/SiO$_2$ support has been used for Suzuki reactions; however, initial results using both conventional heating and microwave irradiation proved disappointing. A thin film of gold (Scheme 4.71) was applied to assist the microwave heating and, when used in conjunction with lower microwave power and a 60 s residence time, the yield of the reaction between 4-bromobenzonitrile and phenylboronic acid improved from 37 to 91%. However, this product contained 5% biphenyl and 12% of the dehalogenated starting material, presumably as a contaminant [147,148].

Our group has published extensively on the use of Pd EnCat as an immobilized source of palladium [149,150], and we have applied this catalyst system to a series of Suzuki reactions in flow using both standard and cooled microwave heating. The heterogeneous catalyst was packed within a glass U-tube insert (Figure 4.5) and placed within a microwave cavity. The use of compressed air to cool the reactor was found to be highly advantageous. Stock solutions of boronic acid, aryl halide and tetrabutylammonium acetate base were constantly fed through the flow reactor (Scheme 4.72). A pulsed cycle of heating was required because continuous heating led to decomposition of the encapsulated catalyst. The optimum conditions for the

Scheme 4.71 Pd/SiO$_2$ catalyzed Suzuki reactions.

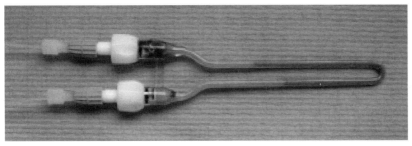

Figure 4.5 Flow reactor microwave insert containing Pd EnCat.

reaction were the use of a 30 s heating phase at 50 W followed by an 18 s cooling period with no power application; a constant stream of compressed air was used throughout. The solution was flowed at a rate of 0.1 ml/min – corresponding to a residence time of 65 s within the microwave cavity – and the outflow stream directed through a column of Amberlyst 15 that acted as a scavenger for the remaining base and boron-containing by-products. The products were isolated simply by evaporation in yields comparable to the batch methods attempted. When higher purity products were required a column of QP-TU was placed immediately after the Amberlyst 15 to scavenge any leached palladium [151,152].

The synthesis of multiple substrates proved possible. Ten different sequential reactions were carried out in an automated fashion with only a wash period between each reagent plug, clearly demonstrating the use of this technology for the high-throughput synthesis of compound libraries in both high purity and yield (Table 4.1).

As a final test of the system a scaled reaction was performed, producing 9.34 g (40.8 mmol) of **31** with a catalyst loading of 182 mg Pd EnCat (equating to 0.2 mol% palladium), demonstrating that this system is applicable to multigram synthesis.

A palladium-coated glass capillary could be prepared by heating a solution of Pd(OAc)$_2$ at 150 °C for 30 min within the tube to deposit a layer of Pd(0). The capillary was further heated to 350 °C for three cycles of 1 min to improve the porosity of the

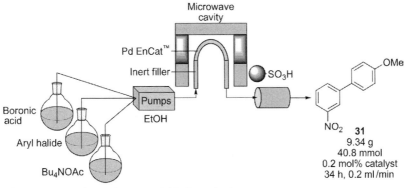

Scheme 4.72 Suzuki coupling using Pd EnCat and microwaves.

Table 4.1 Sequential microwave Suzuki reactions applied in flow.

Input number	Boronic acid	Halide	Yield (purity)
1	Ph-B(OH)₂	4-bromoacetophenone (MeOC, Br)	90 (>98)
2	benzofuran-2-B(OH)₂	1-bromonaphthalene	87 (>98)
3	3-nitrophenyl-B(OH)₂ (NO₂)	methyl 2-bromo-5-methoxybenzoate (MeO, Br, CO₂Me)	94 (>98)
4	3-nitrophenyl-B(OH)₂ (NO₂)	2-bromoanisole (Br, OMe)	92 (92)
5	4-methylphenyl-B(OH)₂ (Me)	2-bromoanisole (Br, OMe)	88 (94)
6	benzofuran-2-B(OH)₂	4-acetylphenyl triflate (MeOC, OTf)	97 (91)
7	thiophen-2-B(OH)₂ (S)	4-bromobenzonitrile (NC, Br)	76 (>98)
8	4-fluoro-3-chlorophenyl-B(OH)₂ (F, Cl)	4-nitrobromobenzene (O₂N, Br)	89 (>98)
9	3,4,5-trimethoxyphenyl-B(OH)₂ (MeO, MeO, OMe)	4-bromoanisole (MeO, Br)	94 (91)
10	pyridin-3-B(OH)₂ (N)	2-bromopyridine (Br, N)	95 (82)

film. The finished capillary could then be used for a series of Suzuki and Heck reactions. These reactions were performed at temperatures between 200 and 225 °C under microwave irradiation (Figure 4.6), resulting in conversions of 58–97% for both reactions. The palladium analysis (by ICP-AES) of the crude reaction was found to be 19.2 ppm, which was removed using silica chromatography [153].

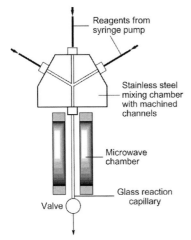

Figure 4.6 Glass capillary microwave reactor.

Salen-like ligands have also been used to complex, and hence immobilize, a number of metals onto polystyrene and silica for use in a variety of cross-coupling reactions. The salen–palladium complex **32** was used in Suzuki reactions (Figure 4.7). The column containing the immobilized catalyst was heated to 100 °C in a water bath and the reagents were recirculated continuously at a flow rate of 6 μl/min for 5 h, resulting in conversions in the range of 65–75% [154].

The use of a functionalized silica-supported salen–nickel complex has allowed Kumada cross-couplings to be performed in flow; the corresponding polystyrene supported complex was shown to be inferior for a number of reasons. Catalyst **33** (Figure 4.7) with the longer tether was found to be more active than the benzyl ether tether used for catalyst **34**. This was postulated to be due to the fact that catalyst **33** resided further away from the silica surface and hence was more available for reaction. Under the conditions used a maximum conversion of 65% was found for the 1 : 1 reaction of 4-bromoanisole and phenylmagnesium chloride, which was found to be comparable to that obtained in batch mode. However, during the reaction catalyst degradation was observed and the conversion reduced from 60% in the first hour to 30% in the fifth hour of the reaction [155,156].

4.2.5.4 Olefin Metathesis
Olefin metathesis has come rightly to occupy a key role in modern organic synthesis, especially in natural product and polymer synthesis, eventually leading

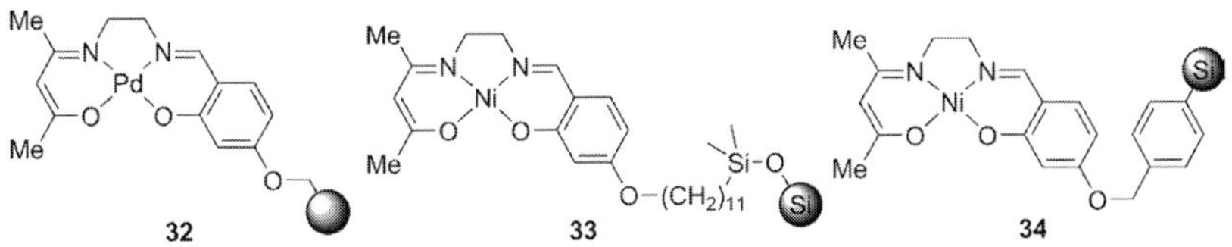

32   **33**   **34**

Figure 4.7 Salen-like catalysts used for flow coupling.

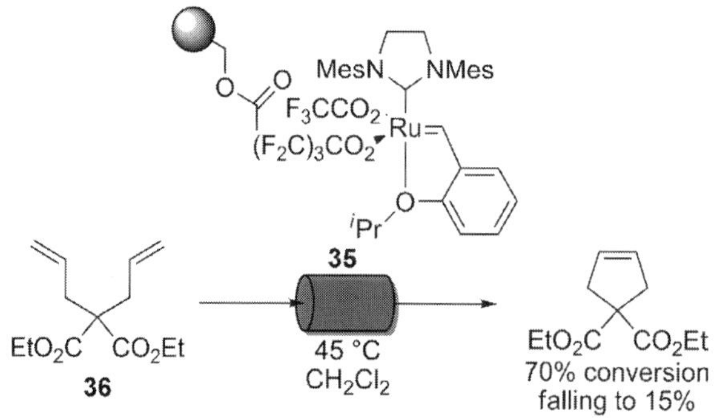

Scheme 4.73 Flow metathesis catalyst.

to the Nobel Prize being presented to Grubbs, Schrock and Chauvin in 2005. Barrett [157,158] and Nolan [159] have developed similar 'Boomerang' catalysts by using a polymer with many free styrene groups. These so-called 'Boomerang' systems are not suitable for continuous-flow processes because the active catalyst is a solution phase species and therefore prone to elution, which has led to the development of two different approaches to the immobilization of metathesis catalysts for flow reactors.

The most successful approach has been to immobilize the catalyst by substitution of the equatorial chloride ligands of the Grubbs–Hoveyda catalyst, by reacting it with a ROMP-derived monolith featuring a terminal silver carboxylate residue. This resulted in a metathesis catalyst **35** suitable for flow chemistry (Scheme 4.73). Initial conversion for the ring closing metathesis of **36** was 70%; however, after 140 min the conversion had reduced to 15%. The turn over numbers (TONs) were comparatively high (~500), and leaching levels of only 1.8 ppm Ru and 0.01 ppm Ag were recorded [160].

A different approach has used a basic tagged ruthenium complex for facile recovery and immobilization using PS-SO₃H [161,162]. A chemically tagged catalyst was immobilized to the solid phase (**37**), and the catalyst was activated because of the ammonium ion's electron-withdrawing properties (Scheme 4.74). The catalyst was used for ring closing metathesis in a continuously recirculated system (5 mol% Ru)

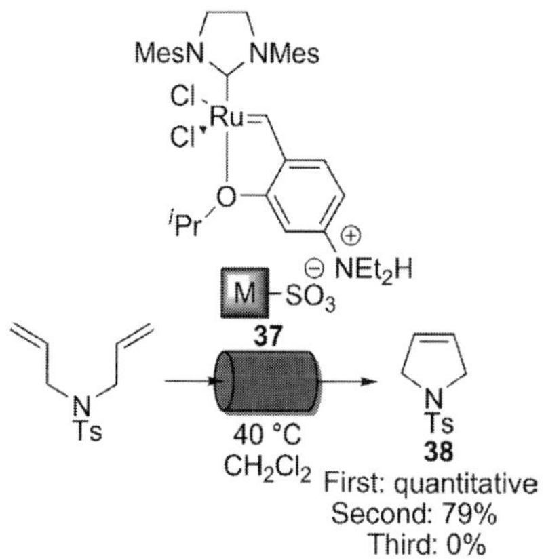

Scheme 4.74 Flow metathesis system.

followed by isolation of **38** 4 h later in quantitative yield. The second run afforded the product in 79% yield; however, no product was isolated on the third run. The column could be cleaned and fresh catalyst attached, but the amount of leaching was much more pronounced than the corresponding batch experiments, as is to be expected when the immobilization mechanism is considered [163].

4.2.6
Enantioselective Reactions

4.2.6.1 Hydrolytic Kinetic Resolution

Salen-based ligands have become an important motif in asymmetric catalysis [164,165], with their metal complexes catalyzing a broad range of reactions such as epoxidations, Diels–Alder reactions and hydrolytic kinetic resolution (HKR) [166]. The silica-supported catalyst **39** has been used in a variety of batch HKR reactions resulting in very high ee, typically >90% for a range of substrates and nucleophiles. This catalyst was used in a flow system for the HKR of epoxide **40** (Scheme 4.75), and in only one pass the product **41** was isolated in 39% yield and 94% ee; the column could also be reused. Considering the previous mechanistic work undertaken using the salen–chromium complex, it was expected that the loading would affect both the yield and the ee because of the importance of neighboring group participation in the postulated transition state of the reaction [167]. By changing the loading of the silica gel, it was found that the ee of the reactions was unaffected; however, there was a correlation observed between the yield of the reaction and the loading, with a minimum loading of catalyst required on the silica gel surface for the reaction to proceed [168].

A salen–cobalt complex has been appended to the PASSflow monolith system to form catalyst **42** and used for the dynamic kinetic resolution of epibromohydrin, **43**. Because **43** undergoes rapid racemization under the conditions used, all the starting materials can theoretically be converted to the desired diol **44** (Scheme 4.75). The

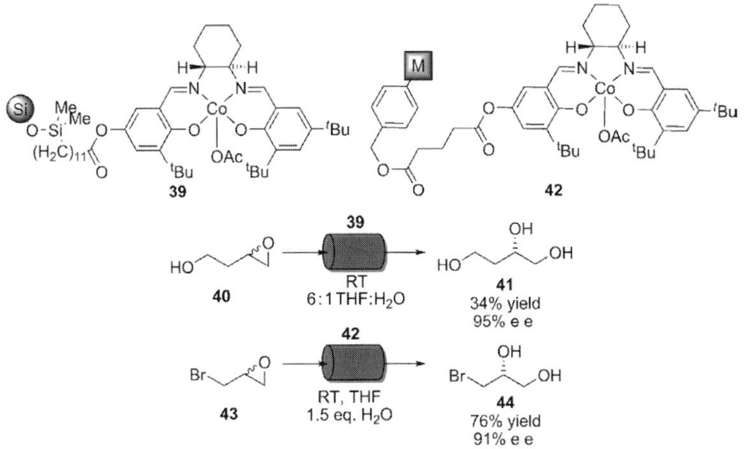

Scheme 4.75 Enantioselective kinetic resolutions.

system was recirculated for 20 h and used in three runs, resulting in yields of 76–83% and 91–93% ee. A fourth attempt was run constantly for 6 days to form 10 mmol of product in 76% yield and 93% ee, illustrating the stability of the catalyst over long reaction times [133,169].

4.2.6.2 Organometallic Additions

The asymmetric addition of diethyl zinc to aryl aldehydes in flow has evoked much academic interest. Supported amino alcohol **6** (Scheme 4.52) was placed in a jacketed column cooled to 0 °C and pretreated with the aldehyde. A 1 : 1.4 aldehyde : diethyl zinc solution was then flowed through the column and collected, and after a simple aqueous workup, the desired product was obtained. In one example, 5 mmol of **6** was used to produce 90 mmol of product in 90% yield and 94% ee, and in a second example 0.7 mmol of **6** produced 58 mmol of product with 92% ee [170].

A further example used the supported amino alcohols **45**, **46** and **47** (Scheme 4.76), where the reagents were pumped up from the bottom of the polymer using a pair of long needles connected to peristaltic pumps. The product was collected from the top using another pump and quenched in a solution of dilute hydrochloric acid. For the first run with catalyst **46**, the yields and ee were excellent (94% yield in 97% ee), but when **46** was recovered and reused, the yield dropped to 75% and the ee to 50%. This was ascribed to degradation of both the chiral and backbone sites of the polymer by diethyl zinc, again demonstrating that not only do the solid supports need to be mechanically sound but both the backbone support and active site must be also chemically resistant to the reaction conditions [171].

A final example of chiral additions to aldehydes used monolithic catalyst **48** (Scheme 4.76), using a chiral moiety derived from a waste material from the synthesis

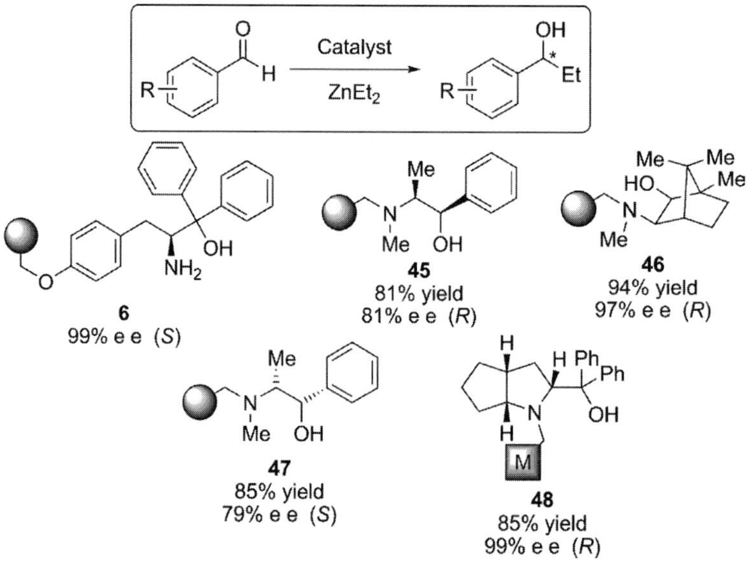

Scheme 4.76 Catalysts used for diethylzinc addition reaction.

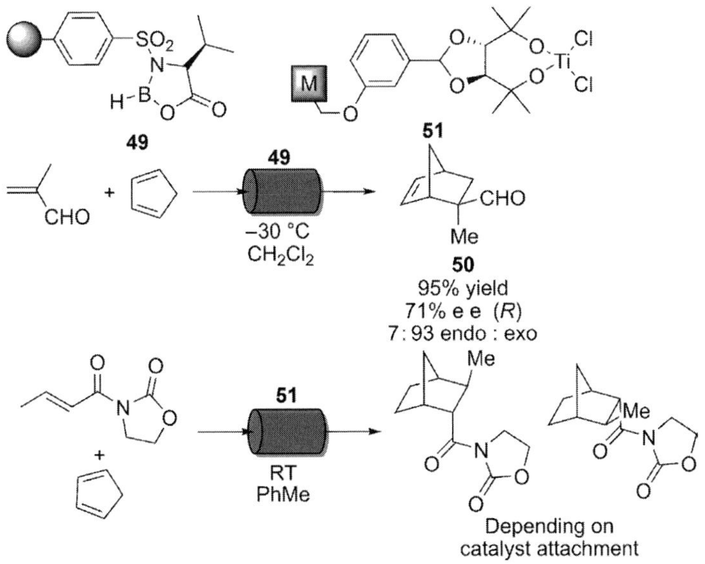

Scheme 4.77 Enantioselective Diels–Alder reactions applied in flow.

of Ramipril. Following preactivation of the column using a diethyl zinc solution at −10 °C, the aldehyde/diethyl zinc solution was continuously recirculated at room temperature through the catalyst column. After 24 h the system was stopped and the solution collected. It was then quenched with 2 M hydrochloric acid, to furnish 85% of the desired product in 99% ee. Three further 24 h cycles were run, and the system gave very similar results (84% yield, 99% ee) [172].

4.2.6.3 Enantioselective Diels–Alder Reactions

Chiral oxazoborolidine systems have been regularly used as catalysts for the enantioselective Diels–Alder reaction [173]. Such a catalyst has been immobilized by copolymerization of a sulfonamide-modified styrene monomer to produce polystyrene beads and subsequently reacted with borane to furnish catalyst **49** (Scheme 4.77). A column of **49** was cooled to −30 °C and a 1 : 1.5 methylacrolein : cyclopentadiene solution was flowed in. Following aqueous workup and column chromatography, the desired product **50** was isolated to yield 138 mmol of product in 95% yield and 71% ee by using only 5.7 mmol of catalyst. This result was found to be comparable with the heterogeneous batch reactions that were also attempted [174].

It has been observed that enantioselective polymer-bound catalysts prepared by copolymerization produce in some cases better asymmetric inductions than systems prepared by grafting [175]. After much optimization, a monolithic polymer catalyst **51** suitable for a titanium-TADDOLate catalyzed Diels–Alder reaction was developed (Scheme 4.77). The monolith was applied in a flow system both under one pass and 24 h recirculation conditions, the latter producing the best yield (55%) and ee (23%); however, this contrasts poorly with the homogeneous batch reaction although the ee is comparable with the heterogeneous batch process. The reversal of topicity was also

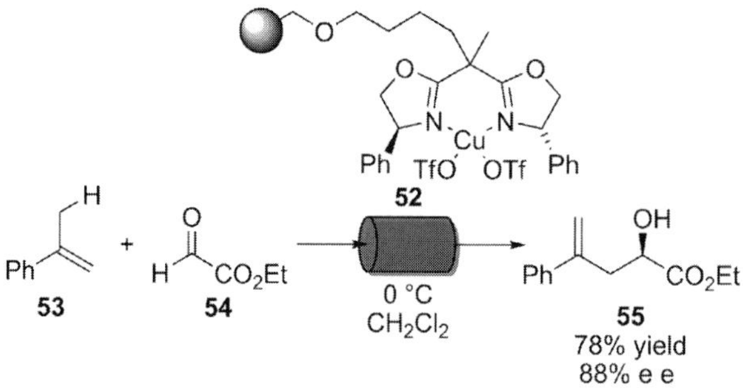

Scheme 4.78 Ene reaction.

discussed when monoliths prepared by different methods (grafting versus copoly-merization) were applied to the Diels–Alder reaction. The monoliths were stable for 6–8 months before a decrease in their activity was observed, and when regenerated they immediately regained their vigor. However, the long-term stability was lost after regeneration, and both activity and selectivity decreased sharply after use. This was believed to be due to titanium complexes that were not bonded to the chiral ligand catalyzing the racemic reaction [176].

4.2.6.4 Ene Reactions

Like the salen ligand, the *bis*(oxazoline) (BOx) motif has become a privileged ligand structure, and the copper complexes of this ligand class have been used to catalyze a variety of transformations [177,165]. The copolymerized catalyst **52** (Scheme 4.78) was milled before being placed in a steel HPLC column cooled to 0 °C. A solution of **53** and **54** was then pumped through the system using a standard HPLC pump. The first pass conversion of 83% was deemed acceptable and was not optimized further. Five batches were run, taking a total of approximately 80 h to complete, and following chromatographic purification, a total of 23 mmol of **55** was synthesized in 78% yield and in 88% ee [178].

4.2.6.5 Cyclopropanation

An immobilized Ru-PyBOx catalyst **56** was synthesized by thermally induced radical copolymerization and used to catalyze the flow cyclopropanation reaction between styrene and ethyldiazoacetate (Scheme 4.79), building on initial work performed in batch [179]. For the flow example, a range of experiments using the starting materials in CH_2Cl_2 solution, neat or in $scCO_2$ were assessed and the results are summarized in Table 4.2 [180].

The results under flow conditions showed a slight improvement in nearly all areas compared with the representative batch reaction – importantly, both the yield and chemoselectivity were significantly improved – but all reactions still produced a signi-ficant amount of the dimerized by-products (ethyl fumarate (**57**) and ethyl maleate (**58**)). The solventless system allowed larger quantities of material to be prepared, whereas the $scCO_2$ system provided products contaminated with only traces of ruthenium (less than 1 ppm).

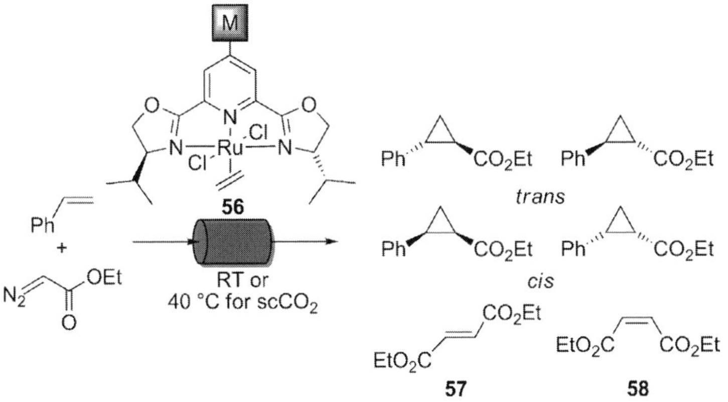

Scheme 4.79 Cyclopropanation in flow.

Table 4.2 Cyclopropanation under various conditions.

	Conversion	Yield (%)[a]	Chemoselectivity (%)[b]	cis/trans	ee cis (%)	ee trans (%)
Ru-PyBOx solution phase batch reaction	81	44	68	87/13	48	77
6 : 1 CH₂Cl₂ (RT)[c]	92	53	73	80/20	48	79
7 : 1 no solvent (RT)[c]	92	72	88	83/17	43	79
scCO₂ (40 °C)	41	22	71	85/15	56	89

[a]Cyclopropanation product isolated (combined *cis* and *trans*).
[b]Proportion of cyclopropanation product.
[c]Styrene : ethyldiazoacetate ratio.

4.2.6.6 Asymmetric Conjugate Addition

Amberlyst 21 (**59**) and solid-supported cinchonidine (**60**) have been used to catalyze the Michael reaction between **61** and methyl vinyl ketone **62** in flow (Scheme 4.80). The reactions using Amberlyst 21 were run at 50 °C and required a residence time of 6 h (120 µl/min) for the reaction to reach completion (99% yield). The asymmetric reactions using **60** under the same conditions formed **63** in 97% yield with 52% ee of the S-isomer; the system could be run continuously for 72 h without any observed loss of activity [181].

4.2.6.7 β-Lactam Synthesis

A series of β-lactams (**64**) have been synthesized through the use of an immobilized cinchona alkaloid catalyst. This is postulated to proceed via the cycloaddition of an imine, and a ketene formed *in situ* through deprotonation of an acid chloride (Scheme 4.81). Different system configurations were described in the paper; however, a column filled with a 5 : 1 mixture of solid K₂CO₃ and immobilized-quinine derivative **65** cooled to −45 °C was found to be the most practical. The solution of the acid chloride and imine was dripped through the column and then directed

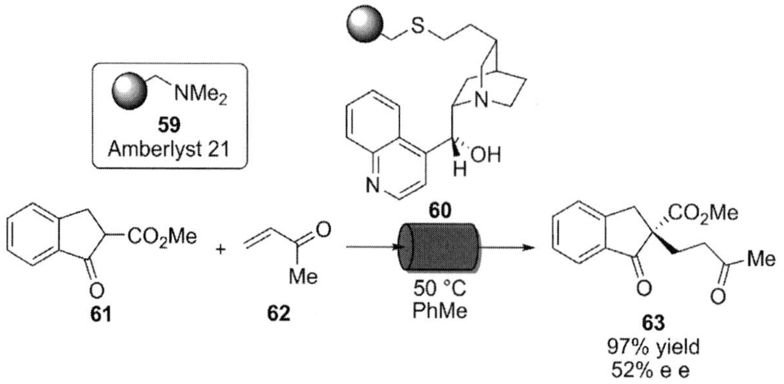

Scheme 4.80 Asymmetric conjugate addition.

through a benzyl amine scavenging resin to remove the excess ketene, affording the β-lactam in 58% yield (3 : 1 d.r., 88% ee). Although this setup did not provide as high yield/selectivity as other reactor setups (65% yield, 10 : 1 d.r., 93% ee), the method was robust enough to form gram quantities of the desired compound [182].

4.2.6.8 Asymmetric Chlorination

The same quinine-derived catalyst **65** could be used to transform acid chlorides into chiral α-chloro-active esters in flow (Scheme 4.82). The authors propose that the reaction proceeds through an ammonium enolate species; however, by analogy to the ketene proposed for the β-lactam synthesis, a pericyclic chloride-transfer ene reaction could also be envisaged. A gravity-fed column packed with **65** was used, acting as both the asymmetric promoter of the reaction and the stoichiometric dehydrohalogenating agent. A plug of the acid chloride and chlorinating agent **66** was trickled through the column cooled to 0 °C and the α-chlorinated ester **67** eluted from the bottom. Once the plug of reagents had passed through the column, the quinine moiety was regenerated using a DIPEA/THF (DIPEA: diisoproylethylamine) solution, and after

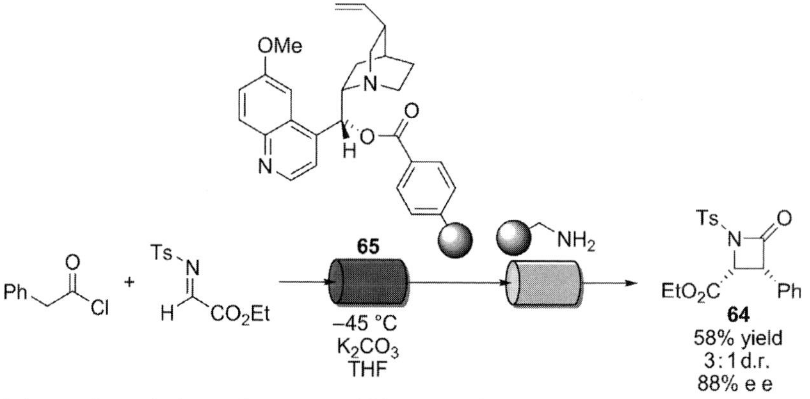

Scheme 4.81 *In situ* ketene formation and β-lactam synthesis.

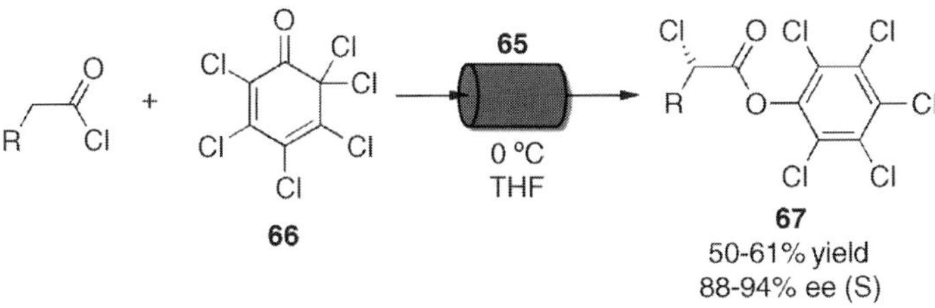

Scheme 4.82 Asymmetric α-chlorination and active ester formation.

further THF washing the column was ready to be reused. The ester products **72** were isolated in 50–61% yield and 88–94% ee [183].

4.2.7
Multistep Synthesis

Solid-supported reagents have been used extensively in multistep organic syntheses [73–77,79,184,185]. Ideally, the use of these reagents should provide clean products without chromatography, crystallization, distillation or any traditional workup procedures. With the rapid growth of single-step transformations within the flow domain, the next obvious challenge is to apply these principles to multistep flow synthesis. Nevertheless, there are obvious difficulties inherent in multistep synthesis in flow, such as (a) the number of reaction steps; (b) compatibility of the solvent with all reaction steps; (c) solvent switching protocols; (d) regeneration of columns; (e) the need for intermediate purification by scavenging or in-line preparative chromatography; (f) dilution effects of solid supports; and (g) flow rate control of HPLC-driven systems. As such, although there are many advantages to be gained from flow synthesis, both experience and engineering are required to fully reap the benefits of flow in multistep transformations.

Lectka used gravity-fed columns for the synthesis of BMS-275291 (**68**), a compound in Stage III clinical trials as a treatment for cancer, in a continuous process (Scheme 4.83). Reagent Stream A underwent the previously described enantioselective α-chlorination/active ester formation (this time using the pseudo-enantiomeric quinidine-derived catalyst **69** [186]) followed by removal of the excess ketene using a piperazine scavenging resin that was found not to react with the active ester. Stream B underwent a PS-DCC (**70**)-mediated peptide coupling, followed by deprotection of the Fmoc-protected amine using PS-Trisamine (**71**). It was the key for Streams A and B to meet each other at a similar concentration to ensure an efficient conversion. The authors worried that the flow rates would have to be timed like a 'ballet dance'; however, at a flow rate of 0.1 ml/min a smooth reaction was found to occur provided that both flow streams started eluting their products at the same time. This combined flow stream was then directed through a column of celite, which allowed the amidation reaction to reach completion. Initial investigations found this step took 18 h; however, by adding a stream of CH_3OH the reaction was completed in 5 h. Finally, the α-chloro group created in Stream A was displaced using

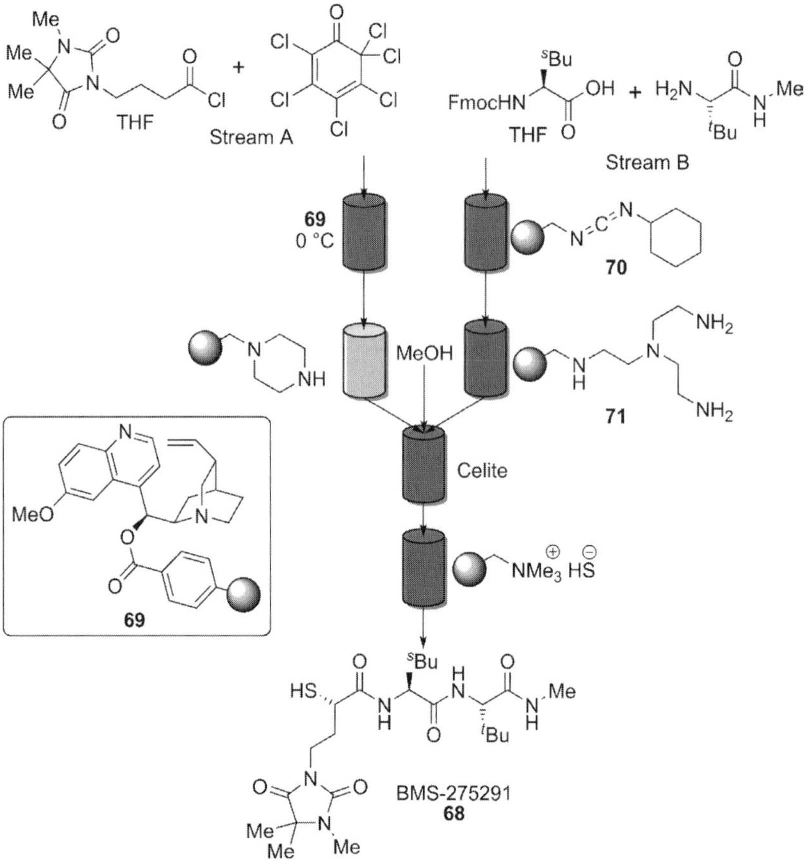

Scheme 4.83 Lectka's flow synthesis of BMS-275291.

HS⁻ immobilized onto an ion-exchange resin, yielding 55 mg of BMS-275291, in 34% yield (83% d.r.). The entire process took approximately 15 h [90].

Our group has developed a method for synthesizing peptides rapidly, reproducibly and in high purity. An N-protected amino acid **72** was pretreated with PyBrOP, **73**, before flowing through a column of PS-HOBt (**74**). This sequesters the amino acid moiety onto the solid phase as the active ester, whereas the by-products of the reaction can be directed to waste (Scheme 4.84). After a suitable washing period, the supported active ester column was then connected in-line to a series of other reagents in the sequence: PS-DMAP (**75**), PS-Bt-OAA and finally PS-SO₃H (**76**). The hydrochloride salt of a second, C-protected, amino acid **77**, was then directed through this series of columns. The PS-DMAP free-based **77** *in situ*, which then flowed through the PS-Bt-OAA forming the peptide bond. The flow stream was then directed through the PS-SO₃H column, scavenging any unreacted amine. Finally, the solvent was evaporated to give the desired product **78** in high purity. Typically 50–200 mg of Boc, Fmoc and Cbz N-protected dipeptides (typical yields 61–81%)

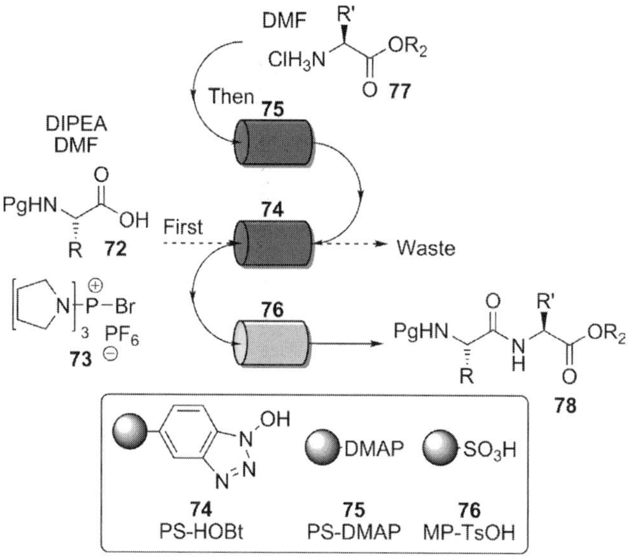

Scheme 4.84 Peptide synthesis in flow.

was generated in only 3–4 h and required no chromatography, which compares well to the overnight reaction times needed in batch [130].

This process could also be applied to the synthesis of tripeptides. A Cbz-protected dipeptide (Cbz-Ala-Gly-OEt) was deprotected in flow using the H-Cube system, and using the above procedure the tripeptide Cbz-Phe-Ala-Gly-OEt was obtained in 59% overall yield and in only 6–7 h, based on the longest linear sequence from glycine.

A more challenging example of multistep synthesis in flow was the synthesis of the natural product grossamide, **79** (Scheme 4.85) [88]. Amide **80** was synthesized according to the protocol described above, and once it was eluted from the PS-SO$_3$H scavenging column the flow stream was diluted (3 : 1) with a second input solution containing a hydrogen peroxide–urea complex and sodium dihydrogen phosphate buffer. This combined flow stream passed through a third column containing horseradish peroxidase on silica to perform the oxidative dimerization of **80** to give grossamide. This paper describes a number of important techniques and procedures that are important for future multistep syntheses: (a) both the input of starting materials and the product elution were controlled using liquid handlers; (b) throughout the initial optimization procedures, the reactions were monitored using in-line HPLC, providing rapid optimization of each reaction step; and (c) the system employed a number of columns and valves, and an in-line UV detector was used to monitor the flow stream's progress. Although the switching was performed manually, a computer could be used to further automate the system using the UV detector as a trigger.

The most important paper to date on multistep synthesis in flow mode is the total synthesis of (±)-oxomaritidine, **81** [89]. This compound has been previously synthesized by our group in five steps and in 62% overall yield using only

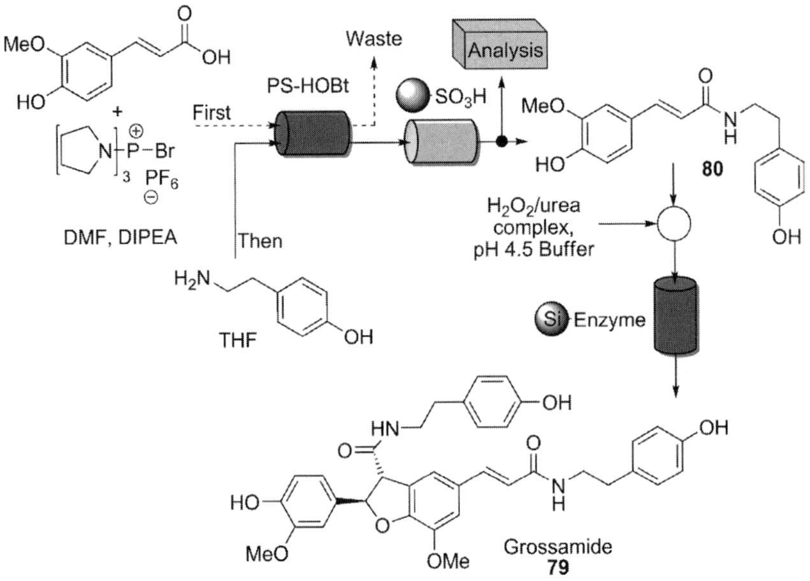

Scheme 4.85 Flow synthesis of grossamide.

solid-supported reagents [187] and was considered an interesting comparison to attempt as a flow synthesis (Scheme 4.86). This synthesis demonstrates many of the advantages of flow chemistry. First, a potentially toxic and explosive azide **82** was generated and immediately reacted with a column of PS-PBu₂ (**83**) furnishing the corresponding aza-Wittig intermediate on polymer support. This prevents exposure of the chemist to the azide and allows the slow and controlled release of nitrogen gas following formation of the iminophosphorane. This column was switched in-line to a second flow stream incorporating the flow oxidation product **26** that reacted to give the desired imine **84**. The solution of imine **84** was subjected to continuous-flow hydrogenation using an H-Cube flow hydrogenator, forming the secondary amine **85**. The THF solvent was removed using a V-10 solvent evaporator [188] and redissolved in CH₂Cl₂, an operation that could be accomplished in less than 10 min.

The last series of reactions consisted of the trifluoroacetylation of **85** using trifluoroacetic anhydride (TFAA) on a glass microfluidic chip heated to 80 °C; the excess TFAA was scavenged using QP-BZA. This reaction stream was directed through a column of PS-PIFA (**86**) that performed the oxidative phenolic coupling, generating the seven-member tricyclic intermediate **87**. This solution then passed into a column of hydroxide ion-exchange resin (**88**), deprotecting the secondary amine that in turn underwent spontaneous cyclization to give the desired compound (±)-oxomaritidine (**81**) in 90% purity.

The entire sequence was completed in around 6 h, which compares favorably to an estimated batch reaction time of 4 days under conventional batch conditions. The overall reproducible yield of the synthesis was found to be 40%; however, it was found that the phenolic oxidation step was yield-limiting (producing only 50% of the desired

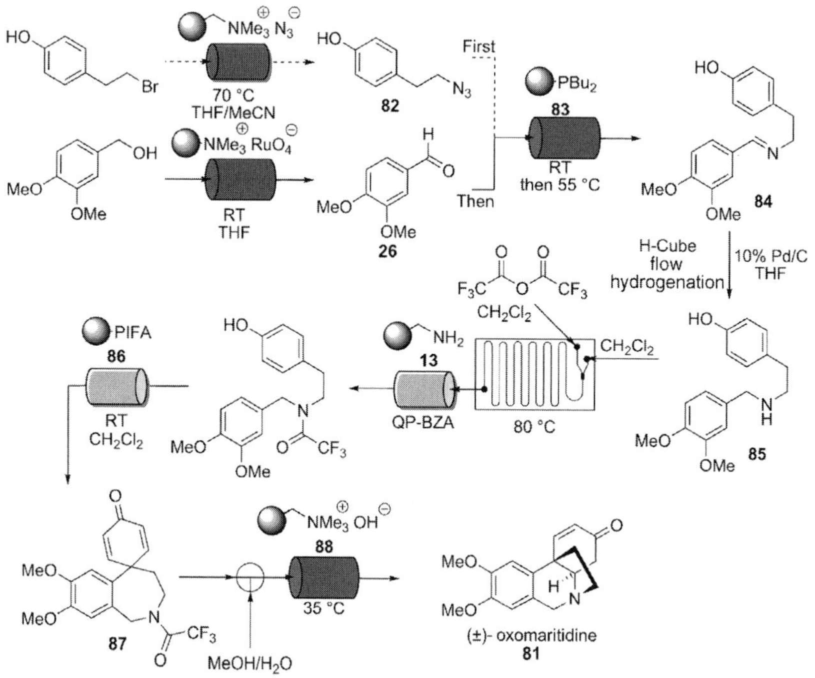

Scheme 4.86 Total synthesis of (±)-oxomaritidine in flow.

product) – the other steps occurred in near quantitative fashion. It is important to note that there are many differences between the solid supported reagent and flow synthesis routes. The previously reported synthesis could not be robotically transferred to flow, illustrating that solution phase, immobilized reagent and flow chemistries all have their own intricacies. It will take time for researchers to optimize and employ these varied technologies effectively.

4.2.8
Conclusions and Outlook

There is no doubt that the incorporation of immobilized species adds enormously to the application of flow-based chemistry for the production of advanced, pure chemical entities. Nevertheless, the area continues to evolve to meet the challenges posed, especially those involved in multistep applications. However, with the commercial availability of modular and readily reconfigured flow reactor components, the field is changing rapidly. Restyled, flexible laboratory arrangements and on-demand synthesis of building blocks are creating new working environments. Minimization of chemical waste and excessive solvent use is forcing a change of attitude. Similarly, the changing demographics stimulate the need for enhanced innovative synthesis technology. Flow chemistry inevitably generates information-rich sequences that aid the discovery process; however, the devices are only effective in the skilled hands of the operators and should not be seen as replacements for manpower. Rather, they are an enhancement to the skills and creativity of the chemist, thereby incentivizing the workforce. Integration of the devices into wireless knowledge capture and evaluation through *in silico* avatars are realistic. Given the opportunities for the more accurate evaluation of reactive surfaces and the massive developments in nanotechnology, the picture is both assuring and exciting.

References

67 Karnatz, F.A. and Whitmore, F.C. (1932) *Journal of the American Chemical Society,* **54**, 3461.

68 Akelah, A. and Sherrington, D.C. (1981) *Chemical Reviews,* **81**, 557–587.

69 Ley, S.V., Baxendale, I.R., Bream, R.N., Jackson, P.S., Leach, A.G., Longbottom, D.A., Nesi, M., Scott, J.S., Storer, R.I. and Taylor, S.J. (2000) *Journal of the Chemical Society, Perkin Transactions 1,* 3815–4195.

70 Kirschning, A., Monenschein, H. and Wittenberg, R. (2001) *Angewandte Chemie-International Edition,* **40**, 650–679.

71 Kirschning, A. (2004) Immobilized catalysts, in *Topics in Current Chemistry,* Springer, Berlin, Heidelberg.

72 Ley, S.V., Baxendale, I.R. and Myers, R.M. (2006) Polymer-supported reagents and scavengers in synthesis, in *Comprehensive Medicinal Chemistry II, Drug Discovery Technologies, vol. 3* (eds D.J. Triggle and J. B. Taylor), Elsevier, Oxford, pp. 791–836.

73 Nicolaou, K.C., Winssinger, N., Pastor, J., Ninkovic, S., Sarabia, F., He, Y., Vourloumis, D., Yang, Z., Li, T., Giannakakou, P. and Hamel, E. (1997) *Nature,* **387**, 268–272.

74 Baxendale, I.R., Ley, S.V., Nessi, M. and Piutti, C. (2002) *Tetrahedron Letters,* **58**, 6285–6304.

75 Baxendale, I.R., Storer, R.I. and Ley, S.V. (2003) Supported reagents and scavengers in multi-step organic synthesis, in *Polymeric Materials in Organic Synthesis and Catalysis* (ed. M.R. Buchmeiser), VCH, Berlin, pp. 53–136.

76 Storer, R.I., Takemoto, T., Jackson, P.S. and Ley, S.V. (2003) *Angewandte Chemie-International Edition,* **42**, 2521–2525.

77 Storer, R.I., Takemoto, T., Jackson, P.S., Brown, D.S., Baxendale, I.R. and Ley, S.V. (2004) *Chemistry – A European Journal,* **10**, 2529–2547.

78 Baxendale, I.R. and Ley, S.V. (2005) *Industrial & Engineering Chemistry Research,* **44**, 8588–8592.

79 Baxendale, I.R. and Ley, S.V. (2005) *Current Organic Chemistry,* **9**, 1521–1534.

80 Ley, S.V., Baxendale, I.R. and Myers, R.M. (2006) The use of polymer supported reagents and scavengers in the synthesis of natural products, in *Combinatorial Synthesis of Natural Product-Based Libraries* (ed. E.A.M. Boldi), CRC Press, Boca-Raton, FL, USA, pp. 131–163.

81 Hodge, P. (1997) *Chemical Society Reviews,* **26**, 417–424.

82 Hodge, P. (2003) *Current Opinion in Chemical Biology,* **7**, 362–373.

83 Jas, G. and Kirschning, A. (2003) *Chemistry – A European Journal,* **9**, 5708–5723.

84 Hodge, P. (2005) *Industrial & Engineering Chemistry Research,* **44**, 8542–8553.

85 Kirschning, A., Solodenko, W. and Mennecke, K. (2006) *Chemistry – A European Journal,* **12**, 5972–5990.

86 Baxendale, I.R. and Ley, S.V. (2007) Solid supported reagents in multi-step flow synthesis, in *New Avenues to Efficient Chemical Synthesis – Emerging Technologies* (eds P.H. Seeberger and T. Blume), Springer, Berlin, Heidelberg, pp. 151–185.

87 Hornung, C.H., Mackley, M.R., Baxendale, I.R. and Ley, S.V. (2007) *Organic Process Research & Development,* **11**, 399–405.

88 Baxendale, I.R., Griffiths-Jones, C.M., Ley, S.V. and Tranmer, G.K. (2006) *Synlett*, **3**, 427–430.

89 Baxendale, I.R., Deeley, J., Griffiths-Jones, C.M., Ley, S.V., Saaby, S. and Tranmer, G.K. (2006) *Chemical Communications*, **24**, 2566–2568.

90 France, S., Bernstein, D., Weatherwax, A. and Lectka, T. (2005) *Organic Letters*, **7**, 3009–3012.

91 Merrifield, R.B. (1963) *Journal of the American Chemical Society*, **85**, 2149–2154.

92 Bayer, E., Jung, G., Halasz, I. and Sebastian, I. (1970) *Tetrahedron Letters*, **51**, 4503–4505.

93 Lukas, T.J., Prystowsky, M.B. and Erickson, B.W. (1981) *Proceedings of the National Academy of Sciences of the United States of America*, **78**, 2791–2795.

94 Atherton, E., Brown, E., Sheppard, R.C. and Rosevear, A. (1981) *Chemical Communications*, **21**, 1151–1152.

95 Dryland, A. and Sheppard, R.C. (1986) *Journal of the Chemical Society, Perkin Transactions 1*, 125–137.

96 Dryland, A. and Sheppard, R.C. (1988) *Tetrahedron*, **44**, 859–876.

97 Harrison, C.R. and Hodge, P. (1976) *Journal of the Chemical Society, Perkin Transactions 1*, 2252–2254.

98 Angeletti, E., Canepa, C., Martinetti, G. and Venturello, P. (1988) *Tetrahedron Letters*, **29**, 2261–2264.

99 Angeletti, E., Canepa, C., Martinetti, G. and Venturello, P. (1989) *Journal of the Chemical Society, Perkin Transactions 1*, 105–107.

100 Itsuno, S., Ito, K., Maruyama, T., Kanda, N., Hirao, A. and Nakahama, S. (1986) *Bulletin of the Chemical Society of Japan*, **59**, 3329–3331.

101 Jönsson, D., Warrington, B.H. and Ladlow, M. (2004) *Journal of Combinatorial Chemistry*, **6**, 584–595.

102 Griffiths-Jones, C.M., Hopkin, M.D., Jonsson, D., Ley, S.V., Tapolczay, D.J., Vickerstaffe, E. and Ladlow, M. (2007) *Journal of Combinatorial Chemistry*, **9**, 422–430.

103 Baumann, M., Baxendale, I.R., Ley, S.V., Smith, C.D. and Tranmer, G.K. (2006) *Organic Letters*, **8**, 5231–5234.

104 Baxendale, I.R., Ley, S.V., Smith, C.D., Tamborini, L. and Voica, F. Manuscript in preparation.

105 Smith, C.J., Iglesias-Sigüenza, F.J., Baxendale, I.R. and Ley, S.V. (2007) *Organic & Biomolecular Chemistry*, 2759–2761.

106 Moorhouse, A.D. and Moses, J.E. (2007) *Chemical Society Reviews*, **36**, 1249–1262.

107 Hinchcliffe, A., Hughes, C., Pears, D.A. and Pitts, M.R. (2007) *Organic Process Research & Development*, **11**, 477–481.

108 Smith, C.D., Baxendale, I.R., Lanners, S., Hayward, J.J., Smith, S.C. and Ley, S.V. (2007) *Organic & Biomolecular Chemistry*, **5**, 1559–1561.

109 Svec, F. and Huber, C.G. (2006) *Analytical Chemistry*, **78**, 2100–2107.

110 Svec, F. (2004) *Journal of Separation Science*, **27**, 747–766.

111 Peters, E.C., Svec, F. and Fréchet, J.M.J. (1999) *Advanced Materials*, **11**, 1169–1181.

112 Tripp, J.A., Stein, J.A., Svec, F. and Fréchet, J.M.J. (2000) *Organic Letters*, **2**, 195–198.

113 Tripp, J.A., Svec, F. and Fréchet, J.M.J. (2001) *Journal of Combinatorial Chemistry*, **3**, 216–223.

114 Svec, F. and Fréchet, J.M.J. (1995) *Macromolecules*, **28**, 7580–7582.

115 Viklund, C., Svec, F., Fréchet, J.M.J. and Irgum, K. (1996) *Chemistry of Materials*, **8**, 744–750.

116 Kirschning, A., Altwicker, C., Dräger, G., Harders, J., Hoffmann, N., Hoffmann, U., Schönfeld, H., Solodenko, W. and Kunz, U. (2001) *Angewandte Chemie-International Edition*, **40**, 3995–3998.

117 Solodenko, W., Kunz, U., Jas, G. and Kirschning, A. (2002) *Bioorganic and Medicinal Chemistry*, **12**, 1833–1835.

118 Hegedus, L.S. (1999) *Transition Metals in the Synthesis of Complex Organic Molecules*, 2nd edn, University Science Books, Sausalito, CA.

119 Smith, M.D., Stepan, A.F., Ramarao, C., Brennan, P.E. and Ley, S.V. (2003) *Chemical Communications*, 2652–2653.

120 Andrews, S.P., Stepan, A.F., Tanaka, H., Ley, S.V. and Smith, M.D. (2005) *Advanced Synthesis & Catalysis*, **347**, 647–654.

121 Lohmann, S., Andrews, S.P., Burke, B.J., Smith, M.D., Attfield, J.P., Tanaka, H., Kaneko, K. and Ley, S.V. (2005) *Synlett*, 1291–1295.

122 Kunz, U., Kirschning, A., Wen, H.-L., Solodenko, W., Cecilia, R., Kappe, C.O. and Turek, T. (2005) *Catalysis Today*, **105**, 318–324.

123 Ley, S.V., Ramarao, C., Gordon, R.S., Holmes, A.B., Morrison, A.J., McConvey, I.F., Shirley, I.M., Smith, S.C. and Smith, M.D. (2002) *Chemical Communications*, 1134–1135.

124 Ley, S.V., Ramarao, C., Lee, A.-L., Østergaard, N., Smith, S.C. and Shirley, I. M. (2003) *Organic Letters*, **5**, 185–187.

125 Broadwater, S.J. and McQuade, D.T. (2006) *Journal of Organic Chemistry*, **71**, 2131–2134.

126 Haag, R. and Roller, S. (2004) Polymeric supports for the immobilisation of catalysts, in *Immobilized Catalysts* (ed. A. Kirschning), Springer, Berlin, Heidelberg, New York, pp.1–42.

127 Desai, B. and Kappe, C.O. (2005) *Journal of Combinatorial Chemistry*, **7**, 641–643.

128 Jones, R.V., Godorhazy, L., Varga, N., Szalay, D., Urge, L. and Darvas, F. (2006) *Journal of Combinatorial Chemistry*, **8**, 110–116.

129 Saaby, S., Knudsen, K.R., Ladlow, M. and Ley, S.V. (2005) *Chemical Communications*, 2909–2911.

130 Baxendale, I.R., Ley, S.V., Smith, C.D. and Tranmer, G.K. (2006) *Chemical Communications*, 4835–4837.

131 Franckevičius, V., Knudsen, K.R., Ladlow, M., Longbottom, D.A. and Ley, S.V. (2006) *Synlett*, 889–892.

132 Knudsen, K.R., Holden, J., Ley, S.V. and Ladlow, M. (2007) *Advanced Synthesis & Catalysis*, **349**, 535–538.

133 Kunz, U., Schonfeld, H., Kirschning, A. and Solodenko, W. (2003) *Journal of Chromatography A*, **1006**, 241–249.

134 Solodenko, W., Wen, H., Leue, S., Stuhlmann, F., Sourkouni-Argirusi, G., Jas, G., Schönfeld, H., Kunz, U. and Kirschning, A. (2004) *European Journal of Organic Chemistry*, 3601–3610.

135 Kobayashi, J., Mori, Y., Okamoto, K., Akiyama, R., Ueno, M., Kitamori, T. and Kobayashi, S. (2004) *Science*, **304**, 1305–1308.

136 Kobayashi, J., Mori, Y. and Kobayashi, S. (2005) *Chemical Communications*, **20**, 2567–2568.

137 Kobayashi, J., Mori, Y. and Kobayashi, S. (2005) *Advanced Synthesis & Catalysis*, **347**, 1889–1892.

138 Künzle, N., Mallat, T. and Baiker, A. (2003) *Applied Catalysis A – General*, **238**, 251–257.

139 Sandee, A.J., Petra, D.G.I., Reek, J.N.H., Kamer, P.C.J. and van Leeuwen, P.W.N. M. (2001) *Chemistry – A European Journal*, **7**, 1202–1208.

140 Laue, S., Greiner, L., Wöltinger, J. and Liese, A. (2001) *Advanced Synthesis & Catalysis*, **343**, 711–720.

141 Kobayashi, S., Miyamura, H., Akiyama, R. and Ishida, T. (2005) *Journal of the American Chemical Society*, **127**, 9251–9254.

142 Wiles, C., Watts, P. and Haswell, S.J. (2006) *Tetrahedron Letters*, **47**, 5261–5264.

143 Kunz, U., Schönfeld, H., Solodenko, W., Jas, G. and Kirschning, A. (2005) *Industrial & Engineering Chemistry Research*, **44**, 8458–8467.

144 Karbass, N., Sans, V., Garcia-Verdugo, E., Burguete, M.I. and Luis, S.V. (2006) *Chemical Communications*, 3095–3097.

145 http://www.vapourtec.co.uk/.

146 Nikbin, N., Ladlow, M. and Ley, S.V. (2007) *Organic Process Research & Development*, **11**, 458–462.

147 He, P., Haswell, S.J. and Fletcher, P.D.I. (2004) *Lab on a Chip*, **4**, 38–41.

148 He, P., Haswell, S.J. and Fletcher, P.D.I. (2004) *Applied Catalysis A – General*, **274**, 111–114.

149 Ley, S.V., Mitchell, C., Pears, D., Ramarao, C., Yu, J.-Q. and Zhou, W. (2003) *Organic Letters*, **5**, 4665–4668.

150 Yu, J.-Q., Wu, H.-C., Ramarao, C., Spencer, J.B. and Ley, S.V. (2003) *Chemical Communications*, 678–679.

151 Lee, C.K.Y., Holmes, A.B., Ley, S.V., McConvey, I.F., Al-Duri, B., Leeke, G.A., Santos, R.C.D. and Seville, J.P.K. (2005) *Chemical Communications*, 2175–2177.

152 Baxendale, I.R., Griffiths-Jones, C.M., Ley, S.V. and Tranmer, G.K. (2006) *Chemistry – A European Journal*, **12**, 4407–4416.

153 Shore, G., Morin, S. and Organ, M.G. (2006) *Angewandte Chemie-International Edition*, **45**, 2761–2766.

154 Phan, N.T.S., Khan, J. and Styring, P. (2005) *Tetrahedron*, **61**, 12065–12073.

155 Haswell, S.J., O'Sullivan, B. and Styring, P. (2001) *Lab on a Chip*, **1**, 164–166.

156 Phan, N.T.S., Brown, D.H. and Styring, P. (2004) *Green Chemistry*, **6**, 526–532.

157 Ahmed, M., Barrett, A.G.M., Braddock, D.C., Cramp, S.M. and Procopiou, P.A. (1999) *Tetrahedron Letters*, **40**, 8657–8662.

158 Ahmed, M., Arnauld, T., Barrett, A.G.M., Braddock, D.C. and Procopiou, P.A. (2000) *Synlett*, 1007–1009.

159 Jafarpour, L. and Nolan, S.P. (2000) *Organic Letters*, **2**, 4075–4078.

160 Krause, J.O., Lubbad, S.H., Nuyken, O. and Buchmeiser, M.R. (2003) *Macromolecular Rapid Communications*, **24**, 875–878.

161 Michrowska, A., Guajski, L. and Grela, K. (2006) *Chemical Communications*, 841–843.

162 Michrowska, A., Guajski, L., Kaczmarska, Z., Mennecke, K., Kirschning, A. and Grela, K. (2006) *Green Chemistry*, **8**, 685–688.

163 Michrowska, A., Mennecke, K., Kunz, U., Kirschning, A. and Grela, K. (2006) *Journal of the American Chemical Society*, **128**, 13261–13267.

164 Katsuki, T. (2003) *Synlett*, 281–297.

165 Yoon, T.P. and Jacobsen, E.N. (2003) *Science*, **299**, 1691–1693.

166 Jacobsen, E.N. (2000) *Accounts of Chemical Research*, **33**, 421–431.

167 Collman, J.P., Belmont, J.A. and Brauman, J.I. (1983) *Journal of the American Chemical Society*, **105**, 7288–7294.

168 Annis, D.A. and Jacobsen, E.N. (1999) *Journal of the American Chemical Society*, **121**, 4147–4154.

169 Solodenko, W., Jas, G., Kunz, U. and Kirschning, A. (2007) *Synthesis*, 583–589.

170 Itsuno, S., Sakurai, Y., Ita, K., Maruyama, T., Nakahama, S. and Fréchet, J.M.J. (1990) *Journal of Organic Chemistry*, **55**, 304–310.

171 Hodge, P., Sung, D.W.L. and Stratford, P.W. (1999) *Journal of the Chemical Society, Perkin Transactions 1*, 2335–2342.

172 Burguete, M.I., Garcá-Verdugo, E., Vicent, M.J., Luis, S.V., Pennemann, H., von Keyserling, N.G. and Martens, J. (2002) *Organic Letters*, **4**, 3947–3950.

173 Corey, E.J. (2002) *Angewandte Chemie-International Edition*, **41**, 1650–1667.

174 Kamahori, K., Ito, K. and Itsuno, S. (1996) *The Journal of Organic Chemistry*, **61**, 8321–8324.

175 Altava, l., Burguete, M.I., Garcá-Verdugo, E., Luis, S.V., and Vicent, M.J. 2006, *Green Chemistry*, **8**, 717–726.

176 Altava, B., Burguete, M.I., Fraile, J.M., Garcá, J.I., Luis, S.V., Mayoral, J.A. and Vicent, M.J. (2000) *Angewandte Chemie-International Edition*, **39**, 15031506.

177 Johnson, J.S. and Evans, D.A. (2000) *Accounts of Chemical Research*, **33**, 325–335.

178 Mandoli, A., Orlandi, S., Pini, D. and Salvadori, P. (2004) *Tetrahedron: Asymmetry*, **15**, 3233–3244.

179 Cornejo, A., Fraile, J.M., Garcá, J.I., Gil, M.J., Luis, S.V., Martínez-Merino, V. and

Mayoral, J.A. (2005) *The Journal of Organic Chemistry*, **70**, 5536–5544.

180 Burguete, M.I., Cornejo, A., Garća-Verdugo, E., Gil, M.J., Luis, S.V., Mayoral, J.A., Martínez-Merino, V. and Sokolova, M. (2007) *The Journal of Organic Chemistry*, **72**, 4344–4350.

181 Bonfils, F., Cazaux, I., Hodge, P. and Caze, C. (2006) *Organic & Biomolecular Chemistry*, **4**, 493–497.

182 Hafez, A.M., Taggi, A.E., Dudding, T. and Lectka, T. (2001) *Journal of the American Chemical Society*, **123**, 10853–10859.

183 Bernstein, D., France, S., Wolfer, J. and Lectka, T. (2005) *Tetrahedron: Asymmetry*, **16**, 3481–3483.

184 Baxendale, I.R., Lee, A.-L. and Ley, S.V. (2002) *Journal of the Chemical Society, Perkin Transactions 1*, 1850–1857.

185 Baxendale, I.R., Ley, S.V., Nessi, M. and Piutti, C. (2002) *Tetrahedron*, **58**, 6285–6304.

186 Personal communication from T. Lectka.

187 Ley, S.V., Schucht, O., Thomas, A.W. and Murray, P.J. (1999) *Journal of the Chemical Society, Perkin Transactions 1*, 1251–1252.

188 http://www.biotage.com/.

4.3
Liquid–Liquid Biphasic Reactions

Batoul Ahmed-Omer, Thomas Wirth

4.3.1
Introduction

Organic solvents play a key role in many chemical processes within the pharmaceutical and chemical industry. Solvent loss during the process is often inevitable because of its volatility causing environmental concerns and adverse health effects. Hence, the search for cleaner chemical processes that reduce the release of harmful chemicals into the environment is a great challenge in organic synthesis. The use of two immiscible solvents in so-called biphasic reactions can be a possible solution to solvent loss or recycling as different solubilities of substrate, reagent/catalyst and product can be used advantageously to setup an economically and environmentally friendly process. There are many applications of biphasic reactions in many different areas of chemistry [189–191].

Examples of applying biphasic systems to catalyzed reactions, such as phase-transfer catalysis, overpower the stoichiometric reactions. In a typical catalytic biphasic system, one phase contains the catalyst, while the other phase contains the substrate. In some systems, the catalyst and substrates are in the same phase, while the product produced is transferred to the second phase. In a typical reaction, when the two phases are mixed during the reaction and after completion, the catalyst remains in one phase ready for recycling while the product can be isolated from the second phase. The most common solvent combination consists of an organic solvent combined with another immiscible solvent that, in most applications, is water. However, there are few examples of suitable water-soluble and stable catalysts, and therefore various applications are limited to some extent [192]. Immiscible solvents other than water are recently becoming more applicable in biphasic catalysis because of the better solubility and stability of various catalysts in such solvents. For example, ionic liquids and fluorous solvents have many successful applications in liquid–liquid

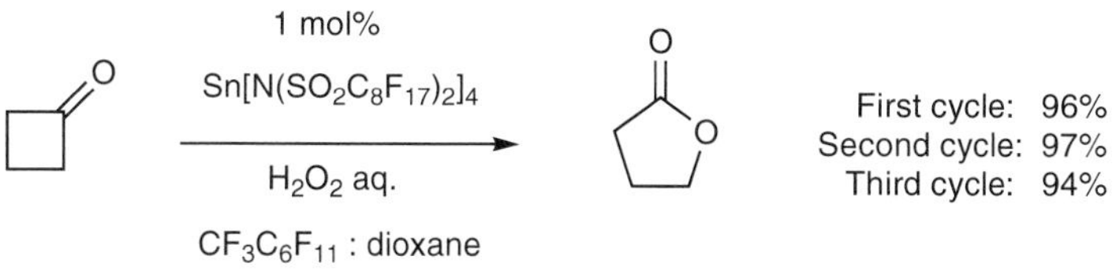

1 mol%

Sn[N(SO$_2$C$_8$F$_{17}$)$_2$]$_4$

H$_2$O$_2$ aq.

CF$_3$C$_6$F$_{11}$: dioxane

First cycle: 96%
Second cycle: 97%
Third cycle: 94%

Figure 4.8 Baeyer–Villiger reaction of cyclobutanone in fluorous media [196].

biphasic syntheses, such as Heck reactions and hydroformylations using ionic liquid media, or Baeyer–Villiger reactions using Lewis acidic catalysts in fluorous biphasic systems as shown as an example in Figure 4.8 [193–196].

4.3.2
Background

In liquid–liquid systems, molecules at the region of contact between the two phases have a different molecular environment compared to those in the bulk of both phases. There are equal cohesive forces in all directions between molecules inside the phase, whereas those at the region of contact have unbalanced cohesive forces because they are not wholly surrounded by the same molecules. Consequently, they are strongly attracted toward the direction of the bulk phase. As a result, a boundary between the two phases, known as the interface or surface area, is formed making it more difficult for one phase mixing with the other. The force at the surface or interface is defined as the surface or interfacial tension. The term interface is used when both phases are liquids, whereas surface is usually used for gas–liquid or solid–liquid systems. An example of boundary formation in nature is the liquid–air surface in which the unbalanced attractive forces result in liquid surface contraction by pulling molecules at the surface to the bulk of the liquid. This contraction is the reason for the formation of spherical liquid droplets in nature. The stronger the cohesive forces in a phase, the higher the interfacial tension. Any decrease in the strength of the interaction will lead to a weaker tension that consequently increases the miscibility of the system. There are many ways of increasing the miscibility; for example, by adding surfactants, mechanical stirring or applying high temperatures. In liquid–liquid systems, the surface in each phase is subject to forces from the other that makes the surface tension weaker than the gas–liquid surface. The formation of a boundary in a liquid–liquid system depends not only on the differences in the molecular environment between the phase and the interface area but also on the degree of phase saturation in certain solvent systems. This is especially the case for partially miscible liquid–liquid systems. Such systems consist of phase α and phase β. The addition of a small amount of phase α to a large amount of phase β leads to the formation of a miscible system. When phase α is continuously added to the system, a single phase is retained until phase β becomes fully saturated with phase α, and then two separated phases are formed. If addition of phase α continues until phase β becomes the minor and completely soluble phase, a miscible system will form again [189,197,198].

4.3.3
Kinetics of Biphasic Systems

Reactions in biphasic systems can take place either at the interface or in the bulk of one of the phases. The reaction at the interface depends on the reactants meeting at the interface boundary. This means, the interface area as well as the diffusion rate across the bulk of the phase plays an important role. On the contrary, in reactions that take place in the bulk phase, the reactants have to be transferred first through the interface before the reactions take place. In this case, the rate of diffusion across the interface is an important factor. Diffusion across the interface is more complicated than the diffusion across a phase, as the mass transfer of the reactant across the interface must be taken into account. Hence, the solubility of the reactants in each phase has to be considered, as this has an effect on diffusion across the interface. In a system where the solubility of a reactant is the same in both phases, the reactant diffuses from the concentrated phase to the less concentrated phase across the interface. This takes into account the mass transfer of the reactant from one phase into the other through the interface. The rate of diffusion J in such systems is described in Equation 4.1, where D is the diffusion coefficient, x is the diffusion distance and l is the interface thickness (Figure 4.9).

$$ J = \frac{D}{l\,d([A]-[B])/\mathrm{d}x}. \tag{4.1} $$

A reactant with different solubilities in both phases will lead to diffusion in the direction of the phase in which the reactant is more soluble. Hence, the diffusion is affected by the concentration relative to the saturation of the solution and not by the absolute concentration of the reactant. The ratio of distribution of solute between the two phases is known as partition coefficient that can be determined by measurements

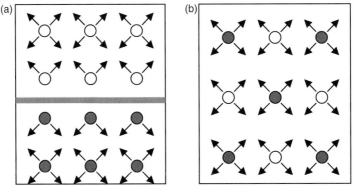

Figure 4.9 (a) Formation of an interface in a biphasic system of two solvents as result of unbalanced forces at the boundary in contrast to (b) a single phase system of two solvents without any boundary as a homogenous mixture.

of the relative solubility of the reagent in each phase of the system under identical physical conditions.

4.3.4
Biphasic Flow in Microchannels

For a biphasic reaction in a flask, the liquid phase with greater density is found at the bottom, whereas the lighter liquid is on top forming an undisturbed flat interface. When the system is agitated, drops of various sizes of one phase form within the other phase. In a microchannel, the two immiscible liquids create various flow patterns, from segregation of phases by each other to stratified parallel flow. These flow patterns are characterized by a number of dimensionless parameters that depend on the channel dimension's channel surface and the liquid properties. Variation of these parameters affects the stability of the flow pattern and can lead to transition from one flow pattern to another. The characteristics of liquid flow in a channel were first studied in 1883 by Osborne Reynolds by pumping a liquid continuously into a glass tube while introducing a fine strand of colored water to the flow. Reynolds observed that at low flow rate, laminar flow behavior was dominated in which the colored strand flew in straight parallel streams along the flow direction. As the flow rate was increased, the colored strands were broken into vortices until a point where turbulent flow behavior was dominant across the tube [199,200] (Figure 4.10).

Transformation from laminar to turbulent flow is characterized by a significant value of a dimensionless quantity known as Reynolds number (Re). Reynolds number relates inertial and viscous forces as represented in Equation 4.2; where ρ is the fluid density, v_s is the fluid velocity, L is the characteristic length (cross section flow in a channel) and μ is the fluid viscosity.

Reynolds number (Re):

$$Re = \rho v_s \frac{L}{\mu} \quad (Re = \text{inertial forces/viscous forces}). \tag{4.2}$$

At low values of the Reynolds number ($Re < 2000$), viscous force dominates resulting in a laminar flow. At a high Reynolds number ($Re > 3000$), inertial forces are dominant resulting in a turbulent flow. However, within a certain range of Reynolds number (Re 2000–3000), the flow is neither laminar nor turbulent because the transformation occurs gradually.

(a) **Laminar flow** (b) **Turbulent flow**

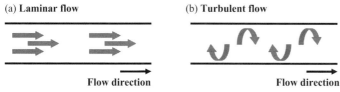

Flow direction **Flow direction**

Figure 4.10 Schematic illustration of the liquid flow in a channel shown by arrows: (a) Laminar flow at low flow rate and (b) turbulent flow at high flow rate.

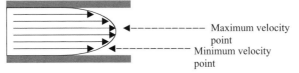

Figure 4.11 Velocity profile of single laminar flow in a microchannel.

The velocity profile of the flow in a channel varies across the diameter of the channel regardless of the flow rate. It has a minimum value ($\sim$0) near the channel walls and a maximum value at the center of the flow. This variation in velocities arises from the adhesive forces between the channel walls and the liquid, causing the liquid layers nearest to the walls to be slower than those in the center. Consequently, the layer region nearest to the wall always exhibits a laminar flow even at high Reynolds numbers as the viscous forces dominate [200,201] (Figure 4.11).

Viscous and inertial forces are related to surface tension by the dimensionless Capillary and Weber numbers. Capillary number (Ca), as shown in Equation 4.3, describes the relative importance of viscosity and surface tension, where μ represents the viscosity, υ is the velocity and σ is the surface tension.

Capillary number (Ca)

$$Ca = \frac{\mu\upsilon}{\sigma}.\tag{4.3}$$

The Weber number (We) is shown in Equation 4.4 and relates the inertial forces to the surface tension, where ρ is the density of the fluid, l represents the characteristic channel length, υ is the velocity of the fluid and σ is the surface tension.

Weber number (We)

$$We = \frac{\rho\upsilon^2 l}{\sigma}.\tag{4.4}$$

In microscale channels, the viscous forces dominate the inertial effect resulting in a low Reynolds numbers. Hence, laminar flow behavior is dominant and mixing occurs via diffusion. However, in a liquid–liquid system, the interfacial forces acting on the interface add complexity to the laminar flow as the relationship between interfacial forces and other forces of inertia and viscous results in a variety of interface and flow patterns. Günther and Jensen [202] illustrated this relationship as a function of the channel dimension and velocity as shown in Figure 4.12. The most regularly shaped flow pattern is achieved when interfacial forces dominate over inertia and viscous forces at low Reynolds numbers, as represented in Figure 4.12 by the area below the yellow plane [202,203].

In a macroscale channel, gravitational force has an effect on the flow pattern of a biphasic system; consequently, the flow pattern varies between vertical and horizontal channels. However, in a microchannel, the gravity effect is dominated by the viscous forces that are expressed by the ratio of gravity force and the surface tension using the Bond number (Bo) as expressed in Equation 4.5; where $\Delta\rho$ is the density difference between two immiscible liquids, g is acceleration due to gravity, d_h is the channel dimension and σ the surface tension.

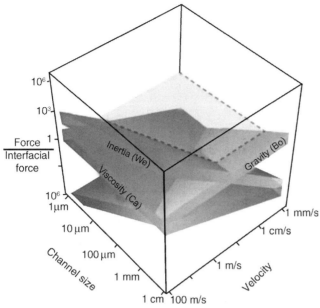

Figure 4.12 Effect of interfacial forces on inertia, viscous, gravity forces with velocity and channel diameter: the balance between these forces where all have a value of 1 is represented by the plane with dotted boundaries (Reproduced by permission of The Royal Society of Chemistry) [202].

Bond number (Bo)

$$Bo = \frac{(\Delta\rho)g d_h^2}{\sigma}. \qquad (4.5)$$

Low Bond numbers indicate the dominance of surface tension over the gravitational force in a system. By applying factors that can affect the strength of the interfacial forces, by using surfactants or applying high temperatures, for example, the gravity force can then dominate resulting in a high Bond number.

4.3.5
Surface–Liquid and Liquid–Liquid Interactions

When cohesive forces of a liquid exceed the attractive forces between the liquid and the surface (adhesive forces), a high-surface tension is formed resulting in a drop of liquid on the surface. On the contrary, when the adhesive forces dominate the cohesive forces, a low-surface tension forms and the liquid will spread out over and wet the surface. The extent of contraction or spreading of the liquid depends on the strength of surface tension or, in other words, the wetting properties of the surface and liquid. The degree of wetting can be expressed with the contact angle θ between

(a) (b)

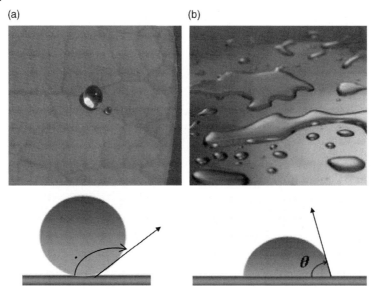

Figure 4.13 Wetting versus nonwetting. (a) The water drop on a
lotus leaf with hydrophobic surface leading to a large contact angle
[204]. (b) The water drop on a glass surface with hydrophilic
properties leading to a small contact angle [205].

the surface and the liquid. For example, a drop of water on a hydrophobic (nonwetting) surface would have a large contact angle going toward 180°, whereas when the water spreads on a hydrophilic surface (wetting), the contact angle is small, approaching zero [203] (Figure 4.13).

Accordingly, on a surface with a medium hydrophobic property, the interface boundary of oil and water mixture, for instance, will contact the wall surface forming a point between the interface and the surface. The relation between the surface tension and the two liquids can be expressed by Young's equation: where $\gamma_{\text{oil–water}}$ is the interface tension between oil and water, while $\gamma_{\text{oil-surface}}/\gamma_{\text{water-surface}}$ is the surface tension between the channel walls and the two phases [202]

$$\gamma_{\text{oil–water}}\cos\theta = \gamma_{\text{oil-surface}} - \gamma_{\text{water-surface}}. \tag{4.6}$$

The effect of surface tension in a microchannel with the same wetting property has on a flow of oil–water will result in a segmented flow pattern given the fact that oil has some wetting properties on the surface. The oil spreads through the channel material, whereas the water forms segments that fill the channel diameter. In this case, the surface tension between the channel surface and the water has to be higher than the interface tension between the water and oil. If the interfacial tension between oil and water is higher than that between the water and surface, the use of surfactants, for example, can help to reduce the interfacial tension. However, the surface tension between the two phases should be high enough to maintain the shape of the segments and avoid their destruction caused by the shear stress between them. In contrast, the interface boundary of oil and water mixture on a hydrophilic surface

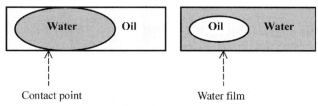

Contact point   Water film

Figure 4.14 (a) Segmented flow of oil and water in a channel with a medium hydrophobic properties – the interface boundary and the channel surface are in contact; (b) droplet flow of oil and water in a channel with hydrophilic properties – the interface boundary and the channel surface are not in contact.

does not form a contact point, as a result of water phase spreading over the surface forming a film that separates the oil phase from the wall-forming monodisperse oil droplets (Figure 4.14).

In the case where both liquids partially and equally wet the surface channel, the interface boundary takes irregular shapes, affecting the reproducibility of flow patterns. This demonstrates the direct effect that surface tension has on shaping the flow pattern of liquid–liquid systems. Hence, it is important to have control over the wetting properties to maintain regular flow patterns. Control can be easily achieved by treatment of the channel surface rather than by varying the properties of the liquids. For example, microchannels made from certain polymers can be hydrophobic, such as polymethyl methyl acrylate (PMMA), making it difficult to flow through water. Because the functional groups on these polymers are causing the hydrophobic character, one solution would be to treat the surface with UV light causing the polymer to break and therefore change the critical surface tension. On the contrary, hydrophilic surfaces such as glass or silicon can be turned into a hydrophobic surface by silanization. In this method, the free hydroxyl groups of the glass surface are deactivated by the coupling reaction with silanes leading to a nonwetting surface [203,206,207].

An alternative way of obtaining control over wetting properties is the use of surfactants to alter the properties of the liquid phases. Dreyfus *et al.* investigated the effect that surfactants have on the flow pattern of an immiscible mixture of water and tetradecane on glass and silicon microchannels with three inlets as shown in Figure 4.15 [208]. When the surfactant concentration is higher than the critical micellar concentration, a complete wetting with respect to oil is achieved and the surface becomes completely hydrophobic. Initially, an experiment was conducted in which water and oil were injected into a microchannel through the three inlet ports, the water from the two side inlets (A and C as shown in Figure 4.15) and the oil from the central inlet (B) in a range of different flow rate ratios, followed by the inverse of the flow order. From that, a flow pattern map could be obtained as shown in Figure 4.15. In the case where oil is injected from the central inlet at high flow rates, well-defined and separated aqueous droplets were obtained and carried through the wetting oil flow as shown in the map. As the flow rate of oil decreases, the aqueous drops begin to extend in length and connect in forming a pearl necklace-like flow.

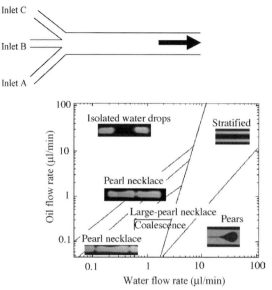

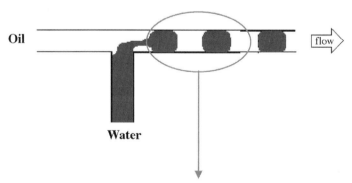

Figure 4.15 Various possible flow patterns in hydrophobic media using a cross inlet junction (reprinted Figure 4.10 with permission from [208]. Copyright 2003 by the American Physical Society).

Increasing the flow rate of water results in its spreading in stratified fashion as the velocity forces dominate the interfacial one. The oil flow in this case forms either a pearlike drop flow when it is at low flow rate or a stratified oil stream at high flow rate. The inlet junction has an influence on the flow pattern regime as well. To prove this, water flow was introduced from the central inlet and the oil from the two external inlets. Under conditions of high flow rates, similar flow patterns were observed as those shown in Figure 4.16 but in a reverse fashion. Well-defined aqueous droplets were attained at high-water flow rate combined with low-oil flow rate and no pear-

Figure 4.16 Dynamical process leading to drop formation of water segments in a main stream of oil flow with internal vortex circulation in each phase segment.

shaped droplets were observed. In addition, when both oil and water were introduced at high flow rate, the oil spreads over the channel walls, whereas the aqueous stream flows in between. This clearly demonstrates that the formation of droplets, segmented or parallel flow, does not only depend on the surface and interfacial forces but also on other factors such as inlet geometry and flow rate. Depending on the conditions, these factors could either generate instabilities at the liquid–liquid interface leading to monodisperse segments or stabilize them to form stratified flow patterns.

Droplets and segmented flow are generated at the inlet section because of the buildup of pressure. The buildup of pressure is a result of increased resistance of the water flow emerging from the inlet against the oil flow. This results in an interfacial instability that favors the interfacial area of the emerging flow head to increase through the mechanism of pinching leading a segment. As the water segment begins to detach itself from the main water flow at the inlet, a tail at the back of the segment attached to the flow forms. The tail breaks eventually as the segment is pulled away by the oil flow to separate the segment from the main flow.

The interaction between the segment and the channel wall causes a shear stress because of the adhesive forces. This shear results in velocity streams in a straightened fashion, from the end of the segment to the start in the direction of the flow. As the flow stream approaches the interface boundary, diversion of the stream occurs causing internal vortices inside the segment that are localized in the front and back parts of the segment. One complete cycle of recirculation occurs when the segment has traveled its length. Factors such as the segment size and the distance traveled by the segment affect the rate of internal mixing. Shorter segments have shorter mixing distances and, hence, the mixing is achieved in less time as shown clearly from the microphotographs produced by Ismagilov and coworkers [209] in their studies on the formation and mixing of droplets. These microphotographs (Figure 4.17) show three aqueous streams combined in a PDMS (poly(dimethylsiloxane)) microchannel; two aqueous streams were separated in the middle by a third stream that continuously segmented into a flow of immiscible fluorinated solvent (PFD: perfluorodecalin). The reagents were mixed inside the segments while transported by the carrier PFD. To help the visualization of mixing inside the aqueous segments, one of the main aqueous streams was colored red using an inorganic iron complex while

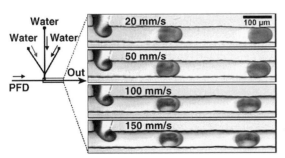

Figure 4.17 Microphotographs illustrating weak dependence of periods, length of plugs and flow patterns inside plugs on total flow velocity (courtesy of the American Chemical Society) [209].

keeping the other streams colorless. The faster the recirculation, the faster is the mixing as the red color is fast distributed evenly around the segment faster. The longer the segment length, the longer it takes for the red color to distribute equally as illustrated in the microphotograph for the larger aqueous segments. In contrast, the experiment without the presence of carrier flow PDF shows laminar flow behavior as the aqueous mixture is not segmented, and no circulation is generated.

As discussed previously, Dreyfus *et al.* found that a stratified flow can be generated when both phases are introduced at high flow rates [208]. The formation of a stable stratified flow depends on the conditions at the inlet junction as well as the channel, affected by the inlet geometry, liquid properties and the flow rate. As the flow rate of both phases increases, the viscous forces dominate the Capillary number (*Ca*) and the interfacial instability reduces as the pressure fluctuation minimizes. One observes the stratification of both phases. Once the parallel streams are formed, mixing occurs via diffusion only between the two phases through the interface and no internal circulation is generated. However, stabilizing parallel flow is not as easy as the formation of droplets or segmented flow because the interfacial stability is very sensitive to viscosity differences between the two phases as uneven velocities at the interface produce vortices leading to turbulent flow. This behavior is known as Kelvin–Helmholtz instability [210].

4.3.6
Liquid–Liquid Microsystems in Organic Synthesis

Exploitation of liquid–liquid microreactor in organic synthesis offers attractive advantages, including the reduction of diffusion path lengths to maximize the rate of mass transfer and reaction rates. Despite the advantages, interest in liquid–liquid microreactors did not take off until recently, perhaps because of the complication of flow pattern manipulation combined with the limited numbers of liquid–liquid reactions. Initial interest focused on the control of parameters responsible for variation in flow patterns to engineer microemulsions or droplets. However, it was soon realized that liquid–liquid microdevices are more than just a tool for controlling flow patterns and further interest developed.

Liquid–liquid chemical reactions that require enhancement in mixing could take advantage of a flow regime in which the two phases were separated to form a segmented flow pattern. Nevertheless, there are only few examples of reactions conducted in parallel flow regime. There is a reasonable interest on segmented flow because of the easy control of the flow regime, the increase in interfacial areas and the efficient mixing inside the segments due to generation of the internal circulations at low Reynolds number compared to parallel flow.

Most examples describe catalyzed biphasic reactions taking the extra advantage of product isolation and catalyst recycling. De Bellefon *et al.* [211] published one of the first examples of biphasic reactions performed in a microreactor. The isomerization of allylic alcohols to carbonyl compounds was conducted in a liquid–liquid system using a micromixer combined with a microchannel tube. As there are limited examples of biphasic isomerization reactions, the authors were interested on

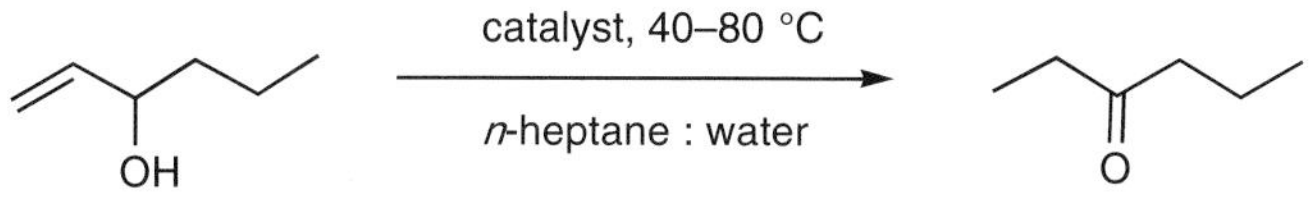

$$\text{(structure with OH)} \xrightarrow[\text{n-heptane : water}]{\text{catalyst, 40–80 °C}} \text{(ketone structure with O)}$$

Figure 4.18 Isomerization of 1-hexene-3-ol to ethyl propyl ketone in a microreactor.

extending the scope by taking advantage of microsystems. To achieve this, various complexes of transition metals with a library of water-soluble ligands were screened for the isomerization of 1-hexene-3-ol to 3-hexanone (Figure 4.18) using an aqueous/hydrocarbon solvent system. The catalysts and substrates were introduced simultaneously in pulsed injection fashion as shown in Figure 4.19. The use of micromixer helped the formation of droplets' flow that was then carried through the integrated microchannel tube. The results of screening different catalysts demonstrated clearly the efficiency of a microsystem for rapidly finding an efficient catalyst. In addition, only a very small amount of catalyst had to be used in the rapid screening of many catalysts. This investigation was one of the first high-throughput screening tests conducted in microreactors (Figure 4.19).

Biphasic catalysis conducted in fluorous media has attracted attention as the interest in environmentally friendly synthetic processes is growing. The advantages of microsystems for biphasic fluorous catalysis were investigated by studying the Mukaiyama aldol reaction to form carbon–carbon bonds between aldehydes and silyl enol ethers in the presence of a Lewis acid catalyst [212]. The reaction was conducted in a biphasic fluorous/organic solvent system using a low concentration of the lanthanide complex $Sc[N(SO_2C_8F_{17})_2]_3$ as a catalyst. The catalyst dissolves only in the fluorous phase, whereas the substrate and product are soluble in the organic phase. A

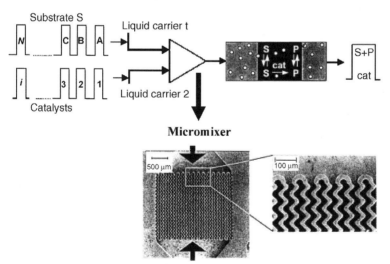

Figure 4.19 Schematic illustration of high-throughput screening used for the isomerization of allylic alcohols to ketones with the illustration of the micromixer used to generate droplet flow (courtesy of Wiley-VCH) [211].

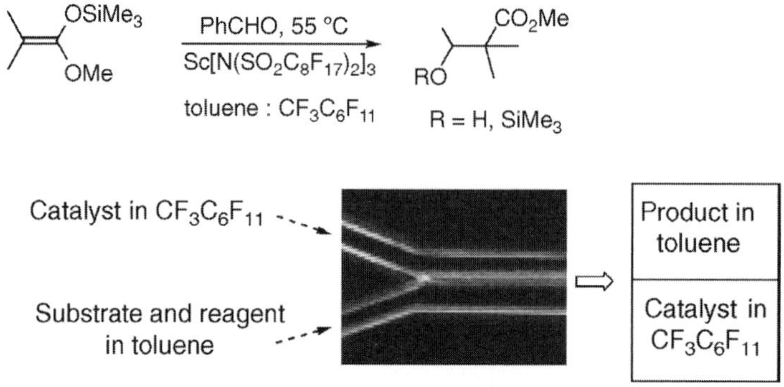

Figure 4.20 Mukaiyama aldol reaction in a borosilicate microreactor under parallel flow (courtesy of Elsevier) [212].

borosilicate microdevice was used, and the delivery of reagents was controlled by a precise syringe delivery system. Both phases were delivered through a Y junction at high pressure. After the reaction, the product was easily isolated from the organic phase, whereas the catalyst remained in the fluorous phase demonstrating the biphasic advantage of product and catalyst separation (Figure 4.20).

Using the same approach, Mikami *et al.* [213] were able to increase the reaction rate of a $Sc[N(SO_2C_8F_{17})_2]_3$ catalyzed Baeyer–Villiger reaction. Only one regioisomeric lactone was obtained in high yields even at very low-catalyst concentrations (0.05 mol%) (Figure 4.21).

A conduction of a reaction under parallel flow in microreactor system was also demonstrated by Kitamori *et al.* where high conversions were achieved for phase-transfer catalyzed reactions [214]. With the use of a glass microchip, the reaction of *p*-nitrobenzene diazonium tetrafluoroborate in water and in ethyl acetate took place in the aqueous layer after rapid phase transfer of 5-methylresorcinol used in excess. Not only higher yields than a conventional reaction in a flask were observed as a result of the large specific interfacial area in the microreactor but also no side products could be detected by fast removal of the main product from the aqueous to the organic phase. The separation of the two phases was easily achieved by splitting the reaction channel into two channels at the end (Figure 4.22).

Benzylations reaction using TBAB (tetrabutylammonium bromide) as a phase-transfer catalyst in glass microchannel reactors has been investigated as well [215]. A dichloromethane/aqueous biphasic system was used in which both the substrate

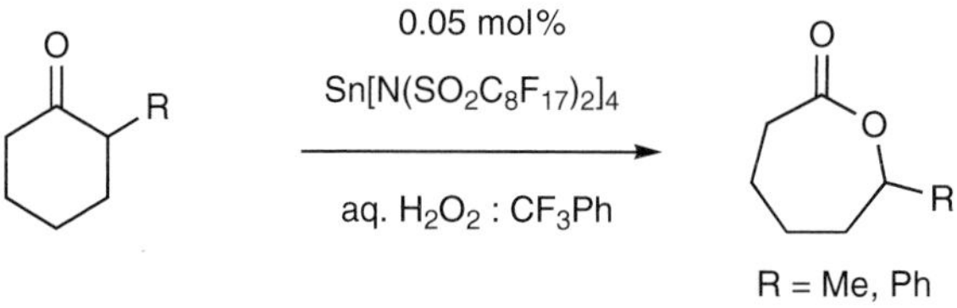

Figure 4.21 Baeyer–Villiger reaction in a microreactor (courtesy of Elsevier) [213].

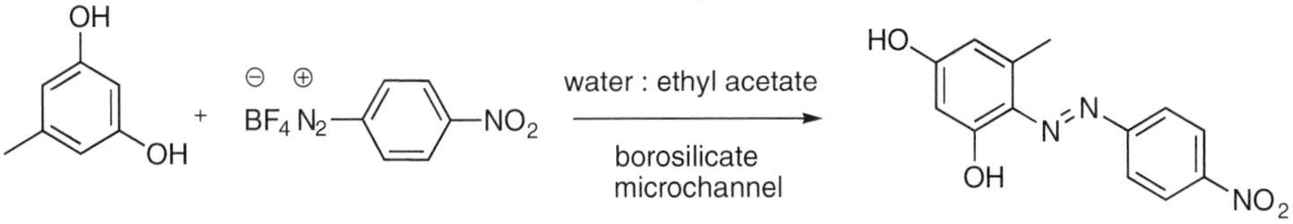

Figure 4.22 Reaction of 5-methylresorcinol with *p*-nitrobenzene diazonium tetrafluoroborate.

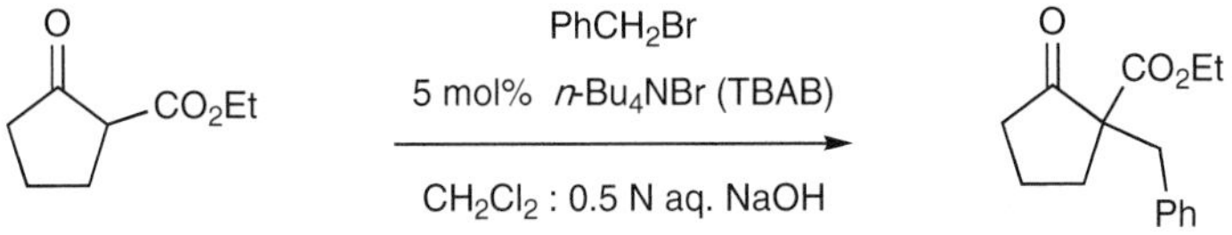

Figure 4.23 Phase-transfer alkylation of ethyl 2-oxocyclopentane carboxylate with benzyl bromide in the presence of TBAB.

(ethyl 2-oxocyclo pentanecarboxylate) and the alkylating agent (benzyl bromide) are dissolved in the organic phase, whereas the transfer catalyst TBAB is dissolved in the NaOH aqueous phase. However, in this case the geometry of the microreactor induced the formation of segmented flow as visualized using an optical microscope. Studies on the effect of the microchannel size on the alkylation reaction were performed. It was found that smaller channels in the microreactor lead to higher rates as the interfacial area increases (Figure 4.23).

Another example of phase-transfer catalysis in biphasic reactions performed in microreactors was reported by Okamoto [216]. Segmented flow was employed using alternating pumping to increase the yield of a biphasic alkylation reaction of malonic ester with iodoethane in the presence of the phase-transfer catalyst TBAHS (tetra-butylammonium hydrogensulfate). Alternate pumping is a technique used to create segmented flow in a microchannel without having to consider the geometry of the inlet junction or the properties of the liquids involved. More detailed studies concerning the principles of this technique without application to chemical reactions have been published by various authors [217–219] (Figures 4.24 and 4.25).

The industrially important nitration of aromatic compounds in a microreactor using two immiscible liquid phases was demonstrated in different studies using either parallel [220] or segmented flow [221]. In all studies, a PTFE capillary microchannel, connected to an inlet junction, was used in which either segmented or parallel flow can be created. The use of PTFE tubing is desirable as it is commercially available and no complicated microfabrication methods are involved.

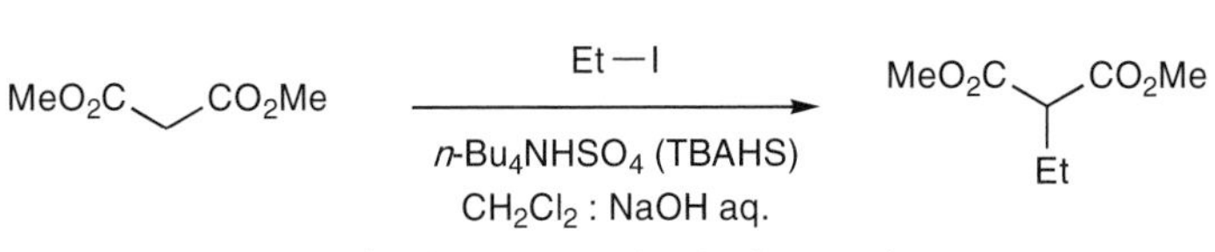

Figure 4.24 Alkylation of malonic ester with iodoethane in the presence of TBAHS as a phase-transfer catalyst.

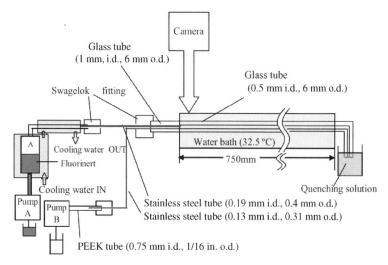

Figure 4.25 Alternate pumping of solution A, containing iodoethane and dimethyl malonate in dichloromethane, and solution B containing TBAHS in a solution of aqueous sodium hydroxide (courtesy of Wiley-VCH) [216].

In the macroscale reaction, the formation of side products such as dinitrobenzene and picric acid is expected as a result of mass transfer limitations. Hence, by using microreactor system, the formation of the side product was reduced and the rate of reaction was increased.

Single-phase reactions can also be conducted in segmented flow fashion in which the reactions take place within a segment while the other immiscible phase is used to form segments. This is a useful way to generate regular turbulence in laminar flow in single-phase reactions [222]. Ismagilov and coworkers [223] demonstrated this technique to optimize deacetylation of ouabain hexaacetate as an example of reaction systems in the microchannel. They have developed a screening system shown in Figure 4.26, in which each segment contains a different reagent, separated by a fluorinated carrier fluid while taking advantage of detection with MALDI-MS.

The segments were introduced in sequence to the carrier fluid flow in which they combine with the substrate solution. A blank solvent segment was introduced between the reagent segments to avoid contamination between them [224]. Following that the flow of the segments was stopped and the microtube system was sealed to keep the segments inside for a specified reaction time. After that the segments were released, collected and then analyzed (Figure 4.26).

The same concept, originally developed by Ismagilov and coworkers [222], was applied to homogeneous catalyzed reactions by introducing an immiscible solvent to the flow generating segmented flow instead of a single flow. As demonstrated by Wirth and coworkers [225], the reaction yield was enhanced for

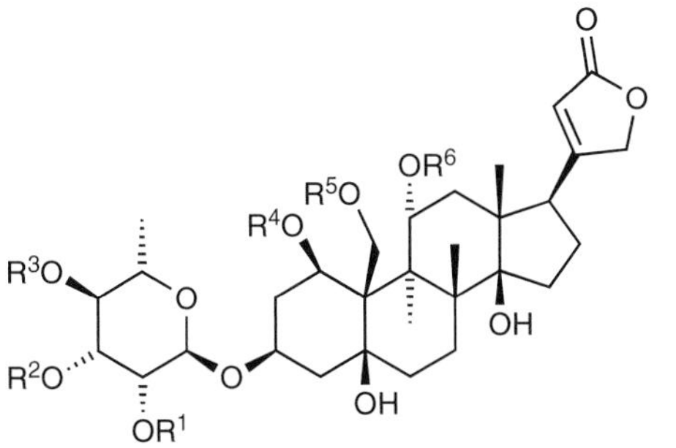

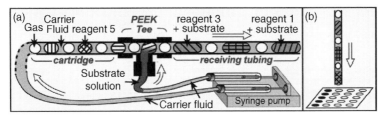

Ouabain $(R^1 - R^6 = H)$

Figure 4.26 (a) Flow screening system for deacetylation of ouabain hexaacetate (R^1–R^6 = Ac) reaction optimizations. (b) After incubation, the segments were deposited onto a sample plate for MALDI-MS (courtesy of the American Chemical Society) [223].

various Heck products compared to conventional methods, and further improving the outcome when using the segmented flow instead of the single flow [226] (Figure 4.27).

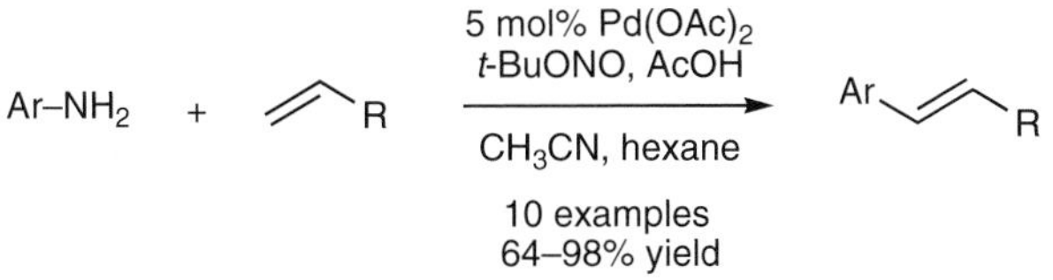

$$Ar–NH_2 \quad + \quad \diagup\!\!\!\diagup R \quad \xrightarrow[\text{CH}_3\text{CN, hexane}]{\substack{5 \text{ mol\% Pd(OAc)}_2 \\ t\text{-BuONO, AcOH}}} \quad Ar\diagdown\!\!\!\diagup R$$

10 examples
64–98% yield

Figure 4.27 Heck reactions accelerated by biphasic flow system.

4.3.7
Conclusions and Outlook

As summarized in this chapter, various synthetic protocols have made advantageous use of biphasic liquid–liquid systems in microreactors. Apart from reaction optimization and acceleration of processes, the application of biphasic flow systems has already led to the development and design of novel protocols, and it is expected that this area of research will grow in future.

References

189 Adams, D.J., Dyson, P.J. and Tavener, S.J. (2004) *Chemistry in Alternative Reaction Media*, Wiley-VCH Verlag GmbH.

190 Ohkouchi, T., Kakutani, T. and Senda, M. (1991) *Bioelectrochemistry and Bioenergetics*, **25**, 81.

191 Lahtinen, R., Fermin, D.J., Kontturi, K. and Girault, H.H. (2000) *Journal of Electroanalytical Chemistry*, **483**, 81.

192 Ahmed-Omer, B., Brandt, J.C. and Wirth, T. (2007) *Organic & Biomolecular Chemistry*, **5**, 733.

193 Earle, M.J. and Seddon, K.R. (2000) *Pure and Applied Chemistry*, **72**, 1391.

194 Zhao, H. and Malhotra, S.V. (2002) *Aldrichimica Acta*, **35**, 75.

195 Tavener, S.J. and Clark, J.H. (2003) *Journal of Fluorine Chemistry*, **123**, 31.

196 Hao, X., Yamazaki, O., Yoshida, A. and Nishikido, J. (2003) *Green Chemistry*, **5**, 524.

197 Levine, I.N. (1995) *Physical Chemistry*, McGraw-Hill.

198 Temperley, H.N.V. and Trevena, D.H. (1978) *Liquids and their Properties*, Ellis Horwood Limited.

199 Holland, F.A. and Bragg, R. (1995) *Fluid Flow for Chemical Engineerings*, 2nd edn, Edward Arnold, a division of Hodder Headline PLC.

200 Coulson, J.M., Richardson, J.F., Backhurst, J.R. and Harker, J.H. (1990) *Chemical Engineering, vol. 1*, 4th edn, Pergamon Press.

201 Janna, W.S. (1983) Introduction to fluid mechanics, Brooks/Cole Engineering Division.

202 Günther, A. and Jensen, K.F. (2006) *Lab on Chip*, **6**, 1487.

203 Baroud, C.N. and Willaime, H. (2004) *Comptes Rendus Physique*, **5**, 547.

204 Lotus Image by Billy Bates, www.Victoria-Adventure.org.

205 Water drops on colored surface photo, photo Publisher: Creatas, Image reference: CR15206048, http://www.inmagine.com/liquid-in-motion-photos/creatas-cr15206-40.

206 Efimenko, K., Wallace, W. and Genzer, J. (2002) *Journal of Colloid and Interface Science*, **254**, 306.

207 Brzoska, J., Azouz, J.B. and Rondelez, F. (1994) *Langmuir*, **10**, 4367.

208 Dreyfus, R., Tabeling, P. and Willaime, H. (2003) *Physical Review Letters*, **90**, 144505.

209 Tice, J.D., Song, H., Lyon, A.D. and Ismagilov, R.F. (2003) *Langmuir*, **19**, 9127.

210 Mikhailovskii, A.B. (1992) *Electromagnetic Instabilities in an Inhomogeneous Plasma*, Institute of Physics Publishing, Bristol.

211 de Bellefon, C., Tanchoux, N., Caravieihes, S., Grenouillet, P. and Hessel, V. (2000) *Angewandte Chemie-International Edition*, **39**, 3442.

212 Mikami, K., Yamanaka, M., Islam, N., Kudo, K., Seino, N. and Shinoda, M. (2003) *Tetrahedron*, **59**, 10593.

213 Mikami, K., Yamanaka, M., Islam, N., Kudo, K., Seino, N. and Shinoda, M. (2004) *Tetrahedron Letters*, **45**, 3681.

214 Hisamoto, H., Kitamori, T., Saito, T., Tokeshi, M. and Hibara, A. (2001) *Chemical Communications*, 2662.

215 Ueno, M., Hisamoto, H., Kitamori, T. and Kobayashi, S. (2003) *Chemical Communications*, 936.

216 Okamoto, H. (2006) *Chemical Engineering & Technology*, **29**, 504.

217 Glasgow, I. and Aubry, N. (2003) *Lab on a Chip*, **3**, 114.

218 MacInnes, J.M., Chen, Z. and Allen, R.W.K. (2005) *Chemical Engineering Science*, **60**, 3453.

219 Stroock, A.D. (2002) *Science*, **295**, 647.

220 Burns, J.R. and Ramshaw, C. (1999) *Transaction of the Institution of Chemical Engineers*, **77**, 206.

221 Dummann, G., Quittmann, U., Groschel, L., Agar, D.W., Worz, O. and Morgenschweis, K. (2003) *Catalysis Today*, **79–80**, 433.

222 Song, H., Chen, D.L. and Ismagilov, R.F. (2006) *Angewandte Chemie-International Edition*, **45**, 7336.

223 Hatakeyama, T., Chen, D.L. and Ismagilov, R.F. (2006) *Journal of the American Chemical Society*, **128**, 2518.

224 Chen, D.L., Li, L., Reyes, S., Adamson, D. N. and Ismagilov, R.F. (2007) *Langmuir*, **23**, 2255.

225 Ahmed, B., Barrow, D. and Wirth, T. (2006) *Advanced Synthesis & Catalysis*, **348**, 1043.

226 Ahmed, B., Barrow, D. and Wirth, T. Unpublished results.

4.4
Gas–Liquid Reactions

Volker Hessel, Patrick Löb, Holger Löwe

4.4.1
Introduction

In traditional chemical engineering and organic synthesis, there are a large variety of reactors that can be used for gas–liquid and gas–liquid–solid reactions. This includes mechanically agitated tanks, slurry reactors, bubble, packed and spray columns, falling-film, loop and trickle-bed reactors and the less widespread used static mixer, venturi and spinning disk reactors [227,228]. Microreaction technology is a new continuous processing concept with different types of reactors comprising microengineered structures [229–247]. These extend the performance of conventional reactors especially in terms of enhanced mass and heat transfer, for example, to be used for fast exothermic reactions and safe operation under extreme processing conditions and with hazardous reagents. The hydrodynamics of gas–liquid microreactors is often characterized by uniform flow patterns such as the Taylor flow (see below and respective reviews [230,248–250] and relevant scientific papers [251–262]). Some gas–liquid microreactors are just miniaturized analogues of their macroscale counterparts, for example, the falling-film microreactor, whereas others offer entirely new multiphase contacting concepts, for example, the Taylor-flow or mesh microreactors.

In the following, some major contacting principles will be given, which will be first accompanied by realized reactor examples. Then, the application of the microreactors for gas–liquid and gas–liquid–solid reactions in the field of organic chemistry will be presented and will cover the major part of this chapter. Information on hydrodynamics, mass transfer, kinetics, modeling and other applications has been given elsewhere (see [230,248–250] and original citations given therein) and does not fall under the umbrella of this book.

4.4.2
Contacting Principles and Microreactors

4.4.2.1 Contacting with Continuous Phases
One way to contact a gas and a liquid is to have both phases continuous, 'side by side'; that is, the fluids are not dispersed into each other. This has the benefit of a known, well-defined and stationary interface. 'Polydispersity' in hydrodynamic

features is absent (for intrinsic reasons), as given typically for a swarm of bubbles flowing in a liquid (e.g. in a bubble column or tube reactor) and also no 'aging'-like coalescence of bubbles can happen. Different from disperse systems, it is only the liquid phase and not the gas phase that needs to be distributed. Phase separation is facile because the phases are never intermixed. Because of the known interfaces and usually also defined liquid and gas layers, numbering up is a valuable concept, as the addition of more channels goes along with the replication of the hydrodynamic conditions. This, however, will only be efficient if flow equipartition is on a high standard and that wetting will be the same for all microchannels. Indeed, continuous-phase microreactors comprise the first reported scaled-out gas–liquid devices that are pilot-scale falling-film and annular-flow microreactors. Two-phase contactors, as well as disperse counterparts, generally need visual control to check the hydrodynamics via transparency of the whole device or inspection windows, which makes construction more complex and limits pressure operation.

4.4.2.2 Falling-Film Microreactor

The falling-film principle uses the wetting of a surface by a liquid stream, governed by gravitational force, which thus spreads to form an expanded thin film (see Figure 4.28), a concept known from macroscale contactors. Typical films have a thickness of a few ten to a few hundred micrometers [230,248,263].

In one version realized, a flow restrictor, typically a slit, creates pressure loss to improve the liquid flow distribution from the incoming stream to the channels of a microstructured plate (see Figure 4.28) [230,248,263]. The liquid streams are recollected via another slit at the end of the plate. The gas flow enters a large gas

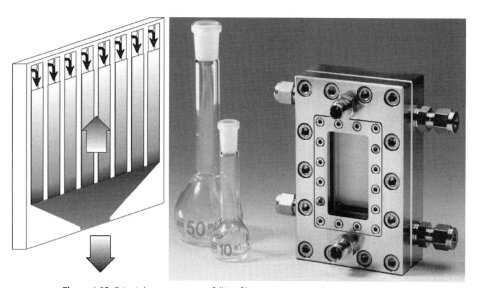

Figure 4.28 Principle to generate a falling film on a structured plate with channels (left). Falling-film microreactor, laboratory version up to 1 l/h (by courtesy of IMM).

chamber, positioned above the microchannels, via a diffuser. Heat exchange is realized by integration of a minichannel heat exchanger, positioned in the rear of the reaction zone. An inspection window above the gas chamber allows to check the falling films. Furthermore, this enables to carry out photochemical gas-liquid reactions [264]. The slit and the gas chamber with its two diffusers are made by microelectrodischarge machining (μEDM). The microchannels are etched in the plate. The heat exchange channels are manufactured by micromilling.

Scaled-out versions of the laboratory falling-film microreactor discussed above were made with about 10 times larger channel surface, which roughly means 10 times larger gas–liquid interface [265]. This was done in two ways. (1) A cylindrical falling-film concept relies largely on internal numbering up, with an increase in the number of microchannels as compared to the laboratory tool by more compact arrangement on a cylindrical metal block (see Figure 4.29). This pilot device has outer dimensions that hardly exceed the laboratory tool. For practical reasons, that is, to facilitate microfabrication, the length of the microchannels is slightly larger in the cylindrical device, so this is not a pure numbering-up approach. (2) Another pilot falling-film device relies on classical scale-up, with both the channel width and the length being enlarged by a factor of 3.3 that gives 10 times larger surface (see Figure 4.29). Although the laboratory tool has an upper flow limit of about 1 l/h, the two pilot reactors come to about 10 l/h. It was announced that as the next generation production, falling-film microstructured reactors are under fabrication with projected flows of about 100 l/h [265].

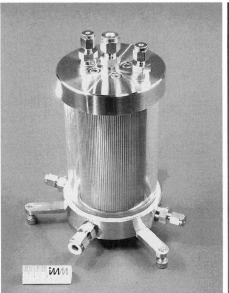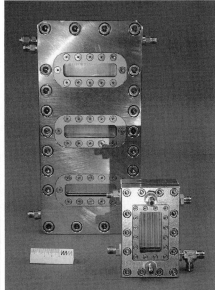

Figure 4.29 Numbered-up, 'cylindrical', (left) and scaled-up, 'large', (right, laboratory version in front) falling-film microreactors (by courtesy of IMM).

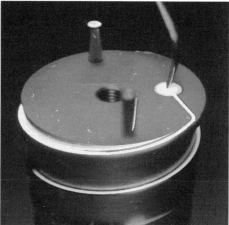

Figure 4.30 Helical falling-film microreactor (right). Flow of the liquid in the helical microchannel, visualized by injection of a fluorescent tracer (by courtesy of Elsevier and IMM) [266].

Both the laboratory and scaled-out falling-film microreactors are limited in residence time [266]. About 1 min is the maximum practical time. An extension to reaction times of several minutes can only be reached by either using vertical reaction channel plates of several 10 cm length exceeding even 1 m, which on its own is neither practical nor a smart solution, or decreasing the falling angle to have less impact of gravity. A new, more compact design based on a helical guiding of the liquid film combines both approaches (see Figure 4.30, left). Owing to the winding of the channel a large length is realized on small footprint and the channel declines at small falling angle. This is, however, done at the expense of throughput so that this device is restricted to analytical investigations.

Initially, taper cone-shaped devices were made from aluminum by computerized numerically controlled (CNC) turning as manufacturing technique using a lathe [266]. These devices are easily amenable to inspection of the complete fluidic path. The empty channel surface and the liquid surface of the filled channel were accessible by white light interferometry. With the use of *n*-butanol as liquid flowing in microchannels of $100\,\mu m \times 300\,\mu m$ cross section, a complete filling of the microchannels without any flooding or undesired wetting of liquid at the walls of the cylinder next to the channel was achieved.

Because the cuts through the taper cone devices showed that structural quality and surface roughness of the microchannels need to be improved, the design of the helical falling-film microreactor was changed and a microstructured cylinder was made of stainless steel, instead of the taper cone (see Figure 4.30, right) [266]. The helical path was set to 7.5. The channel width and depth are set to 300 and $100\,\mu m$, respectively. The diameter of the cylinder amounts to 24.3 mm. Three lengths of the helical path of 540, 1540 and 5390 mm (corresponding to total device lengths of 70, 200 and 700 mm) were made to give residence times of 10, 30 and 105 min, respectively. The gas stream is guided through a cylindrical housing. The residence times of both helical devices, the microstructured taper cone and the cylinder, are about a factor of 50 higher than that of the standard falling-film microreactor [266]. Depending on the viscosity of the solvent, residence times from 3 min for methanol up to 22 min for octanol could be determined.

The housing consists of a capped stainless steel tube [266]. For high-pressure experiments, a stainless steel housing can be used up to 50 bar at 20 °C. Additionally, the stainless steel tube can be replaced by a transparent PMMA housing. This allows a visual inspection of the microchannels in the case of low-pressure experiments (up to 5 bar).

Continuous Contactor with Partly Overlapping Channels   Solute transfer can occur between immiscible phases each flowing in separate adjacent but displaced micro-channels, having only a small conduit in which the fluid interface is stable (partial overlap) [267,268].

This concept of partly overlapping channels was realized by having one plate with a rectangular channel manufactured in silicon by sawing, covered by another plate with a semicircular channel made in glass by wet chemical etching (see Figure 4.31) [267,268]. The glass/silicon plates are joined by anodic bonding. To ensure efficient mass transfer and to stabilize the interface, channel depth (i.e. the diffusion distance to the interface) should be about 100 μm and the channel opening should be about 20 μm [269]. A numbered-up module was developed with 120 partly overlapping microchannels operating in parallel.

Mesh Microcontactor   A mesh microcontactor contains a microstructured plate with regular circular openings through which separate gas and liquid streams come into contact [270,271]. Stability of the interface and prevention of breakthrough are achieved by adjusting the pressure. Gas–liquid operation requires a low gas flow

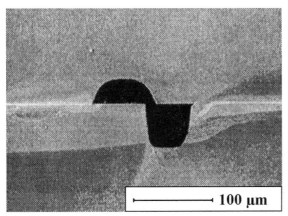

Figure 4.31 Cut through a continuous contactor with partly overlapping channels revealing rectangular and semicircular channels at adjacent but displaced positions (by courtesy of J. Shaw, CRL) [267].

through the microcontactor to achieve the necessary backpressure. Operation is generally possible also in stop-flow mode besides continuous flow.

The first realized mesh microcontactor was shaped to restrict dispersion of sequential samples delivered to and from the reaction zone (see Figure 4.32) [270,271]. A quadrant structure with manifold channel at outlet only and central inlet feed was chosen. Nickel is used for the mesh material because of its robustness, ease of fabrication by pattern plating and its compatibility with a wide range of alkaline to neutral solutions. The mesh is inserted in an enclosure formed from glass and copper. The distance from the mesh to the chamber walls is 100 μm on each side. The reaction chamber volume for each phase is 100 μl. Milling was applied as manufacturing technique for the reactor parts. Nickel mesh fabrication involves photolithography and a two-stage electroplating method.

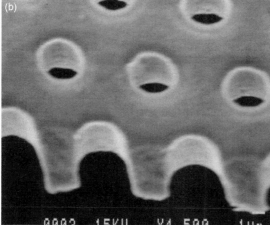

Figure 4.32 Nickel mesh: (a) photograph of complete mesh showing frame and struts. (b) Scanning electron micrograph of mesh pores (by courtesy of Royal Society of Chemistry) [270].

Annular-Flow Microreactors Annular-flow microreactors are generically similar in design to the Taylor-flow reactors, as both flow regimes can be realized in each of these devices. The (small) difference is that usually less attention is paid to the design of the mixing element for gas and liquid streams. No delicate bubble formation but rather a simple split into continuous gas and liquid streams is needed. In the simplest version, just two holes enter in a series into a microchannel, with the liquid stream typically incoming first and surrounding the next opening for the gas that gives a gaseous cylinder with a thin liquid shell. Further, the reaction channel is usually smaller than that of Taylor-flow reactors because annular-flow reactors are used for very fast reactions with high demands for heat transfer.

In one of the very early versions, a microreactor was composed of two plates and a block, made from a special steel, forming the conduits for gas and liquids and a single reaction microchannel, with one of the plates being transparent for visualizing the flow in the latter (see Figure 4.33) [272,273]. The reaction microchannel is cut in the bottom block, and the block is highly polished to ensure gas tightness [242].

A coolant channel is guided through the metal block in a serpentine fashion so that reactant and coolant flows are orthogonal [272]. A thermocouple measures the temperature at the product outlet. This microreactor was developed for the (very fast) fluorination reactions with elemental fluorine. Therefore, the surface of the microchannel was inactivated by exposure to the increasing concentration of fluorine in nitrogen.

After the initial manufacture of a single-microchannel version [272], a numbered-up (scale-out) three-microchannel version followed later [273].

Another annular-flow concept was provided by the so-called dual-channel reactor with two parallel microchannels separated by a wall. In this way, four thin liquid layers in annular flow were created at once (see Figure 4.34, bottom) [274]. In front of this section, the liquid feed enters through a hole directed to the wall, whereas the two gas feeds point to the two reaction channels. Consequently, the liquid flow splits and a larger interface is created than given for single-channel guidance.

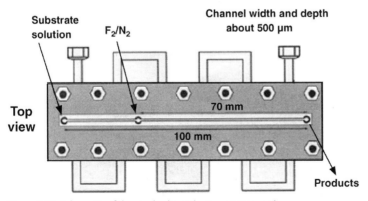

Figure 4.33 Schematic of the single-channel microreactor used for fluorinations with elemental fluorine (by courtesy of the Royal Society of Chemistry) [272].

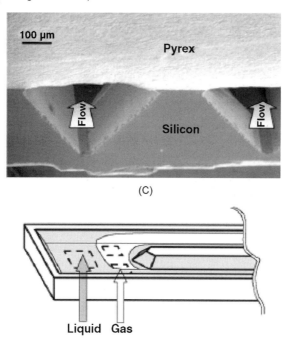

(C)

Liquid Gas

Figure 4.34 Schematic of the inlet section of the dual-channel microreactor (bottom) and scanning electron micrograph of a cut through the dual-channel section (top) (by courtesy of American Chemical Society) [274].

The dual-channel reactor is a silicon chip device and was manufactured by photolithography and potassium hydroxide etching [274]. Silicon oxide was thermally grown on silicon and thin films of nickel were evaporated for passivation because direct fluorination was carried out in this device. Pyrex was bonded anodically to the modified microstructured silicon wafer (see Figure 4.34, top).

A flow-pattern map was derived for nitrogen/acetonitrile flows in the dual-channel microreactor [274]. Bubbly, slug, churn and annular flows as well as wavy annular and wavy annular-dry flows with smaller region of stability were found (see Figure 4.35).

4.4.2.3 Contact with Disperse Phases

Dispersions at micron scale are usually made by merging gas and liquid streams in a mixing element and subsequent decay of the gas stream to a dispersion [251–262]. Mixing elements often have simple shapes such as a mixing tee (dual-feed: gas–liquid) or triple-feed (liquid–gas–liquid) arrangements. The dispersion is passed either in a microchannel (or many of these) or in a larger environment such as a chamber, which, for example, provides volume to fill in porous materials such as catalyst particle beds, foams or artificial structures (microcolumn array). The mechanisms for bubble formation have not been investigated for all of the devices

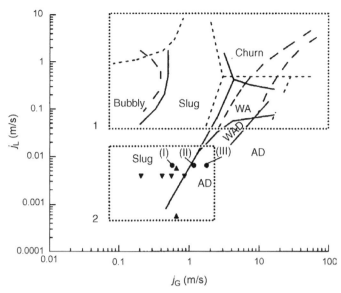

Figure 4.35 Flow pattern map for the nitrogen/ acetonitrile flow in the dual-channel microreactor. Annular flow; wavy annular flow (WA); wavy annular-dry flow, (WAD); slug flow; bubbly flow; annular-dry flow (AD). Transition lines for nitrogen–acetonitrile flows in a triangular channel (224 μm) (solid line). Transition lines for air–water flows in triangular channels (1.097 mm) (dashed lines). Region 2 presents flow conditions in the dual-channel reactor (•) with the acetonitrile–nitrogen system between the limits of channeling (I) and partially dried walls (III). Flow conditions in rectangular channels for a 32-channel reactor (150 μm) (▼) and single-channel reactor (500 μm) (▲) (by courtesy of American Chemical Society) [274].

used for real-case applications, but according to fundamental microfluidic studies shear force or hydrodynamic instability plays a major role.

The different flow patterns largely resemble those known from flow in other continuous flow conduits such as pipes, tubes, capillaries and monoliths [251–262]. Bubbly flow, slug flow (Taylor flow), annular and churn flow and a few more intermediate regimes between the ones mentioned are found. These comprise different gas–liquid configurations such as segmented flow (bubble train), gas core with encompassing stable thin liquid film that wet the channel wall and dynamic wavy liquid films. In case of high gas contents, spray is created with small droplets in continuous gas phase [275,276]. Most often, Taylor and annular flows were used in gas–liquid microreactor engineering, as these two patterns have stable and large known interfaces with good mixing and low axial dispersion. Similar to the falling-film microreactors, inspection windows or, even better, totally transparent devices are almost a must, at least at a prototype and process development stage.

The feed of dispersive systems in many parallel microchannels is not trivial, and mixed flow patterns and even drying of the channels were reported for the first-hour devices [275,276]. Distributor design solutions for phase equipartition were proposed for some devices, for example, for mini-packed reactors [277,278].

Figure 4.36 Design of a Taylor-flow microreactor with feeds and mixing zone (left), and reaction channel and outlet (right). Please note that in the image T-type mixer design is given, which differs somewhat from the one shown in Figure 4.37. Dimensions of the mixers are also given below and the reaction channel is 100 μm wide and 50 μm deep at a length of 2.2 cm. The outer dimensions of this chip are 45 mm × 8 mm × 1.5 mm (by courtesy of M. Warnier, Eindhoven University of Technology).

Taylor-Flow Microreactors Taylor-flow microreactors contain a dispersing mixing element for gas and liquid streams, typically of T and Y shapes, followed by a reaction channel for the segmented gas–liquid flow, often of quite extended length, as the Taylor flow is dominant in typical flow-pattern maps [230,248–250]. In general, all Taylor-flow microreactors can induce other flow patterns as well as the ones mentioned above.

In one version, Taylor-flow microreactors comprised two types of mixer designs followed by a single microchannel (see Figure 4.36) [279].

The gas feed is encompassed by two equal liquid inlets with differing contacting angle of the two liquid streams with respect to the gas stream and different contact region ('T-type and smooth') (see Figure 4.37) [279]. The flow was then introduced into a straight reaction channel in both mixers.

In the T-type mixer, channel width is decreased in two stages, thereby creating an intermediate bubble formation chamber that has the function of assisting in bubble rupture. The idea is to perform bubbles to a specific geometry [279]. Beside bubble rupture, another mode of bubble formation has also been observed. In the smooth mixer, the channel width reduction is performed in a continuous manner, that is, without any stages and edges. Here, gas and liquid flows enter into the reaction channel and remain largely unchanged after initial hydrodynamic decay.

The devices are constructed from two plates, which are irreversibly joined by anodic bonding. Microfabrication was achieved by means of deep reactive ion etching (DRIE) in borosilicate glass.

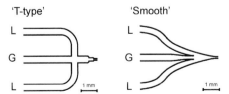

Figure 4.37 Design of two mixing units in Taylor-flow microreactors used for achieving dispersed flow in a downstream channel (100 μm width × 50 μm depth, 2 cm length). L: liquid G: gas (by courtesy of Wiley-VCH Verlag GmbH) [279].

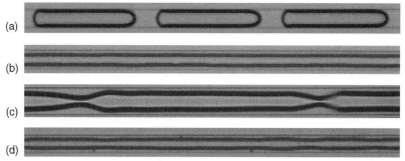

Figure 4.38 Flow patterns of nitrogen-water flows in microchannels of the Taylor-flow microreactor given in Figure 4.36. The superficial gas (U_g) and superficial liquid velocities (U_l) are indicated. The channel has a rectangular cross section of 100 μm × 50 μm. (a) Slug flow: $U_g = 0.5$ m/s, $U_l = 0.1$ m/s; (b) annular flow: $U_g = 5.5$ m/s, $U_l = 0.07$ m/s; (c) ring flow: $U_g = 20$ m/s, $U_l = 0.2$ m/s; (d) churn flow: $U_g = 50$ m/s, $U_l = 0.5$ m/s (by courtesy of Wiley-VCH Verlag GmbH) [279].

With the use of high-speed microscopy imaging, four main flow patterns that are similar to those given in Figure 4.35 [279] were observed. Slug, churn and annular flows were found (see Figure 4.38). Bubbly flow was observed only initially at the channel entrance section that then changed to slug flow because of pressure changes in the microchannel. As another kind of flow pattern, not shown in Figure 4.35, ring flow was observed. This is similar to annular flow but with irregularities of the inner gas core.

With this information, flow-pattern maps were derived (see Figure 4.39) [279]. These maps are schematically in accordance to prior findings of other groups for

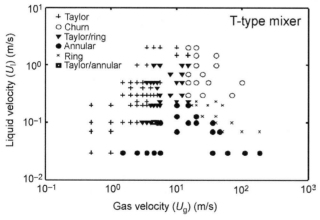

Figure 4.39 Flow patterns for the nitrogen-water system observed for the smooth mixer in the microchannel used at the TU/e. The images were recorded at the indicated superficial gas (U_g) and superficial liquid velocities (U_l). The channel has a rectangular cross section of 100 μm × 50 μm. (a) Slug flow: $U_g = 0.5$ m/s, $U_l = 0.1$ m/s; (b) annular flow: $U_g = 5.5$ m/s, $U_l = 0.07$ m/s; (c) ring flow: $U_g = 20$ m/s, $U_l = 0.2$ m/s; (d) churn flow: $U_g = 50$ m/s, $U_l = 0.5$ m/s (by courtesy of Wiley-VCH Verlag GmbH) [279].

monoliths and also resemble the flow patterns measured for the microbubble column as described below. Details of a significant dependence of these flow patterns (e.g. for the transition from slug to churn flow and the transition from annular to ring flow) on the mixer structure were found. This comes from the high liquid and/or gas velocities used, and the inertia can no longer be neglected with respect to the surface tension.

A transparent microchip reactor with triple mixing element (liquid–gas–liquid) was developed for photochemical gas–liquid synthesis [280]. The mixing element was fed from a liquid inlet port that splits into two channels of equal passage that merge with a third channel, connected to a second port for gas feed. This triple-stream mixer is followed by a long serpentine channel passage that ends in a third outlet port. The channels were micromachined by direct laser lithography and wet chemical etching in the bottom plate. The bottom plate was bonded thermally to a cover.

Like the numbered-up Taylor-flow reactor, the 'microbubble column' achieved dispersion by an interdigital mixing element with many miniaturized mixing tees (see Figures 4.40 and 4.41) [230,248,275,276,279,281,282]. This was one of the very early versions developed that was commercialized later. The dispersion is guided in a microchannel array (for reaction) with exact positioning of each channel of the separate mixing element to the corresponding channel in the reaction array. This is similar to many other Taylor-flow microreactors that, however, usually comprise only one mixing element with two (gas–liquid) or three (liquid–gas–liquid) feed channels connected to one reaction channel. In contrast to the microbubble column, which is made from steel and equipped with an inspection window, many other Taylor-flow microreactors are fully transparent, for example, made from glass, or with a whole transparent cover plate (silicon glass).

To achieve an even higher degree of numbering up, a modular microbubble column reactor was developed that contains a stack of microstructured plates. The construction encompasses five different assembly groups, a cylindrical inner housing, which

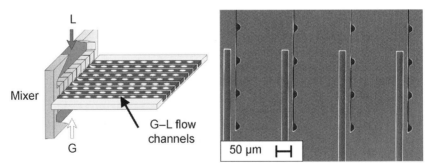

Figure 4.40 Schematic of the 'microbubble column', a numbered-up Taylor-flow microreactor with a mixing for each reaction channel (left) scanning electron micrograph of the mixing element, top view (right). L: liquid; G: gas. The small channels with the semicircular openings are the gas feed and the larger rectangular ones are for liquid feed (by courtesy of VDI Verlag) [275].

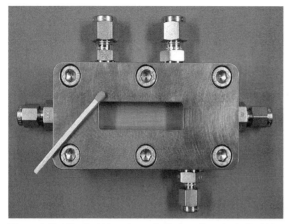

Figure 4.41 Microbubble column (redesigned version). Besides the two inlets for gas and liquid flows and the outlet for the dispersion, two further fluid connectors are there for the incoming and outgoing heat exchange medium (by courtesy of Wiley-VCH Verlag GmbH) [279].

encases the plate stack, two diffusers before and after and a cylindrical outer shell with a flange [283] (see also [284]). The platelets were fabricated by thin-wire μEDM. This construction allows fast plate exchange; for example, only 15–30 min is necessary for cooling from 480 °C to ambient temperature [284]. Because no special inlet manifold was used, some degree of phase separation probably occurred, that is, some micro-channels may have been filled only with liquid, whereas others comprise only gas.

Micromixer–Capillary/Tube Reactors Although the above-mentioned Taylor-flow microreactors are integrated designs of dispersive mixing elements and attached reaction channels, the same type of arrangement can be realized more on a mesoscale by combining micromixers (as single tools) and capillaries or tubes (see [285] for a review on micromixers). Depending on the hydraulic diameter of the latter, the large variety in flow regimes may be lost and rather only bubbly flow and foams may be generated. The most striking advantages of this concept are its simplicity in approach and operation, as well as fast installation, because micromixers are commercially available at comparatively low cost and have the ability to reach high productivity already with one device.

However, coalescence of the foam may occur, as surface forces are less dominant here because of a larger scale. In aqueous systems, this can be prevented by adding surfactants to lower the surface tension. For organic solvents, there is no straightfor-ward solution and only short-term contacting may be realized. In addition, the interface may be not as well defined as for two-phase continuous flow and some of the dispersing microreactors. An exception is given by the formation of the foams that can be quite regular [286].

A large number of micromixers have been used; most often their designs were not oriented on a special use for gas–liquid reactions but rather on mixing of miscible

Figure 4.42 High-pressure interdigital micromixer made of steel (by courtesy of IMM).

liquids [285]. Interdigital micromixers comprise respective feed channel arrays that lead to an alternating arrangement of liquid feed streams generating multilamellae flows, that is, mix by diffusion in the case of miscible liquids [287,288]. In the gas–liquid case considered here, multiple gas jets in a liquid medium are realized that decay into bubbles because of hydrodynamic instability [289,290]. Interdigital micromixers were manufactured as metal/stainless steel devices (see Figure 4.42) [287] or as glass devices (see Figure 4.43) [288].

Split–recombine micromixers with repeated physical separation of fluid streams by branching into separate channels and recombination of channels and streams also perform multilamination when contacting liquid streams only or, probably more correctly expressed, compartmentation of streams. By the reduction of diffusion distances mixing is promoted [291–293]. Owing to the low-pressure loss of this mixing approach, relatively high flows can be realized. One spilt–recombine device was made of silicon with a series of forklike channel segments and was tested, among

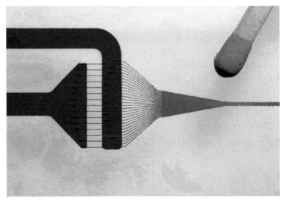

Figure 4.43 Flow focusing interdigital micromixer made of glass (by courtesy of IMM).

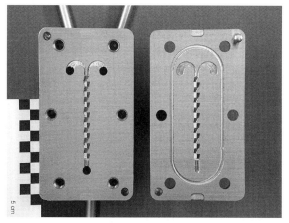

Figure 4.44 Bas-relief micromixer with microstructured ramps in the channel floor and ceiling, termed caterpillar micromixer (by courtesy of IMM).

other uses, for gas–liquid contacting [282]. This plate was joined to a silicon top plate by anodic bonding.

Bas-relief micromixers induce transversal motion, when miscible liquids are considered, to mix by convection [294,295]. In the gas–liquid case to be considered here, the mechanism for bubble formation is yet unclear, but it is probably related to shear forces coming from a similar liquid motion in the dispersed flow. Caterpillar mixers induce such transversal motion by ramplike microstructures, lifted up and down, placed in one channel at the bottom and ceiling [230] (see also [291]). Caterpillar mixers were developed as a family of devices with grouped capacity using smart enlargement of the internal channel and having high volume flows, for example, 100 l/h and more at moderate pressure drops, not exceeding 5 bar (see Figure 4.44).

Mini-Packed-Bed Reactors Packed-bed microreactors have a larger flow-through channel that contains particles brought into contact [274,277,278]. The flow of the gas–liquid mixture goes through the interstices, and in this way a dispersive action is given, that is, continuously renewing the interfaces.

The most striking advantage of the concept is the resemblance of industrially applied principles for gas–liquid and gas–liquid–solid operation. Commercial particles may be employed and exchange is fast and flexible. As a further advantage, relatively large flows are provided, when operating with a single device. Even parallel operation of many devices has been demonstrated [278].

Flow-pattern characterization is more difficult as the particle bed is not transparent and covers most of the flow-through chamber. Owing to the size distribution of the particles and the width distribution of the interstices, one major advantage of microreactors is that the structural and flow regularity is decreased in impact, albeit not lost. Here, the availability of regular particles with uniform size can change the situation.

In one mini-packed-bed reactor, standard porous catalyst particles were inserted in a mini-flow-through chamber (see Figure 4.45) [274,277,278]. Filling of the catalyst

Figure 4.45 A mini-packed-bed reactor (by courtesy of the American Chemical Society) [277].

slurry is achieved via inlet channels, being at both sides at the beginning of the packed bed. An inlet manifold feeds alternately gas and liquid streams into this reaction chamber, thereby achieving a high degree of dispersion. An array of microstructured columns acts as filter at the outlet and retains the catalyst particles.

A multichannel version comprising 10 packed-bed reactors was also produced (see Figure 4.46) [277,278]. The gas flow is distributed by star-type manifolds to the 10 reaction units.

A cartridge heater is inserted in the cover plate of the packed-bed reactor [277]. The base plate provides conduits to the microreactor. The outlets are standard high-pressure fittings. Thermocouples are inserted into the slurry feed channels.

Microfabrication uses photolithography and etching [277,278]. A time-multiplexed inductively coupled plasma etch process was used for making the microchannels. The microstructured plate is covered with a Pyrex wafer by anodic bonding.

A flow-pattern map comprises dispersed flow, annular flow, slug-dispersed flow and slug-annular flow [278]. The highest specific interface measured amounts to $16\,000\,\mathrm{m^2/m^3}$. A porous surface structure ($100\,\mathrm{cm^2}$) in the reaction channel can be generated by a sulfurhexafluoride plasma etch process with silicon nitride masking [278].

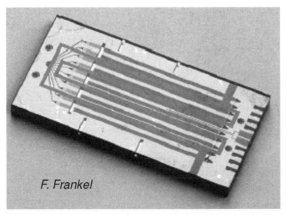

Figure 4.46 Image of a numbered-up 10-channel mini-packed-bed microreactor (by courtesy of IEEE) [278].

4.4.3
Gas–Liquid Reactions

Details on gas–liquid applications in microstructured reactors can be found in the book [230] as well as in the reviews [248–250].

4.4.3.1 **Direct Fluorination of Aromatics**

Direct Fluorination of Aromatics Fluorinated aromatics are particularly important as intermediates in drug synthesis for the pharmaceutical industry. Fluorinations of aromatics are typically carried out as multistep reactions via the Balz–Schiemann route introducing the fluorine moiety through the diazonium BF_4^- precursor. In selected cases, the Halex process that is a nucleophilic aromatic substitution and that only works with selected aromatic compounds having a special substituent pattern is applied. For several decades, attempts were made to use highly reactive elemental fluorine in a direct route and to avoid the circumventing chemistry in solution. However, this route proved to be unselective and harmful [296,297]. This is because of the radical nature of the direct process (under most conditions) that is so fast and exothermic that large amounts of heat are released, which have an autoaccelerating effect on the formation of radicals. As a consequence of this ever-increasing conversion, explosions happened frequently in direct fluorinations [298,299]. Even when carried out safely, fluorine radicals react via nonselective pathways yielding a broad spectrum of side and consecutive products [298,299]. Also chain growth to oligomerization is not uncommon, coating and blocking reactors by precipitation. Furthermore, the insolubility of fluorine in most solvents challenges reactant dosing because the gas–liquid interface, typically not well defined, is now the means to determine and control mass transfer and reaction rate and selectivity.

Operation was studied under extreme dilutions and/or extremely reduced temperatures (cryogenic) [298,300–303], which are the possible ways of reaction control and are not really far-fetching ones because these measures also hamper practical exploitation and are limited to mechanistical and analytical investigations. In addition, a few modern concepts make use of clever combinations of solvents and processing and demonstrate increased selectivity with safe operation at ambient temperature on laboratory scale [303]. Both the low- and the high-temperature studies confirm that an electrophilic pathway is possible with direct fluorination, as given for the other halogenations such as chlorinations and brominations, and being much more selective than radical chemistry. Still, a completely different approach not fixed to certain boundary conditions is desired.

The issues to be solved for direct fluorinations are heat release and mass transfer via the gas–liquid interface. Multiphase microstructured reactors enable process intensification [230,248–250,304–306]. Often geometrically well-defined interfaces are formed with large specific values, for example, up to $20\,000\,m^2/m^3$ and even more. These areas can be easily accessible, as flow conditions are often highly periodic and transparent microreactors are available. For the nondispersing microreactors, the specific interfaces are quasi constant for the whole operation time. This enables to have high local fluorine concentrations at the interfaces and also to have control over them. The heat released at a large scale can be transferred by virtue of the small fluid layers and integration of microheat exchangers. At last, micro-reactors allow to set defined and short reaction times that are necessary as opposing measures to the aggressive conditions deliberately faced.

Direct fluorination of toluene was achieved using the falling-film microreactor and the microbubble column [304,305,307,308] or the dual-channel microreactor [274]. Even extreme conditions were realized, such as operation at high substrate concentration (typically about 0.1 M, but also up to 1.0 M), large fluorine gas contents (up to 50%) and at comparatively high temperatures ($-10\,^\circ$C up to room temperature instead of cryogenic conditions). Monofluorotoluenes were generated with yields up to 28%. Conversions of 44–77% and yields of 60–78% were obtained in a dual-channel microreactor using varying contents of a special solvent mixture (formic acid/acetonitrile) [274].

The distribution of isomers is consistent to an electrophilic mechanism with fluorine cation intermediates.

A small amount of side-chain fluorination was observed [274]. Benzyl fluoride was formed in about the same extent as the *meta* isomer. Small amounts of difluoro-toluenes and trifluorotoluenes as well as some unidentified high-boiling compounds were also found [274]. Additional products were not detected [308].

The reaction cannot be performed in fully fluorinated apolar solvents such as octafluorotoluene, which is probably because of missing polarization of fluorine that is necessary to attack the aromatic system. Polar solvents such as methanol or acetonitrile provide better environments for fluorination [308]. Formic acid as a protic solvent gives even better results and is also quite inert as this small molecule has no labile site for fluorine attack [274].

Conversion rises with the increasing temperature, as expected [308]. For the falling-film microreactor, conversion is increased from 15 to 30% when heating up from -40 to $-15\,^\circ$C. The selectivity varies largely and exhibits no clear trend.

With the use of falling-film microreactor or a microbubble column, yields of up to 28% were obtained with acetonitrile as solvent at conversions ranging from 7 to 76% and selectivities from 31 to 43% with regard to the monofluorinated product [308]. With the use of dual-channel reactor, conversions from 17 to 95% and selectivities from 37 to 10% were achieved using methanol as solvent [274]. The conversion of a laboratory bubble column, taken for comparison, ranged from 6 to 34% with selectivities of 17–50%, which is equivalent to yields of 2–8% [308].

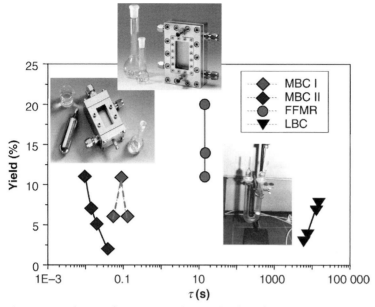

Figure 4.47 Reduction of reaction times by several orders of magnitude using a falling-film microreactor (FFMR) or microbubble columns (MBC I and II, denoting different dimensions, as given in [308]) as compared to standard organic laboratory processing with a laboratory bubble column (LBC). τ: residence time (by courtesy of IMM).

The most striking point about the fluorination results is the high intrinsic speed of the reaction (see Figure 4.47). The falling-film microreactor was operated at seconds scale and the microbubble column even at microseconds scale [308]. This is in contrast to fluorinations in laboratory flasks taking hours.

Accordingly, the respective space–time yields are higher by orders of magnitude [308]. The space–time yields for these microreactors ranged from about 20 000 to 110 000 mol monofluorinated product/$(m^3 h)$. The falling-film microreactor had two times higher space–time yields than the microbubble column. The performance of the laboratory bubble column was in the order of 40–60 mol monofluorinated product/$(m^3 h)$.

The fluorine content in the gas phase of a falling-film microreactor varied between 10, 25 and 50% [308]. A nearly linear increase in conversion results at constant selectivity. The substitution pattern rather than the ratio of *ortho* to *para* isomers is strongly affected.

Direct Fluorination of Aliphatics and Non-C Moieties The basic limitations of direct fluorinations are similar for aliphatic compounds as discussed already using the example of aromatic derivatives and so is the potential of microreactors. The fluorination of ethyl acetoacetate was carried out in an annular-flow microreactor.

71%   12%   3%

The microreactor contained a single channel with two feeds for gas and liquid [309,273]. The gas flows were set so high that an annular-flow regime was reached with a central gas core surrounded by a liquid film wetting the channel. This flow pattern has a very high interface and low liquid-side resistance due to the thin film. Formic acid was used as a solvent.

High yields of the monofluorinated product at short reaction times were obtained that exceeded the performance of standard batch laboratory processing [309]. Yields of 72% were achieved at 99% conversion [309,310]. The metallic construction material interacts with the reaction and impacts the keto/enol equilibrium to the advantage of the enol species that is fluorinated faster.

This microflow processing was also demonstrated using other β-keto esters such as ethyl 2-chloro-3-oxobutanoate [309,273] or ethyl 2-methyl-3-oxobutanoate [273]. Five- and six-ring β-ketoester derivatives such as 3-acetyl-3,4,5-trihydrofuran-2-one **(1)** [273], 2-acetyl cyclohexanone [273] and ethyl 2-oxocyclohexane carboxylate **(2)** [273] were directly fluorinated as well.

(1)

(2)

The performance of a single-channel and a numbered-up three-channel micro-reactor was similar [273]. Some differences were found that probably originate more from fluctuations in the hydrodynamics than from the numbering up.

Hazardous perfluorination processes with high yields were carried out safely in microreactors, such as the perfluorination of tetrahydrofuran and cyclohexane derivatives [309].

A pilot reactor was built for direct fluorination of ethyl acetoacetate and is described in [310] and reported in more detail in this book in Chapter 5 (page 261). A numbered-up reactor with 30 microchannels was used.

Direct Fluorination of Heterocyclic Aromatics Selective fluorination of quinoline aromatics leads to various commercially important products such as 5-fluoroacil, 5-fluoroprimaquine and ciprofloxacin with the fluorine moiety being decisive of their chemical and biological properties.

The fluorination of quinoline was performed in a microstructured reactor operated in the annular-flow regime, which contained one microchannel with two consecutive feeds for gas and liquid [311,312]. The role of the solvent was large. The reaction was totally unselective in acetonitrile and gave only tarlike products. With formic acid, a mixture of mono- and polyfluorinated products besides tar was formed. No tar formation was observed with concentrated sulfuric acid as solvent at 0–5 °C. In this way, a high selectivity of about 91% at medium conversion was achieved. Substitution was effective only in the electron-rich benzenoid core and not in the electron-poor pyridine-type core. The reactivity at the various positions in the quinoline molecule is 5 > 8 > 6 and thus driven by the vicinity to the heteroatom nitrogen that corresponds to the electrophilic reactivity known from proton/deuterium exchange studies in strong acid media.

F_2	Solvent	Conversion	Selectivity			
		(1)	(2)	(3)	(4)	(5)
6 equiv.	conc. H_2SO_4	42%	44	11	23	13
12 equiv.	conc. H_2SO_4	67%	27	8	14	32
7 equiv.	conc. H_2SO_4/PP11	22%	48	11	27	8

Mixtures obtained by diluting sulfuric acid with an inert perfluorocarbon fluid, acting as heat transfer medium, were also used without loss of performance [311,312]. Substituted quinolines allow to study the impact of electron-donating and electron-withdrawing groups on regioselectivity. Electron-donating groups such as $-CH_3$ and $-OCH_3$ direct the fluorine substituent into an *ortho* and *para* pattern, consistent with an electrophilic substitution path and electron-withdrawing groups such as $-NO_2$ leads to a *meta*-substitution pattern.

71% conversion   74%   26%

	solvent	
F_2 (18 equiv.)	conc. H_2SO_4	74%
F_2 (15 equiv.)	Oleum/PP11	59%

4.4.3.2 Oxidations of Alcohols, Diols and Ketones with Fluorine

For some oxidations, toxic heavy metal oxidants are used [313]. This can be circumvented by the use of microreactor technology due to safe handling of elemental fluorine that can be used to mediate oxidation reactions [313]. This is possible in a direct way via fluorine introduction into the substrate and subsequent replacement by an oxygen moiety. In an indirect manner, intermediate oxygen transfer reagents such as HOF·MeCN can be generated by the reaction of aqueous acetonitrile with elemental fluorine that then attacks the substrate. The only by-product is hydrogen fluoride that could be recycled by electrolysis.

The oxidation of cyclohexanol to cyclohexanone with fluorine and aqueous acetonitrile was performed in a single-channel microreactor operated under annular flow at room temperature. A conversion of 84% and a selectivity of 74% were observed [313]. In a similar way, diols such as 1,2-cyclohexanediol were partly or fully oxidized. A 53% selectivity to the monooxidation product was obtained at a conversion of 87%; the dioxidation product was obtained with 30% yield.

The Baeyer–Villiger oxidation of cyclohexanone to the seven-membered lactone used aqueous formic acid (5% water) as medium, and a 60% conversion at 88% selectivity was found [313].

4.4.3.3 Photochlorination of Aromatic Isocyanates

Side-chain photochlorination of toluene isocyanates leads to important industrial intermediates for polyurethane synthesis, one of the most important classes of polymers [264]. Irradiated thin liquid layers in microchannels should have much higher photon efficiency (quantum yield) than given for conventional processing.

The reaction of toluene-2,4-diisocyanate with chlorine to 1-chloromethyl-2,4-diisocyanatobenzene was carried out in a falling-film microstructured reactor with a transparent window for irradiation [264]. There are two modes of reaction. The desired radical process proceeds with the photoinduced homolytic cleavage of the chlorine molecules, and the chlorine radical reacts with the side chain of the aromatic compound. At very high chlorine concentrations radical recombination becomes dominant and consecutive processes such as dichlorination of the side chain may occur as well. Another undesired pathway is the electrophilic ring substitution to toluene-5-chloro-2,4-diisocyanate, promoted by Lewis acidic catalysts in polar solvents at low temperature. Even small metallic impurities probably from corrosion of the reactor material can enhance the formation of electrophilic by-products.

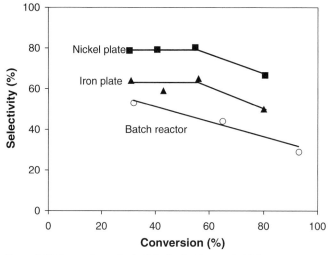

Figure 4.48 Comparison of selectivity for 1-chloromethyl-2,4-
diisocyanatobenzene depending on toluene-2,4-diisocyanate by
using different reactor plate materials (by courtesy of the Swiss
Chemical Society) [264].

The yield of the product 1-chloromethyl-2,4-diisocyanatobenzene ranged from 24
to 54% at 130 °C (conversion: 30–81%; respective selectivity: 79–67%) [264]. The
content of the ring-substituted product decreases with increasing conversion
(12–5%), whereas simultaneously the other by-products were formed in much larger
amount (8–29%) (see Figure 4.48). This is indicative of consecutive reaction steps,
that is, multiple chlorination. The superior performance of the microreactor is
explained by the better photon yield because of the very thin liquid films. Owing to the
low penetration of light into the conventional batch reactor, a large part of the reaction
volume is actually not irradiated. Here, thermal rather than photoinduced pathways
are followed that favor ring chlorination.

Reaction time can also be reduced by means of microprocess technology [264]. A
reaction time of 30 min was necessary for a 30 ml batch reactor to achieve a
conversion of 65% at a selectivity of 45%, whereas with the falling-film microreactor
only ~14 s was required for the same performance. This leads also to a large
space–time yield with 401 mol/(l h) for the falling-film microreactor as compared
to 1.3 mol/(l h) for the batch reactor. The impact of Lewis-acid formation on the
reaction course was investigated by using an iron microreactor instead of the usually
used nickel microreactor. The plate surface is converted to iron chloride under the
reaction conditions. The selectivity of the target product then drops from 67 to 50%.

4.4.3.4 Monochlorination of Acetic Acid

Selectivity is a major issue for the monochlorination of acetic acid to chloroacetic acid
because second chlorination gives dichloroacetic acid and other chlorinated species
including acid chloride formation [314]. The removal of these impurities, especially
of the dichloroacetic acid either by crystallization or by reduction with a palladium
catalyst, is laborious and costly.

Recirculation

A yield of 85% was obtained with a falling-film microstructured reactor that outperforms large-scale bubble column processing [314]. Selectivity was also superior with less than 0.05% dichloroacetic acid formed. Conventional processing has much higher levels of impurity (~3.5%). When both, temperature and pressure, were slightly increased, the yield raised from 85 to 90%. The content of dichloroacetic acid was still negligible (<0.05%).

4.4.3.5 Sulfonation of Toluene

The sulfonation of toluene proceeds via a complex scheme of elemental reactions with numerous side and consecutive reactions. Toluene sulfonic acid and sulfur trioxide can react in a consecutive process to give toluene pyrosulfonic acid, which can undergo further reactions [315,316]. Reaction of this acid with another molecule of toluene yields two molecules toluene sulfonic acid or ditolyl sulfone. In addition, toluene sulfonic anhydride may be formed via reaction of toluene pyrosulfonic acid with toluene sulfonic acid.

Precise control of concentration and residence time can increase the selectivity of the sulfonation of toluene, as this allows to optimally set the interplay between the reactions 4.4.1–4.4.4 [315,316]. The highly exothermic nature of the reaction demands for good temperature control. A single microreactor is not suited to conduct the various reaction steps with all their different needs on temperature and residence time. Thus, a continuously operated plant with many microflow tools was developed. The plant design was based on a fluidic backbone providing unitized ports and plant unit sites to facilitate the connection of microstructured components from different suppliers (see Figure 4.49).

Figure 4.49 Microplant for toluene sulfonation with falling-film microreactor as central apparatus and unitized backbone as fluidic bus system (by courtesy of Elsevier) [315].

The process flow sheet of the microreactor plant is shown in Figure 4.50 [315,316]. Toluene is heated to 40 °C using a microstructured heat exchanger while liquid sulfur trioxide is vaporized by heating to 60 °C. Nitrogen is added to the sulfur trioxide gas in a micromixer for dilution. This stream is then passed into a separator to remove liquid contents before entering the microstructured falling-film reactor where it reacts with the liquid toluene. Following the discussion on the chemistry given above, a delay-loop reactor had to be added to complete

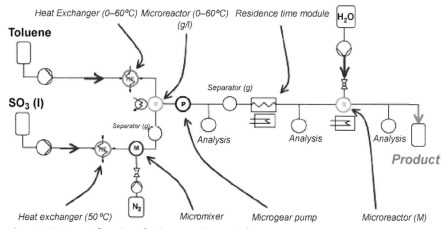

Figure 4.50 Process flow sheet for the microplant used for toluene sulfonation (by courtesy of Elsevier) [315].

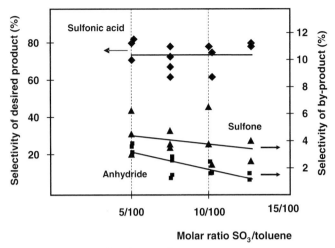

Figure 4.51 Selectivity for toluene sulfonation in a special microplant with falling-film microreactor, when increasing contents of sulfur trioxide are fed (by courtesy of Elsevier) [315].

toluene pyrosulfonic acid conversion. This was achieved in a heated wound tube. The need for anhydride hydrolysis demands for a further processing step that is performed in a tempered liquid–liquid microstructured reactor by mixing with water.

As the distribution of the products depends largely on reaction time and on the type of operations performed, several sampling stations were inserted at various positions of the microreactor plant [315,316]. Analysis was performed directly after the falling-film microreactor, where the ratio of sulfur trioxide to toluene widely varied. The selectivity for the formation of the toluene sulfonic acid is constant at about 73% for molar ratios of sulfur trioxide/toluene ranging from 5/100 to 15/100, whereas the selectivity for anhydride formation decreases from 8 to 2% (see Figure 4.51). The selectivity for the pathway to ditolyl sulfone is low at about 3%. If sampling is done at the stage behind the delay loop, the selectivity of toluene sulfonic acid increases to 82% because of the conversion of toluene pyrosulfonic acid with toluene. The anhydride selectivity is further decreased to 2%, whereas that of ditolyl sulfone is not altered. After hydrolysis with water, a nearly quantitative conversion of the anhydride to the acid is achieved, if mixing conditions and the amount of water were optimized. Totally, a selectivity of 82% for the target product toluene sulfonic acid was achieved at nearly complete conversion. Side products, for example, the sulfones, are given by reaction path (4.4.2) in the scheme above.

4.4.3.6 Photooxidation of α-Terpinene and Cyclopentadiene

A [4 + 2]-cycloaddition of singlet oxygen to α-terpinene in methanol yields ascaridole in the presence of catalytic amounts of Rose Bengal as photosensitizer. The reaction was carried out in one-channel chip microreactor made from glass [280].

This enabled the use of singlet oxygen without the need for preparation of large quantities of oxygenated solutions [280]. Further benefits of microreactor processing are to facilitate scale-up and avoid unwanted sample heating, which results in system simplification by eliminating the need for collimators and refrigeration to control the tungsten lamp power. In addition, safety issues that arise for aerated, oxygenated organic solvents can be addressed differently. Owing to the good mass transfer no pre-saturation of the α-terpinene with oxygen is required, and after reaction the oxygenated solution can be instantly degassed with nitrogen.

The quantum yield should also be high, as the efficiency of the use of light for the reaction can be kept high through the thin liquid layer (the losses by absorption are low, respectively) [280]. Although this is generic to all photochemical applications, it is here of special importance because Rose Bengal has a large extinction coefficient that leads to a substantial increase in absorption. Radiation is not effective anymore after a short path length. In turn, the chip microreactor allowed the use of high sensitizer concentrations of up to 5×10^{-3} M at large optical transmittance of 95%. Conversions of about 80% were obtained at short irradiation times (5 s).

The photooxygenation of cyclopentadiene with Rose Bengal as photosensitizer was performed using a falling-film microreactor in methanol [317]. The endoperoxide is first generated and then reduced to 2-cyclopenten-1,4-diol, which is used as an intermediate in pharmaceutical drug synthesis. This route is not easily possible by batch processing because the explosive endoperoxide intermediate is formed in substantial amounts.

A xenon lamp was used as light source that irradiated the thin falling films flowing through a quartz glass window in the reactor [317]. Thiourea in methanol was used to reduce the labile endoperoxide directly to the stable diol product. The yield of *cis-2-*

cyclopenten-1,4-diol was 20% at 10–15 °C. Other photochemical reactions are described in Chapter 4.1.5 (page 70).

4.4.3.7 Reactive Carbon Dioxide Absorption

This reaction serves as literature-known model reaction to characterize mass transfer efficiency in microreactors [318]. As it is a very fast reaction, solely mass transfer can be analyzed. The analysis can be done simply by titration; the reactants are inexpensive and not toxic.

$$2OH^- + CO_{2(g)} \rightarrow CO_3^{2-} + H_2O. \tag{4.7}$$

The mass transfer efficiency of different gas–liquid contactors as a function of residence time was compared, including an interdigital micromixer, a caterpillar minimixer, a mixing tee and three microbubble columns with microchannels of varying diameter (see Figure 4.52) [318]. The two microbubble columns comprising the smaller microchannels reached almost 100% conversion. The microbubble column with the largest hydraulic diameter reached at best 75% conversion. The respective curve passes over a maximum area due to the antagonistic interplay between residence time and specific interfacial area.

All other devices showed only the increasing part of such dependency; that is, the highest performance was obtained at the longest residence time [318]. The best conversions of interdigital micromixers and caterpillar minimixers of ~78 and ~70%, respectively, still exceed notably the performance of a conventional mixing tee (1 mm inner diameter).

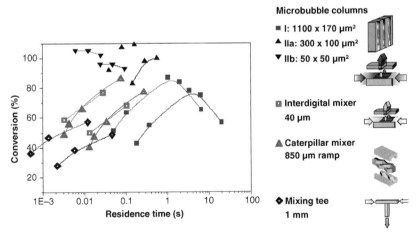

Figure 4.52 Special-type multipurpose microdevices and mixing tee used for investigation of CO$_2$ absorption. Comparison of their reactor performance with their dependence on the residence time. Microbubble column (■) (1100 μm × 170 μm); microbubble column (▲) (300 μm × 100 μm); microbubble column (▼) (50 μm × 50 μm); interdigital mixer (□) (40 μm); caterpillar mixer (△) (850 μm ramp); mixing tee (ε) (1 mm) (by courtesy of IMM) [318].

Table 4.3 Comparison of space–time yields for CO_2 absorption when using microdevices or conventional packed columns.

Reactor type	NaOH (mol/l)	CO_2 (vol%)	Molar ratio CO_2/NaOH	Conversion CO_2 (%)	Space–time yield (mol/m³ s)
Packed column	1.2	12.5	0.41	87	0.61
Packed column	2.0	15.5	0.43	93	0.81
Falling-film microreactor (65 µm)	1.0 (50 ml/h)	8.0	0.40	85	56.1
Falling-film microreactor (65 µm)	2.0 (50 ml/h)	8.0	0.40	61	83.3
Falling-film microreactor (100 µm)	1.0 (200 ml/h)	8.0	0.40	45	83.7
Microbubble column (300 µm × 300 µm)	2.0 (10 ml/h)	8.0 (2400 ml/h)	0.33	100	227
Microbubble column (300 µm × 300 µm)	2.0 (50 ml/h)	8.0 (12 000 ml/h)	0.33	72	816

By courtesy of IMM [318].

The mass transfer efficiency of the falling-film microreactor was determined at various carbon dioxide volume contents (0.1, 1.0 and 2.0 M) [318]. The molar ratio of carbon dioxide to sodium hydroxide was constant at 0.4 for all experiments, that is, the liquid reactant was in light excess. The higher the base concentration, the higher was the conversion of carbon dioxide. For all concentrations, complete absorption was, however, achieved at different carbon dioxide contents in the gas mixture. The results show the interdependency of the carbon dioxide content, the gas flow velocity and the sodium hydroxide concentration.

The mass transfer efficiency of the falling-film microreactor and the microbubble column was compared quantitatively according to the literature reports on conventional packed columns (see Table 4.3) [318]. The process conditions were chosen as similar as possible for the different devices. The conversion of the packed columns was 87–93%; the microdevices had conversions of 45–100%. Furthermore, the space–time yield was compared. Here, the microdevices resulted in larger values by orders of magnitude. The best results for falling-film microreactors and the microbubble columns were 84 and 816 mol/(m³ s), respectively, and are higher than conventional packed-bed reactors by about 0.8 mol/(m³ s).

A more detailed mass transfer study on the carbon dioxide absorption in sodium hydroxide solution was performed using a falling-film microreactor [319]. Experimental investigations were made at a liquid flow of 50 ml/h, with three NaOH concentrations (0.1, 1 and 2 M), at a fixed inlet molar ratio $CO_2 : NaOH$ of 0.4, and for a range of CO_2 concentration of 0.8–100%. A two-dimensional reactor model was developed, and the results are similar to the experimental data at low NaOH concentrations (0.1 and 1 M). The agreement is less pronounced for higher concentrations such as 2 M NaOH, which could be explained by either maldistribution of the liquid through the reactor channels or model simplifications. The model indicates that carbon dioxide is consumed within a short distance from the gas–liquid interface. The variation of the liquid film thickness has no large impact on CO_2 conversion but leads to more efficient consumption of the liquid reactant.

4.4.4
Gas–Liquid–Solid Reactions

4.4.4.1 Cyclohexene Hydrogenation over Pt/Al₂O₃

Cyclohexene hydrogenation is a well-studied process that serves as model reaction to evaluate performance of gas–liquid reactors because it is a fast process causing mass transfer limitations for many reactors [277,278]. Processing at room temperature and atmospheric pressure reduces the technical expenditure for experiments so that the cyclohexene hydrogenation is accepted as a simple and general method for mass transfer evaluation. Flow-pattern maps and kinetics were determined for conventional fixed-bed reactors as well as overall mass transfer coefficients and energy dissipation. In this way, mass transfer can be analyzed quantitatively for new reactor concepts and processing conditions. Besides mass transfer, heat transfer is an issue, as the reaction is exothermic. Hot spot formation should be suppressed as these would decrease selectivity and catalytic activity [277].

Conversions ranged from 2.8 to 16.0% in a packed-bed microreactor [277,278]. Average reaction rate amounted from 8.6×10^{-4} to 1.4×10^{-3} mol/(min g_{cat}) [250] (see also [320]), which is close to the intrinsic reaction rate of 3.4×10^{-3} mol/(min g_{cat}). Cyclohexene was given in excess so that the kinetics was of zero order for this species and also of first order for hydrogen [277]. For this pseudo-first-order reaction with the catalyst surface area of 0.57 m²/g and catalyst loading density of 1 g/cm³, a volumetric rate constant of 16 s^{-1} was determined.

Studies were also performed with an artificial fixed bed composed of an array of microstructured columns made by a plasma etch process. These columns were made porous to increase the surface area to 100 m², which is not far from the porosity of catalyst particles in fixed beds, and then coated with a catalyst [278]. The performance of such catalytic microcolumns was compared with that of a catalytic fixed bed reactor. When normalized to the metal content, the reaction rates of the columnar and the particle-containing reactor are similar with 6.5×10^{-5} and 4.5×10^{-5} mol/(min m²), respectively.

The increased interfacial area in the microreactor led to an increased pressure drop. The energy dissipation factor, the power unit per reactor volume, of the microreactor process was thus higher ($\varepsilon_V = 2$–5 kW/m³) than that of the laboratory trickle-bed reactors ($\varepsilon_V = 0.01$–0.2 kW/m³) [277]. This is, however, outperformed by the still larger gain in mass transfer so that the net performance of the microreactor is better.

4.4.4.2 Hydrogenation of *p*-Nitrotoluene and Nitrobenzene over Pd/C and Pd/Al₂O₃

Nitroaromatics are frequently used in organic synthesis as intermediates for the corresponding aniline derivatives by hydrogenation [283,320]. Among other uses, pharmaceuticals are produced via that route [320]. The high intrinsic reaction rate in hydrogenations of nitrobenzene derivatives cannot be exploited by conventional reactors because of large reaction enthalpies (500–550 kJ/mol), which would release much heat [283,320,321]. Another reason is the loss in selectivity under such conditions by decomposition of the nitroaromatics or by formation of partially hydrogenated intermediates [321]. Therefore, hydrogen supply is restricted to slow down the reaction.

Hydrogenations of nitrobenzene and *p*-nitrotoluene over supported noble metal catalysts are often investigated as model reactions, as they consist of various elemental reactions with different intermediates that can react with each other as depicted in the scheme below for *p*-nitrotoluene [321]. At short reaction times, the intermediates are predominantly formed, whereas complete conversion to *p*-methyl aniline is achieved at long reaction times. Aniline itself can react further to form side products such as cyclohexanol, cyclohexyl amine and other species.

The hydrogenation of *p*-nitrotoluene in the presence of a supported Pd catalyst was carried out in a microreactor with stacked plates with complete selectivity [283,320]. The Pd catalyst was prepared in three different ways. Conversions were 58–98% for an impregnated aluminum oxide wash-coat catalyst, depending on the process conditions. The conversions for an electrodeposited catalyst and an impregnated catalyst on electrooxidized nanoporous substrate were 58 and 89%, respectively. The best latter result is similar to that of a conventional fixed-bed reactor (85%), while the maximum yield of 30% in a microreactor was superior because of the high selectivity.

Computational fluid dynamic (CFD) calculations were performed to give the Pd concentration profile in a nanopore of the oxide catalyst carrier layer [283]. For wet chemical deposition most of the catalyst was deposited in the pore mouth, in the first 4 μm of the pore. Thus, most of the hydrogenation reaction is expected to occur in this location. For electrochemical deposition, large fractions of the catalyst are located both in the pore mouth and in the base. Because the pore base is not expected to contribute to a large extent to hydrogenation, a worse performance was proposed for this case. This was corroborated by experimental evidence. Higher conversions were indeed found for the catalyst prepared by wet chemical deposition.

An even wider variation of preparation procedures for the palladium catalyst was investigated for the hydrogenation of nitrobenzene using a falling-film microreactor [321,323].

$$\text{NO}_2 \xrightarrow{\text{H}_2,\ \text{Pd/Al}_2\text{O}_3} \text{NH}_2$$

A sputtered palladium catalyst exhibited low conversion, and large deactivation of the catalyst was found initially (60 °C; 4 bar) [321]. The corresponding selectivity was also low. A slightly better performance was obtained after an oxidation/reduction cycle. Following a steep initial deactivation, the catalyst activity stabilized at 2–4% conversion and at about 60% selectivity. After reactivation, selectivity initially approached 100%. A palladium catalyst made by UV decomposition had a conversion somewhat higher than the sputtered one. When stopping and restarting microreactor operation, activity changed stepwise. A similar pattern of side products as for the sputtered catalyst was found. An improved performance was found for an impregnated palladium catalyst. Complete conversion was found after 6 h, and selectivity decreased slowly to a level that still was high. For an incipient-wetness palladium catalyst, the best performance of all catalysts investigated was found. Starting from more than 90% conversion, a 75% conversion at selectivity of 80% was reached and maintained for a long time (see Figure 4.53).

It was concluded that the catalyst lifetime is a function of the catalyst loading and does not relate to its way of synthesis [321]. With larger loadings, catalysts are active for a long time before they need reactivation. With regard to lifetime and activity, the four catalysts were ranked as follows: wet impregnation ≫ incipient wetness > UV decomposition of precursors > sputtering. In case of loss of performance, two

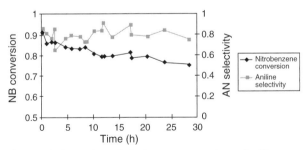

Figure 4.53 Comparison of nitrobenzene conversion and aniline selectivity as a function of reaction time for the incipient wetness catalyst (by courtesy of Elsevier) [321]. NB: nitrobenzene; AN: aniline.

reactivation routes for the catalysts were recommended. First, to try to dissolve the organic residues on the catalysts using dichloromethane or, if this is not successful, to burn these coats by heating in air. In this way, initial activity was gained back showing again complete conversion. Although deactivation of such catalysts still did occur, it could be slowed down by suitable reactivation.

From the list of side products mentioned for the reduction of nitrobenzene, all intermediates except phenylhydroxylamine were identified [321]. Their relative amounts allowed a judgment on the route by which the hydrogenation proceeds. As a result, it was concluded that species containing nitroso, azo and azoxy groups have strong interaction with the catalyst and so are preferably involved in the reaction course. On the contrary, reduction of the hydrazo species was hindered. These assumptions are in agreement with the previous reports.

The hydrodynamics of the falling film governs, as to be expected, the reaction behavior [322,323]. When increasing the flow rate, the reaction time was reduced and the film thickness was increased. For both reasons, the conversion decreased and exhibited in a linear dependence on log Re, Re being the Reynolds number. The hydrogenation conversion had only a weak dependence on hydrogen pressure.

4.4.4.3 Hydrogenation of α-Methylstyrene over Pd/C

The hydrogenation of α-methylstyrene is another standard process for studying mass transfer effects in catalyst pellets and in fixed-bed reactors because the intrinsic kinetics is known [324]. The reaction is relatively fast at room temperature and 1 atm hydrogen pressure.

The hydrogenation of α-methylstyrene was investigated to demonstrate the performance of a packed-bed microreactor with a palladium catalyst supported on activated carbon [324]. The microreactor was operated at 50 °C, and conversions from 20 to 100% were measured. It was determined that the reaction is first order for hydrogen and zero order for α-methylstyrene. Initial reaction rates were close to 0.01 mol/min per reaction channel and were achieved without additional activation of the catalyst. This is in agreement with the literature data on intrinsic kinetics.

Similar studies were made on a mesh microreactor comprising Pd/Al_2O_3 and Pt/Al_2O_3 catalysts coated on a microstructured mesh [270,271]. The global rate constants were 56 and 1.4 1/s for the Pd and Pt catalysts, respectively [271]. An activation energy of 46 ± 5 kJ/mol was found for the Pt catalyst in the mesh reactor, which corresponds to the value of 39 kJ/mol determined for a commercial Pt/Al_2O_3 powder catalyst in a well-behaved batch reactor (see Figure 4.54).

Accordingly, operation was in the chemical regime, that is, governed by kinetics and not controlled by mass transfer [271]. For the Pd catalyst, however, the activation energies measured in the mesh reactor and in a well-behaved batch reactor with commercial Pd/Al_2O_3 powder catalyst were different, being about 0 and 41 kJ/mol,

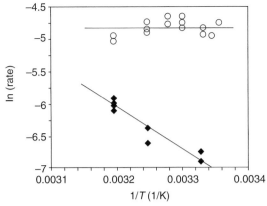

Figure 4.54 Arrhenius plot for the Pd/Al₂O₃ (open circles) and Pt/
Al₂O₃ (filled rhombs) catalysts. Continuous microreactor
operation; 2–3 bar; 1 M reactant solution in methylcyclohexane;
0.1 ml/min liquid flow rate (by courtesy of Royal Society of
Chemistry) [271].

respectively. The activity of the Pd catalyst is much higher so that mass transfer
limitations are given [271].

4.4.4.4 Hydrogen Peroxide Formation from the Elements
The direct route to hydrogen peroxide from hydrogen and oxygen is described in
Chapter 5.4.1 (page 238).

4.4.5
Homogeneously Catalyzed Gas–Liquid Reactions

4.4.5.1 Asymmetric Hydrogenation of Cinnamic Acid Derivatives
The potential of microprocess technology for catalyst and ligand screening for
multiphase flows was evaluated at the asymmetric hydrogenation of Z-methylacet-
amidocinnamate (mac) with enantiomerically pure rhodium diphosphine complexes
[271]. A microstructured mesh reactor was used that separates gas and liquids by a
microstructured porous support (mesh) and has a stable, well-defined gas–liquid
interface through the openings. A test series of 20 enantiomerically pure diphosphines as homogeneous catalysts was carried out. Less than 1 min residence time
in continuous-flow operation was sufficient to complete the reaction for the very
active catalysts. For less active catalysts, longer residence times (up to 30 min) were
needed that could not be realized by continuous operation in a practical manner.
Rather, the flow had to be stopped to allow for batchwise operation. The enantiomeric excesses of the measurements and of published data were in good accordance. The impact of hydrogen pressure on the enantiomeric excesses was
determined as well.

In an earlier investigation, another reactor concept was applied for the transient serial screening of the hydrogenation of a cinnamic methyl ester [324–326]. The microreactor screening setup contained valves for liquid pulse injection (200 μl volume) into the feed stream of an interdigital micromixer. The degree of pulse broadening was characterized in a later study and conditions were optimized [290]. The other feed of the mixer was for the gas stream. The surfactant sodium dodecyl sulfate was added to stabilize the dispersed flow. In this way, rigid foam flows, which were stable for several minutes, were generated. With bubble sizes of about 150–250 μm, large specific gas–liquid interfaces of the foams of up to 50 000 m²/m³ are achieved and improve mass transfer so that operation in a kinetically controlled regime is possible [324–326]. Foam stability increases with hydrogen pressure. By this reaction, foams remain stable up to 12 min at 70 °C, and no coalescence occurs within this period. It has to be, however, mentioned that the range of solvents to be used for a microreactor test unit is limited as a stable foam has to be established. So far, this was only possible for aqueous solutions.

These flows were guided into a wound delay tube for reaction with ethylene glycol/water (60/40 wt%) and hydrogen [324,326]. Efficient mass transfer in a kinetically controlled manner was achieved by the large specific gas–liquid interfaces of the foams of up to 50 000 m²/m³.

Before using the new method for screening, some basic engineering studies were made. The rate of reaction is proportional to increasing catalyst concentration [324,326]. This result suggests operation in a chemical regime. At higher temperatures, larger reaction rates were found. The rate of reaction is proportional to decreasing surfactant (sodium dodecyl sulfate) concentration. No change in the enantiomeric excess was observed, which is another indication for operation in a chemical regime, that is, governed by the kinetics and not by transport issues.

The performance of the new screening method was compared with that of an established method in the same laboratory, using a well-behaved mini-batch reactor (see Table 4.4). The microreactor testing was able to run on average 15 tests per day; at maximum, 40 tests were achieved. The average and the maximum number of the minibatches used routinely were 2 and 3 per day, respectively [327]. Two hundred and fourteen tests were carried out in total during the microreactor campaign [325]. The consumption of noble metal and chiral ligands per test for a microreactor test unit was 5–20 μg and 10 μmol, respectively. This is much lower than that of the minibatch approach with a usage of 500–1000 μg of the noble metal and 0.1 μmol of the chiral ligand [327].

Table 4.4 Benchmarking of figures of merit for the gas–liquid screening using a well-behaved minibatch and a continuous foaming-flow microreactor operation.

Benchmarked property of the g/l screening	Minibatch	Microflow
Reaction volume (ml)	10	0.1
Average amount of Rh/experiment (µg)	500–1000	5–20
Typical amount of ligand/experiment (µmol)	10	0.1
Temperature range (°C)	20–100	20–80
Pressure range (bar)	1–100	1–11
Residence time	>10 min	1–30
Average TTF achieved during study (d^{-1})	2	15
Maximum actual achievable TTF (d^{-1})	3	40
Range of solvents	Large	Restricted
Automation of reagents/catalyst injection	No	Yes
Automation of sample collection	No	Yes

By courtesy of the Swiss Chemical Society [327].

The deviation in the enantiomeric excess (ee) was small, and 90% of all data were within 40 and 48% ee [325]. A kinetic analysis was made by fitting the experimental data to empirical models by parity diagrams (see Figure 4.55). A statistical model with first-order kinetics for hydrogen gave the best fit. 141 from 170 experiments were

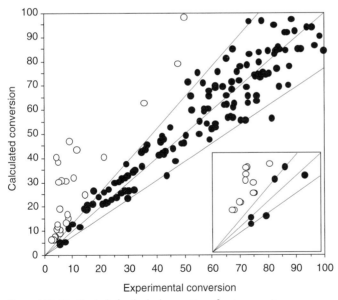

Figure 4.55 Kinetic study for the hydrogenation of a cinnamonic acid derivative: parity plot comparing experimental and modeling data. Open and closed circles correspond to data outside or within a match of experiments and modeling (by courtesy of Elsevier) [325].

properly described within the ranges of deviations set. The few rejected data had higher conversion than theoretically predicted. The kinetic constant and the activation energy were lower for a microreactor test unit than for a batch reactor that was explained by the very low inventory of material [325,327].

4.4.5.2 Asymmetric Hydrogenation of Methylacetamidocinnamate

Enantiomerically pure rhodium phosphine complexes have been evaluated for the asymmetric hydrogenation of the prochiral substrate methylacetamidocinnamate using the helical falling-film microreactor [266]. The ligands in these complexes were mostly diphosphines (P–P), some nitrogen-containing ligands (P–N) and some monophosphines. This library is available from commercial resources. A Diop ligand is the most active under the operating conditions of the test (Diop: 2,3-*o*-diisopropylidene-2,3-dihydroxy-1,4-bis(diphenylphosphino)-butane).

The Rh/Diop catalytic system is one of the fastest catalyzed gas–liquid asymmetric hydrogenations. A (*R*,*S*)-Cy-Cy-Josiphos ligand behaves almost as good as the Diop ligand and provides a better enantioselectivity of 75% (Josiphos: family of ferrocenyl diphosphine ligands; cy: cyclohexyl). The latter is the most active of the Josiphos family (88% conversion). The reproducibility of the data obtained has been checked with the Rh/Diop catalytic system. For more than five tests, the mean deviation was 2% for conversions and less than 1% for the enantiomeric excess that proved the reliability of this new microdevice.

The catalyst precursor complex $[Rh(COD)Diop]BF_4$ has been used for the screening of five substrates containing prochiral C=C double bond (COD = 1,5-cyclooctadiene) [266]. These were methylacetamidoacrylate (**S1**), Z-α-methylacetamidocinnamate (**S2**), dimethylitaconate (**S3**), methone (**S4**) and *rac*-α-pinene (**S5**) (see Figure 4.56). Activated C=C bonds such as those in the two acetamido derivatives were more reactive. The most reactive molecule is the less sterically hindered substrate methylacetamidoacrylate. Reaction was less pronounced for unsubstituted and sterically hindered substrates such as methone. The reduction of C=O bond in α-pinene is more difficult. These results are in agreement with the general trends reported for asymmetric hydrogenations.

Only very low catalyst concentrations down to 5×10^{-5} kmol/m^3 are consumed that keeps also the catalyst inventory very small [266]. Only 0.08 mg of Rh and about 0.2 mg–13 µg of the very expensive chiral ligands (about 300–1000 €/g), depending on their molecular weight, are consumed. Finally, a performance comparison for three different reactors was made for the substrate methylacetamidocinnamate and the two rhodium diphosphine complexes Rh/Josiphos and Rh/Diop (see Figure 4.57). The first reactor was a commercial 'Caroussel' reactor (Radleys

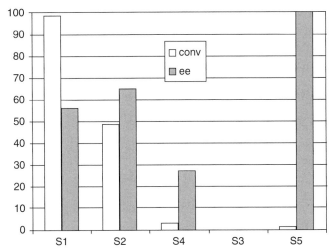

Figure 4.56 Screening of five substrates (see listing in the text) at 8 min and 35 °C (by courtesy of Elsevier) [266].
Conv = conversion; ee = enantiomeric excess.

Discovery Technologies) with a circular arrangement of 12 screw-capped gas-tied glass tubes agitated by magnetic stirring. The second reactor is a commercial pressure batch reactor with a turbine and baffles (Series 4590 from Parr Instruments). The third reactor was the helical falling-film microreactor.

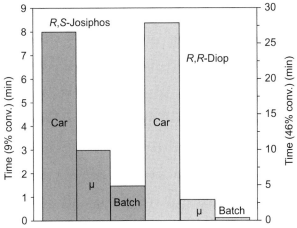

Figure 4.57 Reaction performance comparison of three reactors with the most active catalysts Rh/Josiphos and Rh/Diop. Caroussel (car), helical falling-film microreactor (μ) and Parr (batch) reactor (by courtesy of Elsevier) [266]. 9% conv. and 46% conv. denote a fixed conversion of 9 and 46%, respectively, which have to be achieved.

Enantioselectivity was roughly the same for the three reactors, being 80–90 and 62–65% for the Rh/Josiphos and Rh/Diop catalysts, respectively [266]. Conversion was very different. For fixed reaction time, the batch reactor and the falling-film microreactor had higher conversions than the Caroussel reactor. This was indicative of operation under mass transfer regime in the latter. On the basis of these data, it was concluded that the mass transfer coefficients $k_l a$ of the helical falling-film microreactor are in between the boundaries given by the known $k_l a$ values of $1-2\,s^{-1}$ for small batch reactors and about $0.01\,s^{-1}$ for the Caroussel reactor.

4.4.6
Other Applications

4.4.6.1 Segmented Gas–Liquid Flow for Particle Synthesis

Segmented gas–liquid (Taylor) flow was used for particle synthesis within the liquid slugs. Tetraethylorthosilicate in ethanol was hydrolyzed by a solution of ammonia, water and ethanol (Stöber synthesis) [329]. The resulting silicic acid monomer Si(OH)$_4$ is then converted by polycondensation to colloidal monodisperse silica nanoparticles. These particles have industrial application, for example, in pigments, catalysts, sensors, health care, antireflective coatings and chromatography.

The Taylor-flow microreactor comprised a micromixer for mixing of the precursors for the particle synthesis followed by a gas inlet for separating this continuous mixed liquid stream into segments separated by gas bubbles [328,329].

Monodisperse silica nanoparticles of a diameter of 200–500 nm were obtained [329]. This is explained by the well-defined residence time with reduced axial dispersion of the liquid segments [328,329]. In addition, by moving the liquid segments through a tube recirculation flow sets in, which is very effective in liquid mixing.

4.4.7
Conclusions and Outlook

Two classes of gas–liquid microchannel reactors were developed in the past years – continuous-phase contacting falling film, overlapping channel, mesh and annular-flow approaches and dispersed-phase contacting by Taylor-flow reactors, micromixers for bubble and foam formation and miniaturized packed-bed microreactors that follow classical trickle-bed operation at smaller scale. Gas–liquid–solid processes are possible by using either wall-coated catalyst or mini-packed beds. Numerous applications have demonstrated process intensification in terms of selectivity, space–time yield and safety by use of gas–liquid microreactors, including fluorinations, chlorination, hydrogenations, sulfonations, photooxidations and so on. This is achieved by enhanced mass transfer via the interface and through formation of thin liquid layers, which also give better transfer and allow to conduct photochemical reactions more efficiently. Operation is now possible in new process windows with very aggressive reactants such as elemental fluorine or even under explosive conditions.

The transfer of laboratory process development to pilot scale is not as much practiced so far as compared to the liquid-phase fine-chemical reactions carried out by microprocess technology. This is largely due to open issues on fluid distribution and the fact that throughputs are often not that large as given for liquid-phase devices, for example stemming from the need to make thin films in the falling-film microreactor. However, there may be more reports on transfer to pilot and production for gas–liquid microprocess technology in the coming years, as more laboratory investigations are completed and new reactor technology is available. One example for pilot-scale operation has been already given for the fluorination and the direct synthesis of hydrogen peroxide is on a similar way. Pilot-scale falling-film microreactors have been developed recently that further extend the technological offer for future use.

References

227 Mills, P.L., Ramachandran, P.A. and Chaudhari, R.V. (1992) *Reviews in Chemical Engineering*, **8**, 5.

228 Lee, S.-Y. and Tsui, Y.P. (1999) *Chemical Engineering and Processing*, July, 23–49.

229 Hessel, V., Löwe, H., Müller, A. and Kolb, G. (2005) *Chemical Micro Process Engineering – Processing and Plants*, Wiley-VCH Verlag GmbH, Weinheim.

230 Hessel, V., Hardt, S. and Löwe, H. (2004) *Chemical Micro Process Engineering – Fundamentals, Modeling and Reactions*, Wiley-VCH Verlag GmbH, Weinheim.

231 Kockmann, N. (ed.) (2006) *Micro Process Engineering – Fundamentals, Devices, Fabrication, and Applications*, book of the series: *Advanced Micro and Nanosystems* (eds von Brand, O., Fedder, G.K., Hierold, C., Korvink, J.G. and Tabata, O.), Wiley-VCH Verlag GmbH, Weinheim.

232 Klemm, E., Rudek, M., Markowz, G. and Schütte, R. (2004) *Mikroverfahrenstechnik*, vol. 2 (ed. Küchle, Winnacker), Chemische Technik: Prozesse und Produkte, Neue, Technologien, p. 759.

233 Fletcher, P.D.I., Haswell, S.J., Pombo-Villar, E., Warrington, B.H., Watts, P., Wong, S.Y.F. and Zhang, X. (2002) *Tetrahedron*, **58**, 4735.

234 Kiwi-Minsker, L. and Renken, A. (2005) *Catalysis Today*, **110**, 2.

235 Pennemann, H., Watts, P., Haswell, S., Hessel, V. and Löwe, H. (2004) *Organic Process Research & Development*, **8**, 422.

236 Jähnisch, K., Hessel, V., Löwe, H. and Baerns, M. (2004) *Angewandte Chemie-International Edition*, **43**, 406.

237 Kolb, G. and Hessel, V. (2004) *Chemical Engineering Journal*, **98**, 1.

238 Gavriilidis, A., Angeli, P., Cao, E., Yeong, K.K. and Wan, Y.S.S. (2002) *Transactions of the Institute of Chemical Engineering*, **80 A**, 3.

239 Jensen, K.F. (1999) *AIChE Journal*, **45**, 2051.

240 Jensen, K.F. (1998) *Nature*, **393**, 735.

241 Ehrfeld, W., Hessel, V. and Haverkamp, V. (1999) *Ullmann's Encyclopedia of Industrial Chemistry*, 6th edn, Wiley-VCH Verlag GmbH, Weinheim, p. 1.

242 Hogan, J. (2006) *Nature*, **442**, 351.

243 Thayer, A. (2005) *Chemical & Engineering News*, **83**, 43.

244 Seeberger, P. (2005) *Chemistry Files*, **5**, 2.

245 Geyer, K. *et al.* (2006) *Chemistry – A European Journal*, **12**, 8434.

246 Watts, P. *et al.* (2005) *QSAR Combinatorial Science*, **24**, 701.

247 Wiles, C. *et al.* (2005) *Tetrahedron*, **61**, 701.

248 Hessel, V., Angeli, P., Gavriilidis, A. and Löwe, H. (2005) *Industrial & Engineering Chemistry Research*, **44**, 9750.

249 Günther, A. and Jensen, K.F. (2006) *Lab on a Chip*, **6**, 1487.

250 TeGrotenhuis, W., Stenkamp, S. and Twitchell, A. (2005) *Microreactor Technology and Process Intensification* (eds Y. Wang and J.D. Holladay), American Chemical Society, Washington, DC, p. 360.

251 Bretherton, F.P. (1961) *Journal of Fluid Mechanics*, **10**, 166.

252 Chen, W.L., Twu, M.C. and Pan, C. (2002) *International Journal of Multiphase Flow*, **28**, 1235.

253 Coleman, J.W. and Garimella, S. (1999) *International Journal of Heat and Mass Transfer*, **42**, 2869.

254 Fukano, T. and Kariyasaki, A. (1993) *Nuclear Engineering and Design*, **141**, 59.

255 Hsieh, C.C., Wang, S.B. and Pan, C. (1997) *International Journal of Multiphase Flow*, **23**, 1147.

256 Mishima, K. and Hibiki, T. (1996) *International Journal of Multiphase Flow*, **22**, 703.

257 Peng, X.F. and Wang, B.X. (1993) *International Journal of Heat and Mass Transfer*, **36**, 3421.

258 Thulasidas, T.C., Abraham, M.A. and Cerro, R.L. (1995) *Chemical Engineering Science*, **50**, 183.

259 Thulasidas, T.C., Abraham, M.A. and Cerro, R.L. (1997) *Chemical Engineering Science*, **52**, 2947.

260 Triplett, K.A., Ghiaasiaan, S.M., Abdel-Khalik, S.I. and Sadowski, D.L. (1999) *International Journal of Multiphase Flow*, **25**, 377.

261 Triplett, K.A., Ghiaasiaan, S.M., Abdel-Khalik, S.I., LeMouel, A. and McCord, B.N. (1999) *International Journal of Multiphase Flow*, **25**, 395.

262 Xu, J.L., Cheng, P. and Zhao, T.S. (1999) *International Journal of Multiphase Flow*, **25**, 411.

263 Wille, C. (2002) Entwicklung und Charakterisierung eines Mikrofallfilm-Reaktors für stofftransportlimitierte hochexotherme Gas/Flüssig-Reaktionen, PhD thesis, University Clausthal-Zellerfeld; Fakultät für Bergbau, Hüttenwesen und Maschinenbau.

264 Ehrich, H., Linke, D., Morgenschweis, K., Baerns, M. and Jähnisch, K. (2002) *Chimia*, **56**, 647.

265 Vankayala, B.K., Löb, P., Hessel, V., Menges, G., Hofmann, C., Metzke, D., Kost, H.-J. and Krtschil, U. (2007) 1st International Congress on Green Process Engineering, Toulouse, France.

266 de Bellefon, C., Lamouille, T., Pennemann, H. and Hessel, V. (2005) *Catalysis Today*, **110**, 179.

267 Shaw, J., Turner, C., Miller, B. and Harper, M. (1998) Reaction and transport coupling for liquid and liquid/gas microreactor systems, in Process Miniaturization: 2nd International Conference on Microreaction Technology, IMRET 2; Topical Conference Preprints (eds W. Ehrfeld, I.H. Rinard and R.S. Wegeng), AIChE, New Orleans, USA, pp. 176–180.

268 Bibby, I.P., Harper, M.J. and Shaw, J. (1998) Design and optimisation of micro-fluidic reactors through CFD and analytical modelling, in Process Miniaturization: 2nd International Conference on Microreaction Technology, IMRET2; Topical Conference Preprints (eds W. Ehrfeld, I.H. Rinard and R.S. Wegeng), AIChE, New Orleans, USA, pp. 335–339.

269 Robins, I., Shaw, J., Miller, B., Turner, C. and Harper, M. (1997) Solute transfer by liquid/liquid exchange without mixing in micro-contactor devices, in Microreaction Technology – Proceedings of the 1st International Conference on Microreaction Technology, IMRET 1 (ed. W. Ehrfeld), Springer-Verlag, Berlin, pp. 35–46.

270 Wenn, D.A., Shaw, J.E.A. and Mackenzie, B. (2003) *Lab on a Chip*, **3**, 180.

271 Abdallah, R., Meille, V., Shaw, J., Wenn, D. and de Bellefon, C. (2004) *Chemical Communications*, 372.

272 Chambers, R.D. and Spink, R.C.H. (1999) *Chemical Communications*, 883.

273 Chambers, R.D., Holling, D., Spink, R.C.H. and Sandford, G. (2001) *Lab on a Chip*, **1**, 132.

274 de Mas, N., Günther, A., Schmidt, M.A. and Jensen, K.F. (2003) *Industrial & Engineering Chemistry Research*, **42**, 698.

275 Haverkamp, V. (2002) Charakterisierung einer Mikroblasensäule zur Durchführung stofftransportlimitierter und/oder hochexothermer Gas/Flüssig-Reaktionen (in Fortschritt-Bericht VDI, Reihe 3, Nr. 771), PhD thesis, University Erlangen.

276 Haverkamp, V., Emig, G., Hessel, V., Liauw, M.A. and Löwe, H. (2001) Characterization of a gas/liquid microreactor, the micro bubble column: determination of specific interfacial area, in Microreaction Technology – IMRET 5: Proceedings of the 5th International Conference on Microreaction Technology (eds M. Matlosz, W. Ehrfeld and J.P. Baselt), Springer-Verlag, Berlin, pp. 202–214.

277 Losey, M.W., Schmidt, M.A. and Jensen, K.F. (2001) *Industrial & Engineering Chemistry Research*, **40**, 2555.

278 Losey, M.W., Jackman, R.J., Firebaugh, S.L., Schmidt, M.A. and Jensen, K.F. (2002) *Journal of Microelectromechanical Systems*, **11**, 709.

279 Haverkamp, V., Hessel, V., Löwe, H., Menges, G., Warnier, M.J.F., Rebrov, E.V., de Croon, M.H.J.M., Schouten, J.C. and Liauw, M.A. (2006) *Chemical Engineering & Technology*, **29**, 1015.

280 Wootton, R.C.R., Fortt, R. and de Mello, A.J. (2002) *Organic Process Research & Development*, **60**, 187.

281 Hessel, V., Ehrfeld, W., Golbig, K., Haverkamp, V., Löwe, H., Storz, M., Wille, C., Guber, A., Jähnisch, K. and Baerns, M. (2000) Gas/liquid microreactors for direct fluorination of aromatic compounds using elemental fluorine, in Microreaction Technology: 3rd International Conference on Microreaction Technology, Proceedings of IMRET 3 (ed. W. Ehrfeld), Springer-Verlag, Berlin, pp. 526–540.

282 Haverkamp, V., Hessel, V., Liauw, M.A., Löwe, H. and Menges, M.G. (2007) *Industrial & Engineering Chemistry Research*, **64**, 8558.

283 Födisch, R., Hönicke, D., Xu, Y. and Platzer, B. (2001) Liquid phase hydrogenation of *p*-nitrotoluene in microchannel reactors, in Microreaction Technology – IMRET 5: Proceedings of the 5th International Conference on Microreaction Technology (eds M. Matlosz, W. Ehrfeld and J.P. Baselt), Springer-Verlag, Berlin, pp. 470–478.

284 Kursawe, A. and Hönicke, D. (2000) Epoxidation of ethene with pure oxygen as a model reaction for evaluating the performance of microchannel reactors. Proceedings of the 4th International Conference on Microreaction Technology, IMRET 4, AIChE Topical Conference Proceedings, 5–9 March, Atlanta, USA, pp. 153–166.

285 Hessel, V., Löwe, H. and Schönfeld, F. (2005) *Chemical Engineering Science*, **60**, 2479.

286 Hessel, V., Ehrfeld, W., Golbig, K., Haverkamp, V., Löwe, H. and Richter, T. (1998) Gas/liquid dispersion processes in micromixers: the hexagon flow, in Process Miniaturization: 2nd International Conference on Microreaction Technology, IMRET 2; Topical Conference Preprints (eds W. Ehrfeld, I.H. Rinard and R.S. Wegeng), AIChE, New Orleans, USA, pp. 259–266.

287 Ehrfeld, W., Golbig, K., Hessel, V., Löwe, H. and Richter, T. (1999) *Industrial & Engineering Chemistry Research*, **38** (3), 1075.

288 Hessel, V., Hardt, S., Löwe, H. and Schönfeld, F. (2003) *AIChE Journal*, **49**, 566.

289 Löb, P., Pennemann, H. and Hessel, V. (2004) *Chemical Engineering Journal*, **101**, 75.

290 Pennemann, H., Hessel, V., Kost, H.-J., Löwe, H. and de Bellefon, C. (2004) *AIChE Journal*, **50**, 1814.

291 Schönfeld, F., Hessel, V. and Hofmann, C. (2004) *Lab on a Chip*, **4**, 65.

292 Schwesinger, N., Frank, T. and Wurmus, H. (1996) *Journal of Micromechanics and Microengineering*, **6**, 99.

293 Kim, D.S., Lee, S.H., Kwon, T.H. and Ahn, C.H. (2005) *Lab on a Chip*, **5**, 739.

294 Stroock, A.D., Dertinger, S.K.W., Ajdari, A., Mezic, I., Stone, H.A. and Whitesides, G.M. (2002) *Science*, **295**, 647.

295 Stroock, A.D., Dertinger, S.K.W., Whitesides, G.M. and Ajdari, A. (2002) *Analytical Chemistry*, **74**, 5306.

296 Schiemann, G. and Cornils, B. (1969) *Chemie und Technologie cyclischer Fluorverbindungen*, Ferdinand Enke Verlag, Stuttgart, pp. 188–193.

297 Balz, G. and Schiemann, G. (1927) *Berichte der Deutschen Chemischen Gesellschaft*, **60**, 1186.

298 Grakauskas, V. (1969) *The Journal of Organic Chemistry*, **34**, 2835.

299 Purrington, S.T. and Kagen, B.S. (1986) *Chemical Reviews*, **86**, 997.

300 Grakauskas, V. (1970) *The Journal of Organic Chemistry*, **35**, 723.

301 Cacace, F., Giacomello, P. and Wolf, A.P. (1980) *Journal of the American Chemical Society*, **102**, 3511.

302 Cacace, F. and Wolf, A.P. (1978) *Journal of the American Chemical Society*, **100**, 3639.

303 Conte, L., Gambaretto, G.P., Napoli, M., Fraccaro, C. and Legnaro, E. (1995) *Journal of Fluorine Chemistry*, **70**, 175.

304 Löb, P., Löwe, H. and Hessel, V. (2004) *Journal of Fluorine Chemistry*, **125**, 1677.

305 Hessel, V., Löb, P. and Löwe, H. (2004) *Chimica Oggi – Chemistry Today*, 10.

306 Chambers, R.D. and Sandford, G. (2004) *Chimica Oggi – Chemistry Today*, 6.

307 Hessel, V., Ehrfeld, W., Golbig, K., Haverkamp, V., Löwe, H., Storz, M., Wille, C., Guber, A., Jähnisch, K. and Baerns, M. (1999) Gas–liquid microreactors for direct fluorination of aromatic compounds using elemental fluorine, in Proceedings of the 3rd

International Conference on Microreaction Technology, IMRET 3, Frankfurt, Germany, pp. 526–540.

308 Jähnisch, K., Baerns, M., Hessel, V., Ehrfeld, W., Haverkamp, V., Löwe, H., Wille, C. and Guber, A. (2000) *Journal of Fluorine Chemistry*, **105**, 117.

309 Chambers, R.D. and Spink, R.C.H. (1999) *Chemical Communications*, 883.

310 Chambers, R.D., Fox, M.A., Holling, D., Nakano, T., Okazoe, T. and Sandford, G. (2005) *Lab on a Chip*, **5**, 191.

311 Chambers, R.D., Holling, D., Sandford, G., Batsanov, A.S. and Howard, J.A.K. (2004) *Journal of Fluorine Chemistry*, **125**, 661.

312 Chambers, R.D., Holling, D., Sandford, G., Puschmann, H. and Howard, J.A.K. (2002) *Journal of Fluorine Chemistry*, **117**, 99.

313 Chambers, R.D., Holling, D., Rees, A.J. and Sandford, G. (2003) *Journal of Fluorine Chemistry*, **119**, 81.

314 Wehle, D., Dejmek, M., Rosenthal, J., Ernst, H., Kampmann, D., Trautschold, S. and Pechatschek, R. (2000) Verfahren zur Herstellung von Monochloressigsäure in Mikroreaktoren, Patent 27.07.2000.

315 Müller, A., Cominos, V., Horn, B., Ziogas, A., Jähnisch, K., Grosser, V., Hillmann, V., Jam, K.A., Bazzanella, A., Rinke, G. and Kraut, M. (2005) *Chemical Engineering Journal*, **107**, 205.

316 Müller, A., Cominos, V., Hessel, V., Horn, B., Schürer, J., Ziogas, A., Jähnisch, K., Hillmann, V., Großer, V., Jam, K.A., Bazzanella, A., Rinke, G. and Kraut, M. (2004) *Chemie Ingenieur Technik*, **76**, 641.

317 Jähnisch, K. (2004) *Chemie Ingenieur Technik*, **76**, 630.

318 Hessel, V., Ehrfeld, W., Herweck, T., Haverkamp, V., Löwe, H., Schiewe, J., Wille, C., Kern, T. and Lutz, N. (2000) Gas–liquid microreactors: hydrodynamics and mass transfer, in Proceedings of the 4th International Conference on Microreaction Technology, IMRET 4, Atlanta, USA, pp. 174–186.

319 Zanfir, M., Gavriilidis, A., Wille, C. and Hessel, V. (2005) *Industrial & Engineering Chemistry Research*, **44**, 1742.

320 Födisch, R., Reschetilowski, W. and
Hönicke, D. (1999) Heterogeneously
catalyzed liquid-phase hydrogenation of
nitro-aromatics using microchannel
reactors, in DGMK-Conference on the
Future Role of Aromatics in Refining and
Petrochemistry, Erlangen, Germany, pp.
231–238.

321 Yeong, K.K., Gavriilidis, A., Zapf, R. and
Hessel, V. (2003) *Catalysis Today*, **81**, 641.

322 Yeong, K.K., Gavriilidis, A., Zapf, R., Kost,
H.J., Hessel, V. and Boyde, A. (2005)
Experimental Thermal and Fluid Science,
80, 463.

323 Yeong, K.K., Gavriilidis, A., Zapf, R. and
Hessel, V. (2004) *Chemical Engineering
Science*, **59**, 3491.

324 de Bellefon, C., Tanchoux, N.,
Caravieilhes, S., Grenouillet, P. and

Hessel, V. (2000) *Angewandte Chemie-
International Edition*, **112**, 3584.

325 de Bellefon, C., Pestre, N., Lamouille, T.
and Grenouillet, P. (2003) *Advanced
Synthesis & Catalysis*, **345**, 190.

326 de Bellefon, C., Caraveilhes, S. and
Grenouillet, P. (2001) *Application of a
Micro Mixer for the High Throughput
Screening of Fluid–Liquid Molecular
Catalysts*, Springer-Verlag, Berlin, pp.
408–413.

327 de Bellefon, C., Abdallah, R., Lamouille,
T., Pestre, N., Caraveilhes, S. and
Grenouillet, P. (2002) *Chimia*, **56**, 621.

328 Günther, A., Khan, S.A., Thalmann, M.,
Trachsel, F. and Jensen, K.F. (2004) *Lab on
a Chip*, **4**, 278.

329 Khan, S.A., Günther, A., Schmidt, M.A. and
Jensen, K.F. (2004) *Langmuir*, **20**, 8604.

4.5
Bioorganic Reactions

Kaspar Koch, Floris P.J.T. Rutjes, Jan C.M. van Hest

4.5.1
General Introduction

Bioorganic chemistry can be defined as the discipline that deals with all chemical processes involving biomolecules. This broad area covers the use of biocatalysts for synthesis and screening, the *in vitro* production of biomolecules such as DNA and the performance of organic reactions in a living environment, such as the cell. The developments in micro total analysis systems (µTAS) and microreactor technology have also had their impact on bioorganic chemistry. As a first application, which started in the 1990s, device miniaturization was envisaged to be very useful for all bioorganic chemistry processes that are set up for diagnostics. This includes enzyme-linked immunosorbent assays (ELISAs) as well as the more recently developed enzyme-mediated protein fragmentation and subsequent mass spectrometric analysis, as applied in the area of proteomics. The advantage of the use of µTAS in these applications is clear because in these cases it is crucial to use as few analytes as necessary.

More recently, microreactor technology has entered the field of biocatalysis; enzymes are used for synthesis rather than for diagnostics. The concept behind the use of biocatalytic microreactor systems is in fact twofold. First, a miniaturized reactor allows an efficient use of small amounts of enzyme, when enzyme kinetics determination is involved. Second, the classical advantages of micro-reactors in synthesis, namely, better control over heat- and mass-transfer processes as well as the high surface-to-volume ratio, allow the development of optimal protocols using biocatalytic species, in particular, when dealing with fast reactions.

The latest development in bioorganic chemistry concerns the study of chemical processes within a cell. The incorporation of cells in a microreactor enables the investigation of the effect of all kinds of molecules on the behavior and the metabolic processes of a single cell. Although still in its infancy, this line of research holds much promise for more effective pharmaceutical screening processes and biotechnology applications.

In this chapter, we will focus on those bioorganic reactions in which biocatalysts, in particular, play a crucial role. We will not discuss peptide or natural product synthesis, as conventional organic chemistry will be covered by other chapters in this book. Also the discussion of the development of DNA chips, certainly one of the most exciting developments in the field of µTAS, is beyond the scope of this chapter, and the interested reader is referred to some excellent reviews [330,331]. First, the application of bioorganic chemistry in diagnostics will be discussed. This will be followed by a discussion on biocatalysis in microreactors. Finally, the recent development of cells on a chip is highlighted.

4.5.2
Diagnostic Applications

One of the most successful applications of microsystem technology is the use of µTAS in diagnostics [332–335]. Microreactors have been integrated into automated analytical systems, which eliminate errors associated with manual protocols. Furthermore microreactors can be coupled with numerous detection techniques and pretreatment of samples can be carried out on the chip. In addition, analytical systems that comprise microreactors are expected to display outstanding reproducibility by replacing batch iterative steps and discrete sample treatment by flow injection systems. The possibility of performing similar analyses in parallel is an attractive feature for screening and routine use.

The first enzymatic microreactors have been developed in the field of diagnostics to facilitate routine work in biochemical analysis. Using these microreactors requires only small sample volumes to obtain the same experimental results as the more traditional techniques. Nowadays, enzyme microreactors are used for a broad range of diagnostic applications, from a simple one-step digestion of proteins to very complex multiple-step reaction modules. An extensive review about enzyme microreactors for chemical analysis and kinetic studies has been published by Urban *et al.* [336]. The most common diagnostic applications for enzyme microreactors will now be highlighted starting with the digestion of proteins and DNA.

4.5.2.1 Protein and DNA Analysis
Protein digestion refers to the method in which the protein of interest is first digested by a proteolytic enzyme and then analyzed by mass spectrometry. This process affords a peptide map that is unique for each protein and allows identification

through a search of existing databases. Peptide mapping is routinely used in proteomics. A comprehensive review about miniaturization in peptide mapping has been published by Foret and Preisler [337]. Protein digestion is currently one of the most important applications of enzyme-based microsystem diagnostics because the use of a miniaturized digestion setup allows the analysis of minute amounts of proteins, with limits of detection in the low femtomoles per microliter range [338–340]. The most frequently used enzyme for peptide mapping is trypsin. This enzyme catalyzes the process of protein digestion through C-terminal hydrolysis of peptide bonds at the basic residues arginine and lysine.

The other advantages of the use of microreactors are the high surface-to-volume ratio and the possibilities for enzyme immobilization within a microchannel to create the so-called immobilized enzyme reactors (IMERS). When soluble trypsin is used, it partially results in autodigestion with the undesired formation of additional peptide fragments, which may complicate the unambiguous assignment of the studied protein [341]. Such unwanted autodigestion can be eliminated by site isolation of the proteolytic enzyme by immobilization on a solid support such as monoliths, which are favored because of their large surface availability. In most cases, immobilization is achieved by nonspecific adsorption. The other, more controlled methods are also possible, such as the well-known biotin–avidin-mediated binding of trypsin onto the surface of fused capillaries [342,343]. The main advantage of this immobilization is the good steric accessibility of the active binding site.

Another immobilization method was described by Maeda and coworkers [344]. They developed a facile and inexpensive preparation method for the formation of an enzyme-polymeric membrane on the inner wall of the microchannel (PTFE) through cross-linking polymerization in a laminar flow. With this approach, α-chymotrypsin was immobilized successfully. The activity of the immobilized enzyme was tested using N-glutaryl-L-phenylalanine p-nitroanilide as substrate, and the reaction products were analyzed offline by HPLC. There was no significant difference in the hydrolysis efficiency compared to solution-phase batchwise reactions using the same enzyme/substrate molar ratio (Scheme 4.87).

Besides the benefits of scale reduction and trypsin immobilization, microsystem technology has other advantages to offer. For example, Ekstrom et al. [345] described a device that integrated an enzyme microreactor with a sample pretreatment robot and

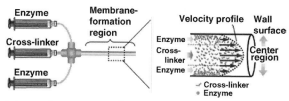

Scheme 4.87 Formation of enzyme–polymer membranes in a microchannel. Reprinted with permission from [344]. Copyright 2005 The Royal Society of Chemistry.

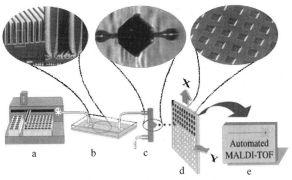

Scheme 4.88 Microfluidic system for MALDI protein analysis. (a) automated sample pretreatment and injection; (b) microreactor; (c) microdispenser used to deposit sample into nanovials; (d) shallow nanovials on the MALDI target plate; and (e) automated MALDI–ToF–MS analysis. Reprinted with permission from [345]. Copyright 2000 American Chemical Society.

a microfabricated microdispenser to transfer digested protein directly to a MALDI target plate for automated MS analysis (Scheme 4.88), thereby enabling a rapid and automated protein identification. An overview of protein digestion in microreactors can be found in Table 4.5.

Besides proteins, DNA can also be digested by enzymes for analytical purposes. These so-called restriction enzymes are isolated from bacteria that recognize specific sequences in DNA and then cut the DNA to produce the so-called restriction fragments. These oligonucleotides are subsequently analyzed, providing information with regard to locations of restriction sites in DNA routinely required in many research protocols in molecular biology. DNA digestion also plays an important role through DNA identification, as performed in genomics research. The first on-chip enzymatic restriction reaction with DNA followed by electrophoretic sizing of the mixture was performed by Jacobson and Ramsey [372] using a monolithic device that executed the biochemical procedure automatically. This device mixed a DNA sample with a restriction enzyme in a 0.7 nl reaction chamber and, after a digestion period, injected the fragments onto a capillary electrophoresis channel for sizing. Materials could be precisely manipulated by computer control within the channel structure using electrokinetic transport. Digestion of a plasmid by an enzyme and fragment analysis were completed in 5 min using only 30 amol of DNA and 2.8×10^{-3} units of enzyme per run.

Burns *et al.* [373] developed a DNA analyzer that provided significant improvement in the speed, portability and cost of DNA analysis. This device was capable of handling aqueous reagent and DNA-containing solutions, mixing the solutions together, amplifying or digesting the DNA to form discrete products and separating and detecting those products.

A heterogeneous assay was developed by Kartalov and Quake for DNA sequencing-by-synthesis to reliably sequence up to four consecutive base pairs [374]. Briefly, the

Table 4.5 Application of enzymatic microreactors in the analysis of proteins.

Application	Enzyme	Method	References
Analysis of proteins	Trypsin	Monolith	[346]
		Fused-silica capillary	[347]
Digestion of proteins	Trypsin	Monolithic capillary column	[348]
		Fused-silica capillary	[349]
		Gel beads	[350]
		Controlled pore glass	[351]
		Fused-silica capillary	[342]
		Injected	[352]
High-speed online protein digestion	Trypsin, glucose oxidase	Porous silicon	[353]
Rapid protein digestion, peptide separation and protein identification	Trypsin	Poly(vinylidene fluoride) in poly(dimethylsiloxane) channel	[354]
Extraction of proteins from 2D gels and digestion	Trypsin	PVDF membrane disk	[355]
Peptide mapping	Trypsin and pepsin	Fused-silica capillary	[356]
	Trypsin	Controlled pore glass	[357]
		Monolith	[358]
		Glycidyl methacrylate-modified cellulose membrane	[359,360]
		Porous polymer monolith	[361]
		Porous polymer monolith in fused-silica capillary	[340]
Protein mapping	Protease	Porous wall of a capillary	[362]
	Trypsin	Porous polymer monolith	[363]
		Reversed-phase beads	[364]
		Poroszyme cartridge	[365]
Peptide mapping analysis of proteins	Protease	Fused-silica capillary	[366]
Protein digestion, peptide separation and protein identification	Pepsin	Gel on a photopolymerized porous silica monolith	[367]
Protein identification	Chymotrypsin	Silicon	[345]
Protein patterning	Trypsin	Micropatterned sol–gel structures in polydimethylsiloxane microchannels	[368]
Proteomic research	Trypsin	Silica gel microchannels	[369]
Online frontal analysis	Trypsin	Glycidyl methacrylate-modified cellulose	[370]
Specific fragmentation of high molecular mass and heterogeneous glycoproteins	Chymotrypsin, trypsin and papain	Magnetic microparticles	[371]

method involved exposing a primed DNA template to a mixture of a known type of standard nucleotide, its fluorescently tagged analogue and DNA polymerase. If the tagged nucleotide was complementary to the template base next to the primer's end, the polymerase could perform a one-nucleobase extension and the partly incorporated fluorescence signal could be detected after a washing step. Iteration with each type of nucleotide revealed the DNA sequence. All these steps were integrated into one-PDMS microreactor. The average read length was three base pairs, which constitute the proof of principle for this sequencing approach.

4.5.2.2 DNA Amplifications

The previous example shows that instead of analyzing DNA by biocatalytic digestion, enzymes can also be used to replicate specific DNA sequences. To not only replicate but also multiply a desired DNA fragment for further research purposes, a process referred to as DNA amplification, the polymerase chain reaction (PCR), can be executed, reported first by Saiki *et al.* in 1985 [375]. PCR is assisted by DNA polymerase. The amplification process consists of a series of cycles. Each cycle can be divided into three steps, each with its own specific reaction temperature: denaturing, extension and annealing. Within one temperature cycle the amount of DNA can be doubled, and 20–35 cycles can therefore produce millions of DNA copies. Conventional PCR instruments usually achieve a heating or cooling rate of about 1–2 °C/s in the temperature range relevant for PCR, leading to a complete PCR analysis time of approximately 1–2 h. This low ramping rate is due to the high thermal capacity of the material from which the PCR reaction system is constructed. Current improved PCR instruments can perform somewhat faster heating cycles and therefore enable faster DNA amplification.

Microsystem technology, however, has created the possibilities for strongly improved PCR procedures. Ever since the first silicon-based stationary PCR chip was described by Northrup *et al.* [376], a variety of PCR microfluidic technologies have facilitated DNA amplification with much faster rates [377,378], as a result of the smaller thermal capacity of the micro-PCR device and a larger heat-transfer rate between the PCR sample and the temperature-controlled microreactor. Nowadays, miniaturized DNA amplification can be performed in three different classes: in stationary chambers [379], in flow-through channels [380,381] and in microchannels using Rayleigh–Bérnard thermal convection [382]. Stationary chambers are miniaturized wells and temperature control is achieved by placing the microchip containing the stationary chambers on special plates that can rapidly heat and cool the specimen. In contrast to this batch process, flow-through channels are more often used (Scheme 4.89).

Other extensions that have been made include the construction of a biochip in which a reverse transcription PCR process can be carried out [384] and the integration of PCR with capillary electrophoresis [385], DNA microarray hybridization [386] and sample preparation [387,388]. The speed of analysis, the ease of integrating different functions and the relative low costs of micro-PCR production are important advantages of the miniaturized PCR technique that suggest that this bioorganic microreactor device will be thoroughly implemented in many analytical labs.

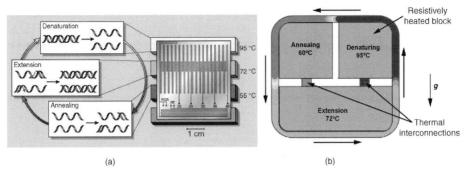

(a)   (b)

Scheme 4.89 Two examples of microreactor PCR devices (a) Sample is continuously pumped through the flow cell and encounters different temperature regions (resp. 55, 72 and 95 °C). Reprinted with permission from [377]. Copyright 2003 Nature Publishing group. (b) Sample is loaded into a dedicated microchip and circulates multiple times through the temperature zones by heat convection and gravity (g). Reprinted with permission from [383]. Copyright 2007 Wiley-VCH Verlag GmbH.

4.5.2.3 Enzyme-Linked Immunoassays

Immunoassays are currently the predominant analytical technique for the quantitative determination of a broad variety of analytes of clinical, medical, biotechnological and environmental significance. The high specificity of the analysis is provided by the antibody molecule, which recognizes the corresponding antigen (analyte). Among the most important advantages of immunoassays are their speed, sensitivity, selectivity and cost effectiveness. Immunoassays can be divided roughly into three classes involving (1) redox labeling, (2) optical labeling and (3) enzyme labeling. Enzyme labeling is very valuable as it can be used to efficiently catalyze a specific reaction to convert a nondetectable substrate into a detectable product. The catalytic reaction by the enzyme amplifies the signal and increases the sensitivity of the assay because the number of detectable molecules can be exponentially higher than the number of antigens. ELISA is the most commonly used method of various immunoassays and is usually performed in 96-well microtiter plates. The whole ELISA process requires a series of mixing, reaction and washing steps involving a tedious and laborious protocol. It may often take many hours to days to perform one assay because of the long incubation time required for each individual step. These long incubation times are mostly attributed to inefficient mass transport from the solution to the surface, although the immunoreaction itself is a rapid process. Furthermore, the antibodies and reagents used in ELISA are expensive. Using miniaturized ELISA tests can help overcome these drawbacks because of their intrinsic microscale dimensions, leading to enhanced mass transfer and hence improved reaction efficiency, reduction of assay time and sample or reagent consumption. Furthermore, it enables the development of portable systems. The most commonly used enzymes and substrates in immunoassays are mentioned in Table 4.6 (Scheme 4.90).

Many different methods have been developed in recent years for miniaturized ELISA, most of them dealing with improving the immobilization of antibodies that

Table 4.6 Representative enzyme labels for immunoassays.

Enzyme labels	Substrate	Reference
Alkaline phosphatase	p-Aminophenyl phosphate	[391]
	p-Nitrophenyl phosphate	[392]
	4-Methylumbelliferyl phosphate	[393]
HRP	Luminol	[394]
	3-(p-Hydroxyphenyl)propionic acid	[395]

From Bange *et al.* [390].

can be immobilized either on channel walls [395–397], on a membrane [398] or on microparticles such as polystyrene beads [399–401].

An example of the immobilization of antibodies on channel surfaces was presented by Eteshola and Leckband [395]. A microfluidic sensor chip was developed to quantify a model analyte (sheep IgM) with sensitivities down to 17 nM. This was achieved by first immobilizing a layer of bovine serum albumine (BSA) onto the channel wall, followed by specific adsorption of protein A to which the primary antibody for IgM was coupled covalently. This antibody could capture IgM, which was detected with the secondary antibody, labeled with horseradish peroxidase (Scheme 4.91). This enzyme catalyzes the conversion of the fluorogenic substrate 3-(p-hydroxyphenyl)propionic acid into a fluorophore, which was quantified off-chip with a spectrofluorometer. The measured fluorescence signal was proportional to the analyte concentration in the test sample.

An example of the immobilization on polystyrene beads is the bead–bed micro-ELISA system of Kitamori *et al.* [403]. The beads were coated with a capture antibody and loaded into the reaction channel of the ELISA microchip (Scheme 4.92a). Next, solutions containing antigen (Scheme 4.92b), a biotinylated secondary antibody (Scheme 4.92c) and a streptavidin–peroxidase conjugate were introduced into the channel (Scheme 4.92d). Finally, the substrate for peroxidase was continuously

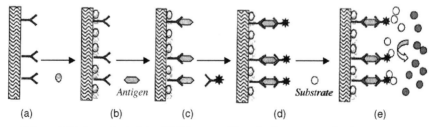

(a) (b) (c) (d) (e)

Scheme 4.90 General principle of ELISA. (a) Antibodies are adsorbed/immobilized onto the surface of the microreactor; (b) remaining available surface is blocked with a nonspecific binding protein such as BSA; (c) antigens are added and bound to the antibodies; (d) antibodies conjugated with enzymes are added and bound to the antigens; (e) substrate is added and is converted by the enzyme into a detectable product. Reprinted with permission from [389]. Copyright 2006 Elsevier.

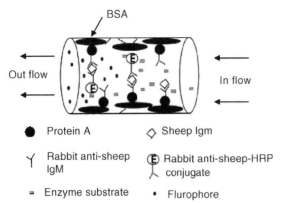

Protein A   Sheep Igm

Rabbit anti-sheep IgM   Rabbit anti-sheep-HRP conjugate

Enzyme substrate   Flurophore

Scheme 4.91 Schematic representation of the ELISA assay in a microchannel in which the primary antibody is immobilized on the channel wall. Reprinted with permission from [395]. Copyright 2001 Elsevier.

pumped into the channel (Scheme 4.92e), while the reaction product was monitored with a thermal lens microscope (TLM) [402] positioned downstream (Scheme 4.92f).

Microfluidic devices have been developed for carrying out various ELISAs for detecting and quantifying biological agents that are of interest in medical diagnostics, food safety surveillance and environmental monitoring. However, there are still few commercial applications of such devices. More robust manufacturing processes and integrated assay systems, including sample agent loading, microfluidics and signal detection, still need to be developed [389].

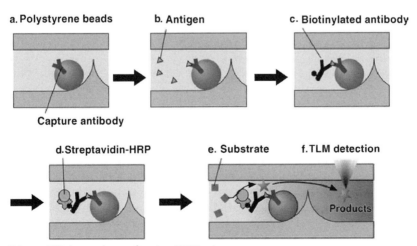

Scheme 4.92 Assay scheme of a micro-ELISA using antibodies immobilized on PS beads. Reprinted with permission from [403]. Copyright 2004 The Royal Society of Chemistry.

4.5.2.4 **Other Diagnostic Applications**

Most diagnostic tests involve blood because it contains large amount of information with respect to different biological functions such as oxygen transport, glucose levels, accumulation of waste products and, obviously, blood coagulation efficiency. One of the essential enzymes in the coagulation process is thrombin – a serine protease that converts soluble fibrinogen into insoluble strands of fibrin, causing the blood to clot. The anticoagulant drug argatroban directly inhibits the active site of thrombin and is an effective treatment for the blood clotting disease heparin-induced thrombocytopenia. Ismagilov and coworkers [404] developed a microfluidic system in which this anticoagulant could be titrated into blood samples to study the clotting time using a droplet-based microfluidic system, which allowed fibrin clots within the droplets to be transported through the channel without contacting or contaminating the channel walls.

L'Hostis *et al.* [405] developed a system in which the oxidation of glucose by glucose oxidase (GOx) was followed by measuring chemiluminiscence of luminol, which was oxidized by hydrogen peroxide formed in the primary reaction (Scheme 4.93). This system allowed the detection of glucose within the biologically relevant range 50–500 μM. In another glucose analysis setup, a tandem enzyme reaction was used by Seong and Crooks [406]. Glucose oxidase and horseradish peroxidase (HRP) were immobilized on polystyrene beads in different parts of the microchannel (Schemes 4.94 and 4.95). The hydrogen peroxide produced by GOx upon oxidation of glucose was used by HRP to oxidize Amplex Red to resorufin. The product resorufin was analyzed online using fluorescence microscopy. Both systems were applied on buffer solutions rather than on blood samples. Normally, biochemical blood tests are performed on serum or plasma, rather than whole blood, to enhance accuracy and lower the detection limit. The microfluidic toolbox offers many tools to separate serum from whole blood [407]. Combining on-chip assays with on-chip separation of plasma from whole blood would dramatically reduce the amount of blood and reagents needed for tests. Although not many examples of the integration

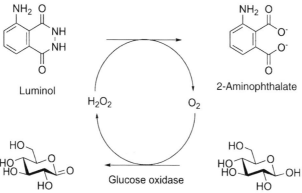

Scheme 4.93 Reaction scheme of the oxidation of luminol to the chemiluminescent product 2-aminophthalate [405].

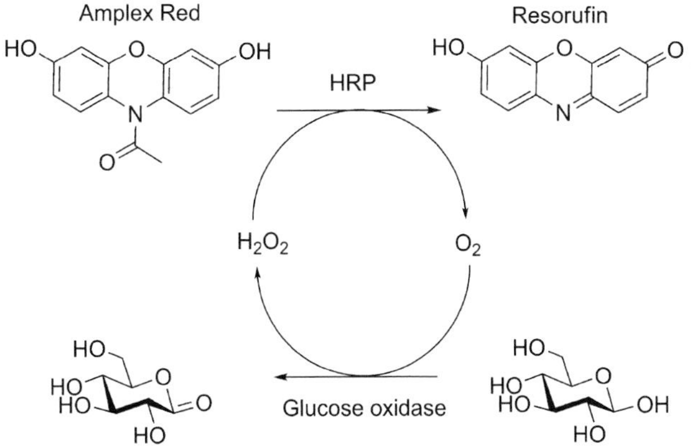

Amplex Red

Resorufin

HRP

H$_2$O$_2$

O$_2$

Glucose oxidase

Scheme 4.94 Reaction scheme of the oxidation of Amplex Red to the fluorescent product resorufin [406].

of both techniques are developed, they are expected to become readily available in coming years.

4.5.3
Biocatalysis

The examples of bioorganic chemistry in the previous paragraph are all concerned with the known biocatalytic assays, in which the effects of miniaturization on the efficiency of the analytical method were investigated. Recently, a new development has started in which the biocatalytic process itself has become the center of attention. Biocatalysis in microreactors, as described in here, deal with the investigation of the use of enzymes for the production of molecules. Two different approaches can be identified. In one line of investigation, the miniaturized reaction environment is used to screen the efficiency of an enzyme. In this case, only small amounts of

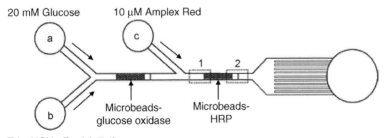

20 mM Glucose   10 μM Amplex Red

a

c

1   2

b

Microbeads-
glucose oxidase

Microbeads-
HRP

Tris–HCl buffer (ph 7.4)

Scheme 4.95 Schematic view of the microreactor in which two enzymes (glucose oxidase and horseradish peroxidase) immobilized on beads are located in different parts of the microchannel. (Reprinted with permission from [406]. Copyright 2002 American Chemical Society.)

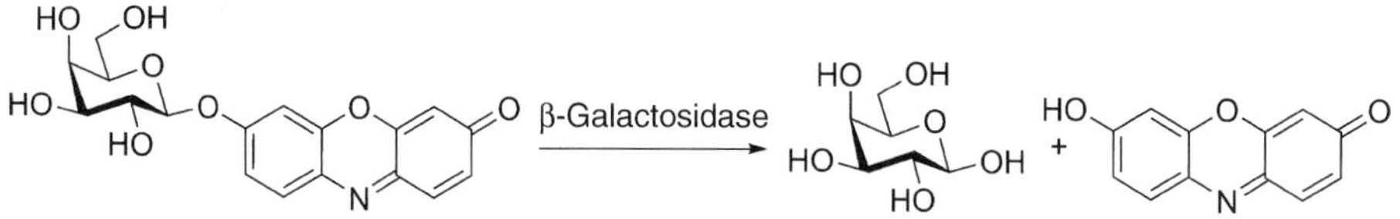

Scheme 4.96 Biocatalytic hydrolysis of resorufin β-D-galactopyranoside to the fluorescent product resorufin [408].

enzyme are required to effectively determine enzyme kinetics. In the second approach, the known advantages of mass and heat transfer in microchannels are employed to improve enzyme-catalyzed product synthesis. Biocatalytic synthesis in microreactors is still a relatively new field and most examples therefore describe the proof of concept with known enzymatic transformations. However, because of the advantages that have been clearly demonstrated, biocatalytic microreactors are also expected to be used for explorative research purposes. Here, we start with a description of enzyme screening, followed by biocatalytic synthesis on a chip.

4.5.3.1 Enzyme Kinetics in Microreactors

One of the first developments in the field of enzyme screening in microreactors was to establish whether there were pronounced differences in kinetics on a micro- and batch scale. The first report of using a microfluidic system for analyzing the kinetics of an enzymatic reaction was provided by Hadd *et al.* [408]. The enzyme β-galactosidase (β-Gal) was assayed using resorufin β-D-galactopyranoside (RBG), a substrate that is hydrolyzed to resorufin, a fluorescent product, which was detected by laser-induced fluorescence (Scheme 4.96). The Michaelis–Menten constants obtained on chip compared well with a conventional enzyme assay. An enzyme assay performed on the microchip within a 20-min period required only 120 pg of enzyme and 7.5 ng of substrate, reducing the amount of reagent consumed by four orders of magnitude over a conventional assay.

The same enzyme was used for the hydrolysis of *p*-nitrophenyl-β-D-galactopyranoside to D-galactose (Scheme 4.97) by Kanno *et al.* [409,410] in a PMMA microreactor. Quantitative hydrolysis was reported and the reaction was about five times faster than the batch reaction. This unexpected rate enhancement is one of the few examples in which a difference was observed in kinetics between batch and microscale. A second example was reported by Maeda and coworkers [411]. They described the trypsin-catalyzed hydrolysis of benzoylarginine-*p*-nitroanilide and found that the rate of reaction seemed to be 20 times greater than the batchwise system (Scheme 4.98).

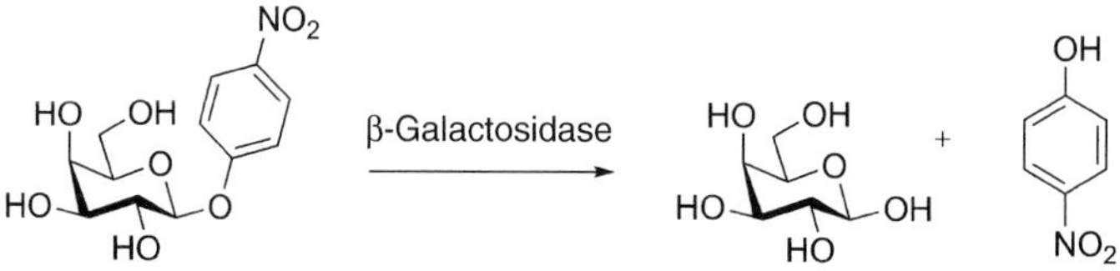

Scheme 4.97 Biocatalytic hydrolysis of *p*-nitrophenyl-β-D-galactopyranoside [409,410].

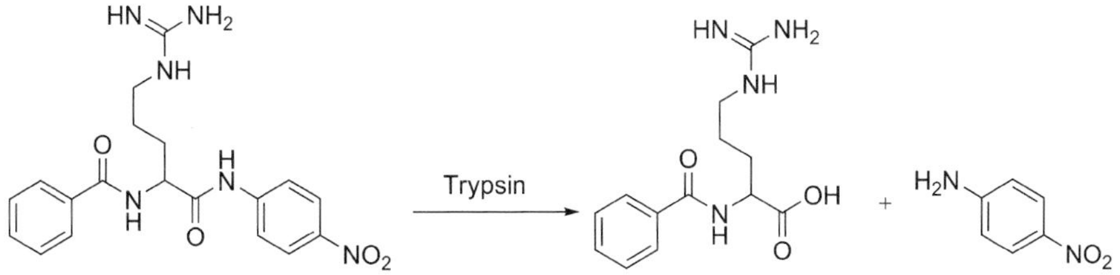

Scheme 4.98 Trypsin-catalyzed hydrolysis of benzoylarginine-*p*-nitroanilide [411].

In most cases, however, rates of biocatalytic reactions in microreactors and batch processes are fairly comparable. Seong *et al.* [412] showed that the Michaelis–Menten kinetics constant determined by a microfluidic device with immobilized horseradish peroxidase was similar to the value obtained during homogeneous catalysis in batch mode (also found by Hadd *et al.* [408]). Jiang *et al.* [370] published a method for determining K_m and v_{max} by applying online frontal analysis of peptides originating from the digestion by trypsin and observed that these constants were very close to the static and off-line detection methods.

Another enzyme that was studied extensively in microreactors to determine kinetic parameters is the model enzyme alkaline phosphatase. Many reports have appeared that differ mainly on the types of enzyme immobilization, such as on glass [413], PDMS [393], beads [414] and in hydrogels [415]. Kerby *et al.* [414], for example, evaluated the difference between mass-transfer effects and reduced efficiencies of the immobilized enzyme in a packed bead glass microreactor. In the absence of mass-transfer resistance, the Michaelis–Menten kinetic parameters were shown to be flow-independent and could be appropriately predicted using low substrate conversion data.

Another development in enzyme screening is to provide systems that combine a high resolution with fast analysis time, which makes it possible to effectively screen multiple enzymes. One elegant method was developed by Ismagilov and coworkers [416]. They created a flow of small nanoliter plugs in a microchannel, separated by an inert liquid. The plugs contained enzymes and substrates. They managed to measure single-turnover kinetics of the enzyme ribonuclease A (RNase A) with millisecond resolution within plugs. Also, the conversion of alkaline phosphatase was studied within plugs on a millisecond timescale [417]. In both cases, the products were analyzed by online fluorescence microscopy.

In a further extension, they demonstrated screening of submicroliter volumes with fluorescein diphosphate (FDP) as substrate against multiple enzymes (lysozyme, RNase, catalase and alkaline phosphatase) using preformed arrays of nanoliter plugs in a three-phase (liquid–liquid–gas) plug flow (Scheme 4.99) [418].

Moore *et al.* [419] used surface-enhanced resonance Raman scattering to detect the activity of hydrolases at ultralow levels. The method was used to rapidly screen the relative activities and enantioselectivities of 14 enzymes including lipases, esterases and proteases. In the current format, the sensitivity of this technique was sufficient to detect 500 enzyme molecules, thus offering the potential to

(a)

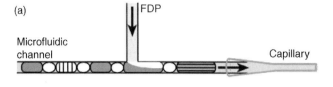

(b)

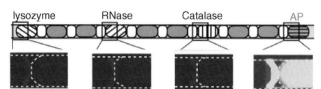

Scheme 4.99 (a) A schematic illustration of the assay of multiple enzymes. (b) The result of the enzymatic assay. The drawing at the top illustrates the array of plugs. The plugs of buffer are depicted in gray, the air bubbles introduced to separate the liquid plugs are in white and the plugs of enzymes are hatched. The images at the bottom are fluorescence micrographs of the corresponding plugs. AP = alkaline phosphatase, FDP = fluorescein diphosphate. Reprinted with permission from [418]. Copyright 2005 Wiley-VCH Verlag GmbH.

detect multiple enzyme activities simultaneously and at levels found within single cells.

Because in a number of cases biocatalytic transformations are not very fast, extended reaction times in microreactors are desired. This is sometimes difficult to accomplish by lowering the flow or increasing the channel length. As an alternative, a stopped-flow approach can be used in which the reaction mixture is pumped through the microchannel and is then halted for a certain reaction time. Kitamori and coworkers [420,421] reported stopped-flow microreactor devices using glass microchips. The products were detected by a thermal lens microscope [402]. An acceleration of a peroxidase-catalyzed (horseradish peroxidase) reaction was observed. However, the reasons for this acceleration are not clear and further studies are necessary. The same enzyme was also used in a microreactor system, developed by the same group, combined with infrared heating of the reaction fluid. The rate of the enzyme reaction, which was initially inhibited owing to the cooling of the chip, was increased by noncontact heating using the photothermal effect produced by the diode laser. Their findings suggest the possibility to control nanoscale reactions and to synthesize substances using photothermal stimulation.

4.5.3.2 Biocatalyzed Synthesis in Microreactors

Besides using microreactor devices for efficient screening of biocatalysts, these can also be applied for enzymatic synthesis. Biocatalyzed reactions have recently received much attention because of the efficiency and selectivity that enzymes have to offer, combined with their ability to act under mild conditions [422,423]. Microreactors offer in this respect the advantage that enzymes can be more optimally used as a result of the higher mass-transfer properties of microchannels and the high surface-

Table 4.7 Enzymatic reactions performed in microreactors.

Reaction	Enzyme	Technique	References
Transglycosylation	β-Galactosidase	Solution phase	[409,410]
Enantioselective epoxidation	Epoxide hydrolase	Solution phase	[424]
Enantioselective reduction of lactic acid	Lactate dehydrogenase	Surface modification	[425–427]
Enantioselective reduction of acetophenone	Alcohol dehydrogenase	Solution phase	[428]
Enantioselective formation of cyanohydrins	Hydroxynitrile lyase	Solution phase	[429]
Transesterification	Protease P	Monolith	[430]
Oxidation of p-chlorophenol	Laccase	Solution phase	[431]
Oxidative polymerization of phenols	Soybean peroxidase	Surface immobilization	[432]
Hydroxylation	PikC hydroxylase	Immobilized	[433]
Polyketide synthesis	THNS	Immobilized	[434]
Tandem hydrolysis and oxidative coupling	Lipase and peroxidase	Immobilized	[435]

to-volume ratio, which becomes relevant when enzymes are immobilized within the reactor. Miniaturized biocatalyzed synthetic reactors, therefore, require only relatively small amounts of catalysts. In this paragraph, some typical examples of the different biocatalytic transformations aimed at product synthesis are discussed, with a distinction between solution phase and immobilized biocatalysts. An overview of the different enzymatic reactions is given in Table 4.7.

The most straightforward method to perform a biocatalyzed reaction in a microreactor is to employ the enzyme in the solution phase. The fluids can be pumped either by syringe pumps or by electric osmotic flow (EOF) through the microchannel. Microreactors made from PMMA, PDMS and glass are mostly used with channel dimensions varying between 50 and 250 μm in width.

One of the early examples was published by Kanno et al. [409,410] in which the enzymatic transglycosylation of p-nitrophenyl-β-D-galactopyranoside (PNPGal) and p-nitrophenyl-2-acetamide-2-deoxy-β-D-glucopyranoside (PNPGlcNAc) was studied (Scheme 4.100). Transglycosylation is a useful method for oligosaccharide synthesis. The authors observed that by maintaining the microreactor at 37 °C, achieved by submerging the reactor in a heated bath, the reaction rate was enhanced compared to a batch reaction executed at the same temperature. This enhancement was attributed to the efficient diffusive mixing obtained within the microchannel compared to a traditional batch setup.

Although enzymes have the advantage of selectivity and efficiency, they are, as previously mentioned, sometimes not very fast. To increase the total reaction time in microspace, Müller et al. [428] constructed a circulating continuous flow enzyme membrane reactor and investigated the reduction of acetophenone to (S)-phenylethanol in the presence of the enzyme alcohol dehydrogenase (ADH)

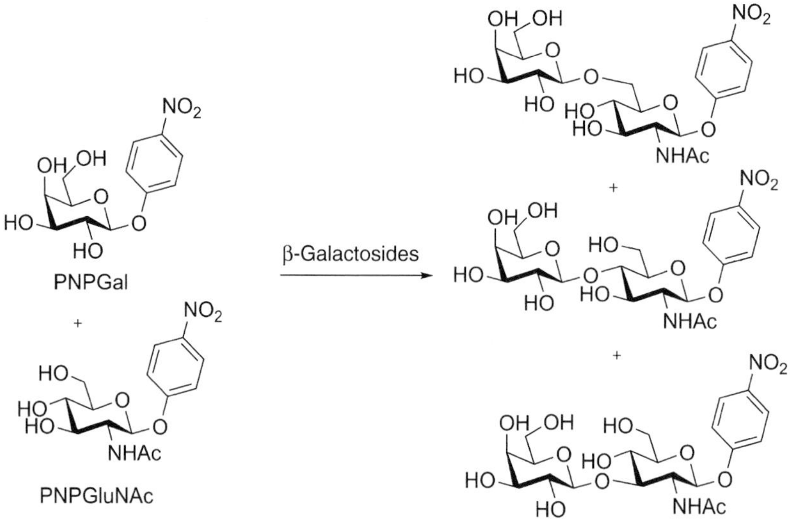

Scheme 4.100 Synthesis of a disaccharide by transglycosylation [409,410].

(Scheme 4.101). The cofactor NADH was regenerated by the simultaneous oxidation of isopropanol to acetone.

Another method for cofactor regeneration was demonstrated by Kenis and coworkers [425]. They recently reported an efficient technique for the electrochemical regeneration of nicotinamide cofactors (NAD) in a PDMS microreactor. The authors found that by employing a multistream laminar flow, comprising a buffer and a reagent stream, efficient regeneration of the cofactor NADH could be achieved at the surface of a gold electrode and a mediator. The mediator flavin adenine dinucleotide (FAD) was chosen because of its known stability over numerous oxidation and reaction cycles. With this approach, enzyme/cofactor regeneration was demonstrated for the conversion of a chiral pyruvate to L-lactate in the presence of lactate dehydrogenase (Scheme 4.102), whereby yields of 41% L-lactate were achieved (turnover number $= 75.6\,h^{-1}$). This result clearly illustrates enhancements over conventional approaches, providing a feasible, cost-effective route to the biocatalytic synthesis of large quantities of chiral products.

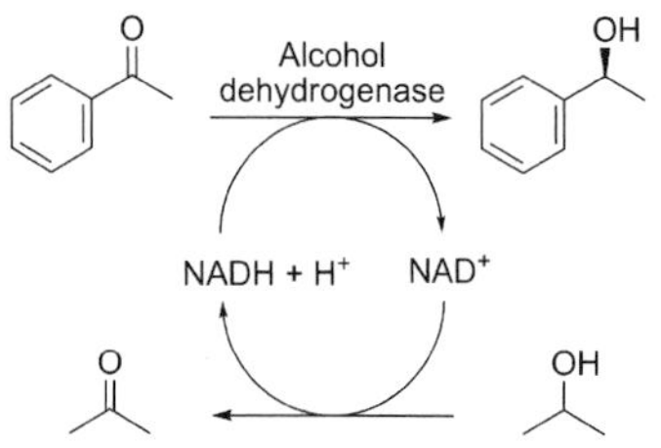

Scheme 4.101 Reaction scheme of reduction of acetophenone by ADH to (S)-phenylethanol with simultaneous cofactor regeneration [428].

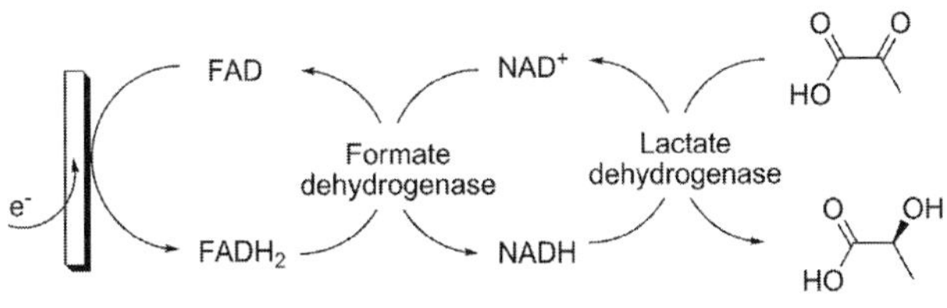

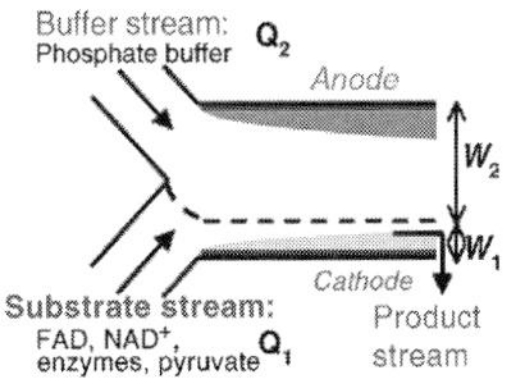

Scheme 4.102 Indirect chemical regeneration of NADH using FAD/FADH$_2$ as the mediator and subsequent biocatalytic conversion [425]. Reprinted with permission from [425]. Copyright 2005 American Chemical Society.

Most biocatalytic conversions are performed with the enzyme immobilized in the microreactor. Miyazaki *et al.* [426] developed a simple noncovalent immobilization method for His-tagged enzymes on a microchannel surface. These enzymes contain a polyhistidine-tag motif that consists of at least six histidine residues, often located at the N- or C-terminus. The His-tag has a strong affinity for nickel and can be reversibly immobilized by a nickel–nitrilotriacetic acid (Ni–NTA) complex (Scheme 4.103), a strategy commonly used in affinity chromatography.

With the help of this method, His-tagged L-lactate dehydrogenase was immobilized. By pumping pyruvic acid as substrate with NADH as cofactor, it was demonstrated that the enzyme was still active in the microchannel. In this case, cofactor was used up. Srinivasan *et al.* [433] incorporated PikC hydroxylase from *Streptomyces venezuelae* into a PDMS-based microfluidic channel with a similar approach. The enzyme was immobilized to Ni–NTA agarose beads with an *in situ* attachment, following the addition of the beads to the microchannel. This enabled the rapid hydroxylation of the macrolide YC-17 to methymycin and neomethymycin (Scheme 4.104) in about equal amounts with a conversion of >90% at a flow rate of 70 nl/min.

Kawakami *et al.* [430] covalently immobilized protease P (from *Aspergillus melleus*) onto a capillary (PEEK) monolith with an inner diameter varying between 0.1 and 2.0 mm using an *in situ* sol–gel method. The enzyme protease P can be used for transesterification. To determine the reaction rate of the system, (*S*)-glycidol and vinyl *n*-butyrate (Scheme 4.105) were used as substrates to be converted by the enzyme into

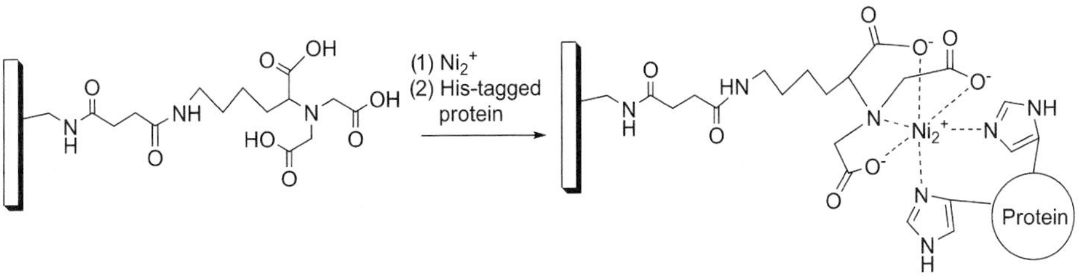

Scheme 4.103 Immobilization of His-tagged enzymes by the Ni–NTA complex (Ni–NTA = nickel–nitrilotriacetic acid) [426].

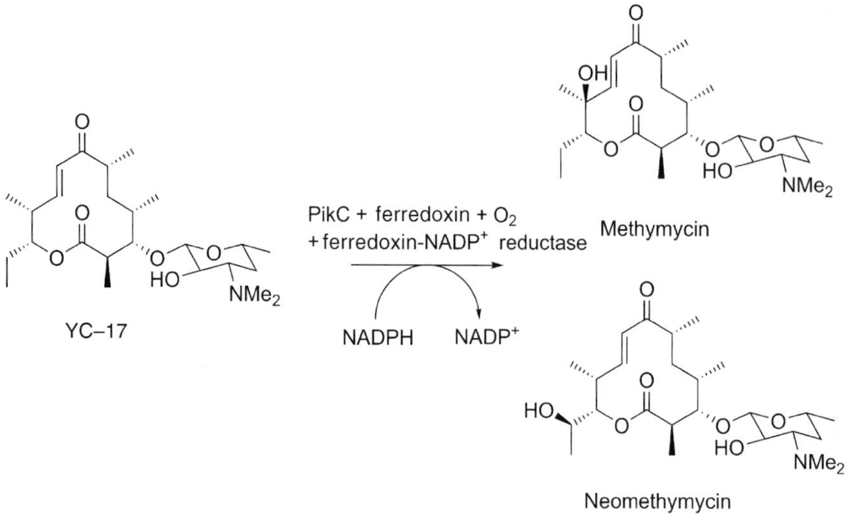

Scheme 4.104 PikC-catalyzed hydroxylation of YC-17 to methymycin and neomethymycin in the presence of NADPH, ferredoxin and ferredoxin–NADP$^+$ reductase [433].

the corresponding ester. Changes in the tube diameter had no influence on the conversion when the linear flow was kept constant. Three monoliths of the same diameter were butt-connected and the conversion rate increased by as much as 50%. This protease P microbioreactor performed better at high flow rate than the control batch experiment, whereas it did not reach the same conversions of the batch experiment at lower flow rate. It suggests that convective flow took place in the enzyme reactor at high flow rates, which thereby enhanced the digestion rates.

In some cases, substrates and enzymes are not soluble in the same solvent. To achieve efficient substrate conversion, a large interface between the immiscible fluids has to be established, by the formation of microemulsions or multiple-phase flow that can be conveniently obtained in microfluidic devices. Until now only a couple of examples are published in which a two-phase flow is used for biocatalysis. Goto and coworkers [431] were first to study an enzymatic reaction in a two-phase flow in a microfluidic device, in which the oxidation of *p*-chlorophenol by the enzyme laccase (lignin peroxidase) was analyzed (Scheme 4.106). The surface-active enzyme was solubilized in a succinic acid aqueous buffer and the substrate (*p*-chlorophenol) was dissolved in isooctane. The transformation of *p*-chlorophenol occurred mainly at

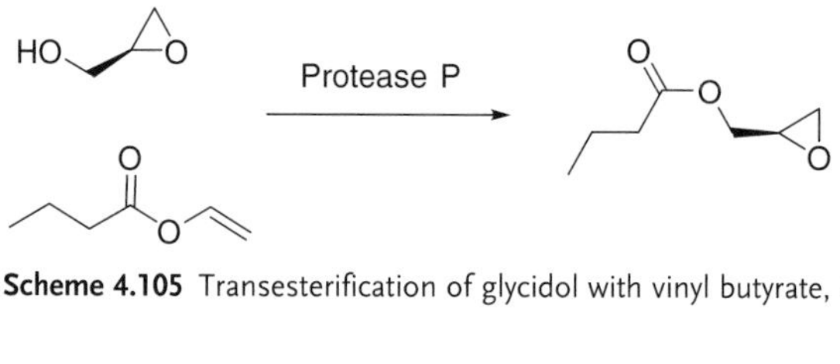

Scheme 4.105 Transesterification of glycidol with vinyl butyrate, catalyzed by protease P [430].

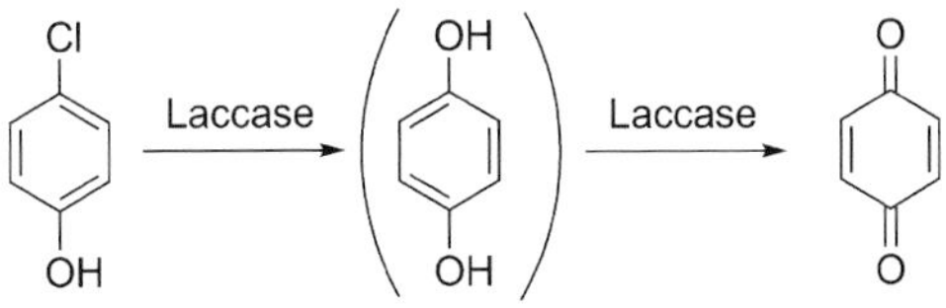

Scheme 4.106 Oxidation of *p*-chlorophenol by laccase [431].

the aqueous–organic interface in the microchannel. The effects of flow velocity and microchannel shape on this enzymatic reaction were investigated. On the basis of a simple theoretical model for oxidation in the microchannel, they concluded that diffusion of the substrate (*p*-chlorophenol) was the rate-limiting step.

Rutjes and coworkers [429] reported an example of a two-phase system in which two crude cell lysates, containing the enzyme hydroxynitrile lyase (HNL) were applied (Scheme 4.107). One crude lysate contained HNL originating from bitter almonds (*Prunus amygdalus*) and the other lysate from a rubber tree (*Hevea brasiliensis*). HNLs catalyze enantioselective C–C bond formation with the addition of hydrogen cyanide to aldehydes or ketones to yield the corresponding optically active cyanohydrins. The enzymes and hydrogen cyanide are solubilized in aqueous medium, whereas the substrates are normally soluble in water-immiscible organic solvents. Typically this reaction has to be carried out by thoroughly mixing the two phases. With the use of a microreactor setup, efficient mixing was directly accomplished. Two parameters (time and ratio of both phases) were screened consuming only minute amounts of the catalyst. The enantiomeric ratio (er) of the cyanohydrins **1** was >98 : 2 for all products except **1c** (er 93 : 7). More importantly, results from the continuous flow reaction were fully consistent with those obtained from the larger batchwise process.

Ku *et al.* [434] demonstrated that microreactors can also be used for tandem reactions. A microreactor containing beads functionalized with polyketide synthase [1,3,6,8-tetrahydroxynaphthalene synthase (THNS)] was coupled to one that contained soybean peroxidase (SBP). THNS was immobilized using the Ni–NTA method

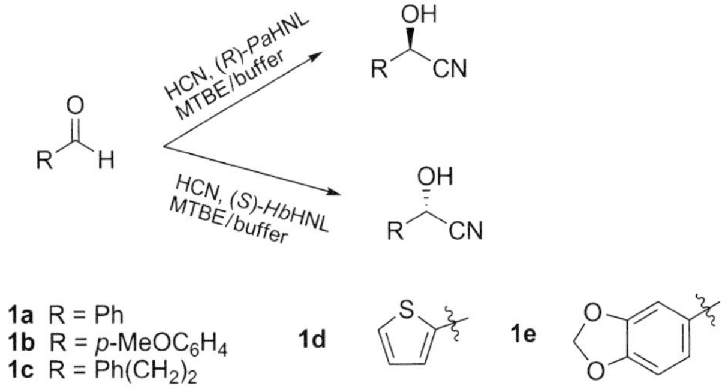

1a R = Ph
1b R = *p*-MeOC$_6$H$_4$
1c R = Ph(CH$_2$)$_2$ **1d**   **1e**

Scheme 4.107 Enzymatic cyanohydrin formation catalyzed by HNL [429] (MTBE: methyl-*tert*-butyl ether; (*R*)-*Pa*HNL: (*R*)-hydroxynitrile lyase originating from *Prunus amygdalus*; (*S*)-*Hb*HNL: (*S*)-hydroxynitrile lyase originating from *Hevea brasiliensis*).

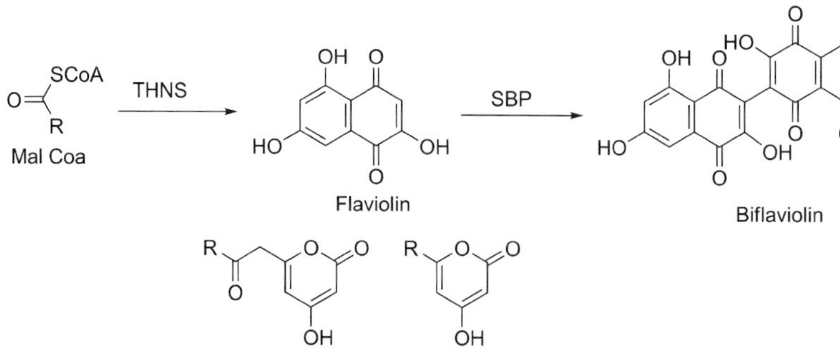

Scheme 4.108 Tandem reaction with THNS and SBP [434].

and SBP was immobilized covalently. The result was a tandem, two-step biochip that enabled the synthesis of novel polyketide derivatives. The first microchannel, consisting of THNS, resulted in the conversion of malonyl–CoA (Mal CoA) to flaviolin (Scheme 4.108) in yields up to 40% with a residence time of 6 min. This conversion is similar to that obtained in batch reactions with a volume of several milliliters after 2 h. Linking this microchannel to the SBP microchannel resulted in the synthesis of biflaviolin (Scheme 4.108).

A second example of a tandem reaction was shown by Srinivasan and coworkers [432,435]. First, they developed an enzyme-containing microfluidic biochip for the oxidative polymerization of phenols. Hydrogen peroxide was used together with soybean peroxidase as phenol oxidizing catalyst of five substrates (**2a–e**) leading to polyphenolic species (Scheme 4.109). The enzyme soybean peroxidase was immobilized covalently on the microchannel wall. The substrate **2b** turned out to be completely converted in the microchannel at a flow rate of about 290 nl/min and the kinetic properties of the SBP-catalyzed oxidation of this substrate were nearly identical to those of the SBP catalysis in solution at larger volumes. Second, two tandem reactions were achieved: (1) a bienzymatic reaction using the combination of an additional lipase-catalyzed hydrolysis of a phenolate before the polymerization reaction (Scheme 4.110) and (2) a trienzymatic reaction using the combination of invertase to generate glucose from sucrose, glucose oxidase to produce *in situ* H_2O_2 and SBP for the subsequent oxidative polymerization (Scheme 4.111). In both bienzymatic and trienzymatic reactions, the microchannel reactions were much

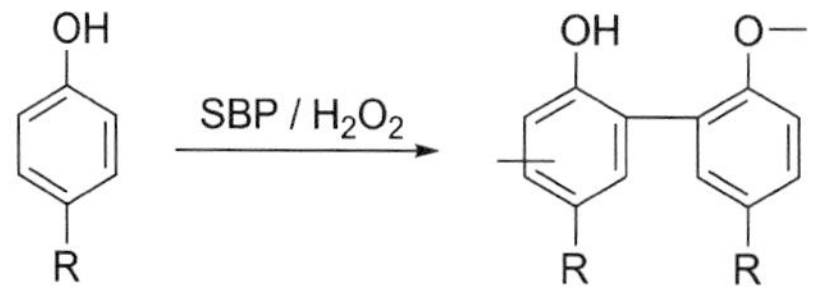

2a R = H
2b R = CH$_3$
2c R = OCH$_3$
2d R = (CH$_2$)$_2$OH
2e R = CH$_2$COOH

Scheme 4.109 Oxidative polymerization of phenols by SBP [432].

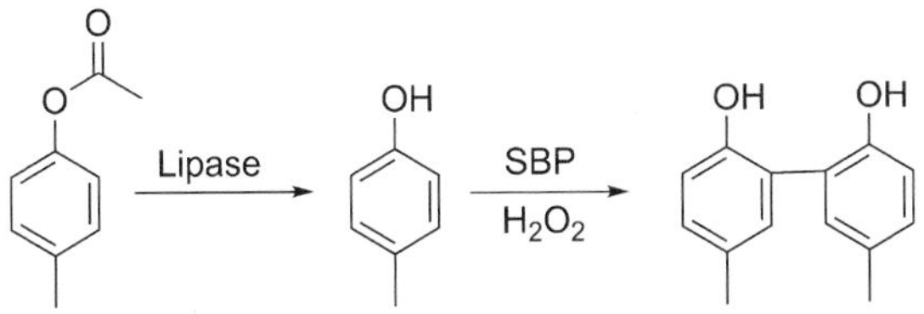

Scheme 4.110 Bienzymatic tandem reaction with lipase and SBP [435].

faster than the solution-phase reaction, probably because of the higher concentrations of the invertase and glucose oxidase in the biochip than in free solution.

Microreactor technology offers the possibility to combine synthesis and analysis on one microfluidic chip. A combination of enantioselective biocatalysis and on-chip analysis has recently been reported by Belder *et al.* [424]. The combination of very fast separations (<1 s) of enantiomers using microchip electrophoresis with enantioselective catalysis allows high-throughput screening of enantioselective catalysts. Various epoxide–hydrolase mutants were screened for the hydrolysis of a specific epoxide to the diol product with direct on-chip analysis of the enantiomeric excess (Scheme 4.112).

Another example in which biocatalysis is combined with analysis is the system reported by Honda *et al.* [436]. A microreaction system, consisting of an enzyme-immobilized microreactor, for optical resolution of racemic amino acids was devel-

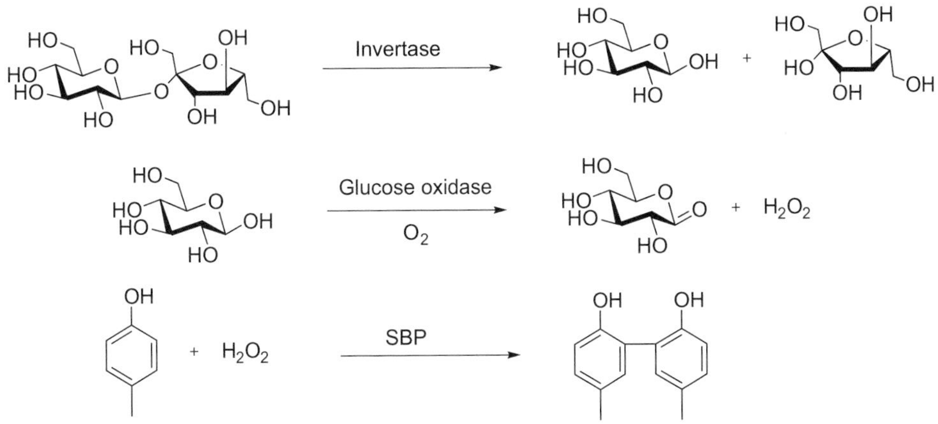

Scheme 4.111 Trienzymatic tandem reaction with invertase, glucose oxidase and soybean peroxidase [435].

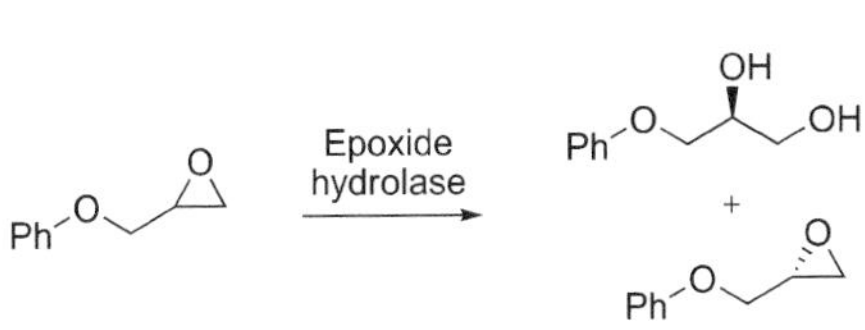

Scheme 4.112 Enantioselective hydrolysis of the epoxide with on-chip analysis [424].

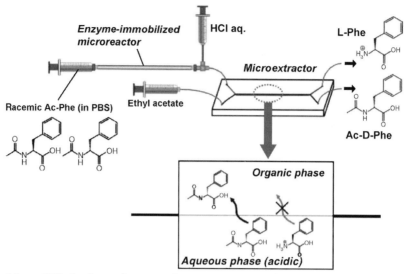

Scheme 4.113 Continuous flow system for racemic resolution of amino acids. Reprinted with permission from [436]. Copyright 2007 The Royal Society of Chemistry.

oped, in which an aminoacylase was immobilized onto the wall and a microextractor. Using an enzyme microreactor, which was prepared by membrane formation on the microchannel surface, enabled a highly enantioselective hydrolysis on the racemic amino acid derivative. The microextractor provided a laminar flow of two immiscible solutions, which enabled selective extraction of the product (Scheme 4.113). With the use of this integrated device, efficient continuous production of optically pure natural and unnatural amino acids was achieved.

4.5.3.3 Complex Catalysis in Microbioreactors

A chemostat or a bioreactor is an environment in which microbes and microbial communities can be controlled very efficiently by continually substituting a fraction of a bacterial culture with sterile nutrients [437,438]. Much effort has been undertaken in miniaturizing these chemostats to reduce the necessity for large quantities of growth media and reagents [439–442]. Kim and Lee developed a silicon microfermenter chip in which the cell density, dissolved oxygen (DO) and glucose levels could be measured. Following up on that, Jensen and coworkers [439] developed a membrane-aerated microbioreactor with a volume as low as 5 μl in which the optical density (OD), DO and pH could be measured online (Scheme 4.114). Furthermore, they showed that the behavior of the bacteria in the microbioreactor was similar to that in the larger (500 ml) bioreactor. This similarity includes growth kinetics, dissolved oxygen profile within the vessel over time, pH profile over time, final number of cells and cell morphology.

In microfluidic chips, small (nano- and picoliter) liquid volumes can be generated by the formation of aqueous droplets in a carrier medium such as a hydrophobic

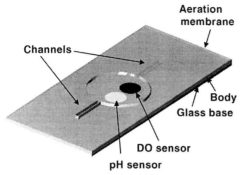

Scheme 4.114 Microbioreactor built of three layers of PDMS on top of a layer of glass. Reprinted with permission from [439]. Copyright 2004 John Wiley & Sons Inc.

compound or a gas. The ability to precisely control the supply of reagents, to handle small liquid volumes without fast evaporation and to form droplets at high speed with homogeneous diameters of a few micrometers makes this a valuable tool for screening experiments that rely on high reproducibility. He *et al.* [443] captured individual (mouse mast) cells within a segmented water–oil flow and performed (laser-induced) photolysis to study the activity of an intracellular enzyme (β-galactosidase). Also *in vitro* expression of green fluorescent protein (GFP) to directly analyze the contents of individual droplets was achieved on a microfluidic chip [444]. This ability to entrap individual selected cells or subcellular organelles opens new possibilities for carrying out single-cell studies and single-organelle measurements.

In principle, it is nowadays possible that all the separate steps in the whole process from cell culturing to biochemical analysis can be miniaturized [445]. Although there is still no system or platform available, inclusive of all these separate steps, the rapid developments in this field indicate such a platform will be available within next few years.

4.5.4
Conclusions

In this chapter, we have described the wide variety of bioorganic reactions, more particularly the biocatalyzed processes that have currently been investigated in microreactors. The use of enzymes in diagnostics, such as protein digestion and ELISA, has drawn most of the attention till date and the development of a ready-to-use technology is within reach. The use of enzymes for compound synthesis is still relatively unexplored and is mostly limited to proof of concept model reactions. This also counts for whole- and single-cell analyses. The first demonstrations of the benefits of microreactor technology in these applications, however, are highly encouraging and will lead to an increased research effort and exciting new opportunities for bioorganic chemistry on a chip in the near future.

References

330 Ramsay, G. (1998) *Nature Biotechnology*, **16**, 40.

331 Heller, M.J. (2002) *Annual Review of Biomedical Engineering*, **4**, 129.

332 Whitesides, G.M. (2006) *Nature*, **442**, 368.

333 Reyes, D.R., Iossifidis, D., Auroux, P.A. and Manz, A. (2002) *Analytical Chemistry*, **74**, 2623.

334 Lee, S.J. and Lee, S.Y. (2004) *Applied Microbiology and Biotechnology*, **64**, 289.

335 Manz, A., Harrison, D.J., Verpoorte, E.M. J., Fettinger, J.C., Paulus, A., Ludi, H. and Widmer, H.M. (1992) *Journal of Chromatography*, **593**, 253.

336 Urban, P.L., Goodall, D.M. and Bruce, N. C. (2006) *Biotechnology Advances*, **24**, 42.

337 Foret, F. and Preisler, J. (2002) *Proteomics*, **2**, 360.

338 Figeys, D., Ning, Y.B. and Aebersold, R. (1997) *Analytical Chemistry*, **69**, 3153.

339 Krenkova, J. and Foret, F. (2004) *Electrophoresis*, **25**, 3550.

340 Peterson, D.S., Rohr, T., Svec, F. and Frechet, J.M.J. (2002) *Analytical Chemistry*, **74**, 4081.

341 Lazar, I.M., Ramsey, R.S. and Ramsey, J. M. (2001) *Analytical Chemistry*, **73**, 1733.

342 Amankwa, L.N. and Kuhr, W.G. (1992) *Analytical Chemistry*, **64**, 1610.

343 Nouaimi, M., Moschel, K. and Bisswanger, H. (2001) *Enzyme and Microbial Technology*, **29**, 567.

344 Honda, T., Miyazaki, M., Nakamura, H. and Maeda, H. (2005) *Chemical Communications*, 5062.

345 Ekstrom, S., Onnerfjord, P., Nilsson, J., Bengtsson, M., Laurell, T. and Marko-Varga, G. (2000) *Analytical Chemistry*, **72**, 286.

346 Xie, S.F., Svec, F. and Frechet, J.M.J. (1999) *Biotechnology and Bioengineering*, **62**, 30.

347 Licklider, L., Kuhr, W.G., Lacey, M.P., Keough, T., Purdon, M.P. and Takigiku, R. (1995) *Analytical Chemistry*, **67**, 4170.

348 Ye, M.L., Hu, S., Schoenherr, R.M. and Dovichi, N.J. (2004) *Electrophoresis*, **25**, 1319.

349 Licklider, L. and Kuhr, W.G. (1998) *Analytical Chemistry*, **70**, 1902.

350 Jin, L.J., Ferrance, J., Sanders, J.C. and Landers, J.P. (2003) *Lab on a Chip*, **3**, 11.

351 Bonneil, E. and Waldron, K.C. (2000) *Talanta*, **53**, 687.

352 Gottschlich, N., Culbertson, C.T., McKnight, T.E., Jacobson, S.C. and Ramsey, J.M. (2000) *Journal of Chromatography B*, **745**, 243.

353 Bengtsson, M., Ekstrom, S., Marko-Varga, G. and Laurell, T. (2002) *Talanta*, **56**, 341.

354 Jiang, Y. and Lee, C.S. (2001) *Journal of Chromatography A*, **924**, 315.

355 Cooper, J.W. and Lee, C.S. (2004) *Analytical Chemistry*, **76**, 2196.

356 Licklider, L. and Kuhr, W.G. (1994) *Analytical Chemistry*, **66**, 4400.

357 Bonneil, E., Mercier, M. and Waldron, K. C. (2000) *Analytica Chimica Acta*, **404**, 29.

358 Peterson, D.S., Rohr, T., Svec, F. and Frechet, J.M.J. (2003) *Analytical Chemistry*, **75**, 5328.

359 Jiang, H.H., Zou, H.F., Wang, H.L., Ni, J. Y. and Zhang, Q. (2000) *Chemical Journal of Chinese Universities-Chinese*, **21**, 702.

360 Jiang, H.H., Zou, H.F., Wang, H.L., Zhang, Q., Ni, J.Y., Zhang, Q.C., Guo, Z. and Chen, X.M. (2000) *Science in China Series B-Chemistry*, **43**, 625.

361 Palm, A.K. and Novotny, M.V. (2004) *Rapid Communications in Mass Spectrometry*, **18**, 1374.

362 Guo, Z., Zhang, Q.C., Lei, Z.D., Kong, L., Mao, X.Q. and Zou, H.F. (2002) *Chemical Journal of Chinese Universities-Chinese*, **23**, 1277.

363 Peterson, D.S., Rohr, T., Svec, F. and Frechet, J.M.J. (2002) *Journal of Proteome Research*, **1**, 563.

364 Ekstrom, S., Malmstrom, J., Wallman, L., Lofgren, M., Nilsson, J., Laurell, T. and Marko-Varga, G. (2002) *Proteomics*, **2**, 413.

365 Samskog, J., Bylund, D., Jacobsson, S.P. and Markides, K.E. (2003) *Journal of Chromatography A*, **998**, 83.

366 Guo, Z., Xu, S.Y., Lei, Z.D., Zou, H.F. and Guo, B.C. (2003) *Electrophoresis*, **24**, 3633.

367 Kato, M., Sakai-Kato, K., Jin, H.M., Kubota, K., Miyano, H., Toyo'oka, T., Dulay, M.T. and Zare, R.N. (2004) *Analytical Chemistry*, **76**, 1896.

368 Kim, Y.D., Park, C.B. and Clark, D.S. (2001) *Biotechnology and Bioengineering*, **73**, 331.

369 Qu, H.Y., Wang, H.T., Huang, Y., Zhong, W., Lu, H.J., Kong, J.L., Yang, P.Y. and Liu, B.H. (2004) *Analytical Chemistry*, **76**, 6426.

370 Jiang, H.H., Zou, H.F., Wang, H.L., Ni, J.Y., Zhang, Q. and Zhang, Y.K. (2000) *Journal of Chromatography A*, **903**, 77.

371 Korecka, L., Bilkova, Z., Holeapek, M., Kralovsky, J., Benes, M., Lenfeld, J., Minc, N., Cecal, R., Viovy, J.L. and Przybylski, M. (2004) *Journal of Chromatography B*, **808**, 15.

372 Jacobson, S.C. and Ramsey, J.M. (1996) *Analytical Chemistry*, **68**, 720.

373 Burns, M.A., Johnson, B.N., Brahmasandra, S.N., Handique, K., Webster, J.R., Krishnan, M., Sammarco, T.S., Man, P.M., Jones, D., Heldsinger, D., Mastrangelo, C.H. and Burke, D.T. (1998) *Science*, **282**, 484.

374 Kartalov, E.P. and Quake, S.R. (2004) *Nucleic Acids Research*, **32**, 2873.

375 Saiki, R.K., Scharf, S., Faloona, F., Mullis, K.B., Horn, G.T., Erlich, H.A. and Arnheim, N. (1985) *Science*, **230**, 1350.

376 Northrup, M.A., Benett, B., Hadley, D., Landre, P., Lehew, S., Richards, J. and Stratton, P. (1998) *Analytical Chemistry*, **70**, 918.

377 de Mello, A.J. (2003) *Nature*, **422**, 28.

378 Zhang, C.S., Xu, J.L., Ma, W.L. and Zheng, W.L. (2006) *Biotechnology Advances*, **24**, 243.

379 Nagai, H., Murakami, Y., Yokoyama, K. and Tamiya, E. (2001) *Biosensors & Bioelectronics*, **16**, 1015.

380 Schneegaß, I., Bräutigam, R. and Köhler, R.M. (2001) *Lab on a Chip*, **1**, 42.

381 Kopp, M.U., de Mello, A.J. and Manz, A. (1998) *Science*, **280**, 1046.

382 Krishnan, M., Ugaz, V.M. and Burns, M.A. (2002) *Science*, **298**, 793.

383 Agrawal, N., Hassan, Y.A. and Ugaz, V.M. (2007) *Angewandte Chemie-International Edition*, **46**, 4316.

384 Liao, C.S., Lee, G.B., Liu, H.S., Hsieh, T.M. and Luo, C.H. (2005) *Nucleic Acids Research*, **33**. e156, doi:10.1093/new/gni157.

385 Woolley, A.T., Hadley, D., Landre, P., de Mello, A.J., Mathies, R.A. and Northrup, M.A. (1996) *Analytical Chemistry*, **68**, 4081.

386 Trau, D., Lee, T.M.H., Lao, A.I.K., Lenigk, R., Hsing, I.M., Ip, N.Y., Carles, M.C. and Sucher, N.J. (2002) *Analytical Chemistry*, **74**, 3168.

387 Zhang, N.Y., Tan, H.D. and Yeung, E.S. (1999) *Analytical Chemistry*, **71**, 1138.

388 Cady, N.C., Stelick, S., Kunnavakkam, M.V. and Batt, C.A. (2005) *Sensors and Actuators B: Chemical*, **107**, 332.

389 Lee, L.J., Yang, S.T., Lai, S.Y., Bai, Y.L., Huang, W.C. and Juang, Y.J. (2006) *Advances in Clinical Chemistry*, **42**, 255.

390 Bange, A., Halsall, H.B. and Heineman, W.R. (2005) *Biosensors & Bioelectronics*, **20**, 2488.

391 Choi, J.-W., Oh, K.W., Han, A., Okulan, N., Wijayawardhana, C.A., Lannes, C., Bhansali, S., Schuleter, K.T., Heineman W.R., Halsall, H.B., Nevin, J.H., Helmicki, A.J., Henderson, H.T. and Ahn, C.H. (2001) *Biomedical Microdevices*, **3**, 191.

392 Wang, J., Ibanez, A. and Chatrathi, M.P. (2003) *Journal of the American Chemical Society*, **125**, 8444.

393 Mao, H.B., Yang, T.L. and Cremer, P.S. (2002) *Analytical Chemistry*, **74**, 379.

394 Yakovleva, J., Davidsson, R., Lobanova, A., Bengtsson, M., Eremin, S., Laurell, T. and Emneus, J. (2002) *Analytical Chemistry*, **74**, 2994.

395 Eteshola, E. and Leckband, D. (2001) *Sensors and Actuators B: Chemical*, **72**, 129.

396 Dodge, A., Fluri, K., Verpoorte, E., de Rooij, N.F. (2001) *Analytical Chemistry*, **73**, 3400.

397 Linder, V., Verpoorte, E., Thormann, W., de Rooij, N.F. and Sigrist, M. (2001) *Analytical Chemistry*, **73**, 4181.

398 Hisamoto, H., Shimizu, Y., Uchiyama, K., Tokeshi, M., Kikutani, Y., Hibara, A. and Kitamori, T. (2003) *Analytical Chemistry*, **75**, 350.

399 Sato, K., Yamanaka, M., Takahashi, H., Tokeshi, M., Kimura, H. and Kitamori, T. (2002) *Electrophoresis*, **23**, 734.

400 Sato, K., Tokeshi, M., Odake, T., Kimura, H., Ooi, T., Nakao, M. and Kitamori, T. (2000) *Analytical Chemistry*, **72**, 1144.

401 Sato, K., Tokeshi, M., Kimura, H. and Kitamori, T. (2001) *Analytical Chemistry*, **73**, 1213.

402 Kitamori, T., Tokeshi, M., Hibara, A. and Sato, K. (2004) *Analytical Chemistry*, **76**, 52A.

403 Sato, K., Yamanaka, M., Hagino, T., Tokeshi, M., Kimura, H. and Kitamori, T. (2004) *Lab on a Chip*, **4**, 570.

404 Song, H., Li, H.W., Munson, M.S., Van Ha, T.G. and Ismagilov, R.F. (2006) *Analytical Chemistry*, **78**, 4839.

405 L'Hostis, E., Michel, P.E., Fiaccabrino, G. C., Strike, D.J., de Rooij, N.F. and Koudelka-Hep, M. (2000) *Sensors and Actuators B: Chemical*, **64**, 156.

406 Seong, G.H. and Crooks, R.M. (2002) *Journal of the American Chemical Society*, **124**, 13360.

407 VanDelinder, V. and Groisman, A. (2006) *Analytical Chemistry*, **78**, 3765.

408 Hadd, A.G., Raymond, D.E., Halliwell, J.W., Jacobson, S.C. and Ramsey, J.M. (1997) *Analytical Chemistry*, **69**, 3407.

409 Kanno, K., Maeda, H., Izumo, S., Ikuno, M., Takeshita, K., Tashiro, A. and Fujii, M. (2002) *Lab on a Chip*, **2**, 15.

410 Kanno, K., Kawazumi, H., Miyazaki, M., Maeda, H. and Fujii, M. (2002) *Australian Journal of Chemistry*, **55**, 687.

411 Miyazaki, M., Nakamura, H. and Maeda, H. (2001) *Chemistry Letters*, **30**, 442.

412 Seong, G.H., Heo, J. and Crooks, R.M. (2003) *Analytical Chemistry*, **75**, 3161.

413 Gleason, N.J. and Carbeck, J.D. (2004) *Langmuir*, **20**, 6374.

414 Kerby, M.B., Legge, R.S. and Tripathi, A. (2006) *Analytical Chemistry*, **78**, 8273.

415 Koh, W.G. and Pishko, M. (2005) *Sensors and Actuators B: Chemical*, **106**, 335.

416 Song, H. and Ismagilov, R.F. (2003) *Journal of the American Chemical Society*, **125**, 14613.

417 Roach, L.S., Song, H. and Ismagilov, R.F. (2005) *Analytical Chemistry*, **77**, 785.

418 Zheng, B. and Ismagilov, R.F. (2005) *Angewandte Chemie-International Edition*, **44**, 2520.

419 Moore, B.D., Stevenson, L., Watt, A., Flitsch, S., Turner, N.J., Cassidy, C. and Graham, D. (2004) *Nature Biotechnology*, **22**, 1133.

420 Tanaka, Y., Slyadnev, M.N., Hibara, A., Tokeshi, M. and Kitamori, T. (2000) *Journal of Chromatography A*, **894**, 45.

421 Tanaka, Y., Slyadnev, M.N., Sato, K., Tokeshi, M., Kim, H.B. and Kitamori, T. (2001) *Analytical Sciences*, **17**, 809.

422 Schoemaker, H.E., Mink, D. and Wubbolts, M.G. (2003) *Science*, **299**, 1694.

423 Schmid, A., Dordick, J.S., Hauer, B., Kiener, A., Wubbolts, M. and Witholt, B. (2001) *Nature*, **409**, 258.

424 Belder, D., Ludwig, M., Wang, L.W. and Reetz, M.T. (2006) *Angewandte Chemie-International Edition*, **45**, 2463.

425 Yoon, S.K., Choban, E.R., Kane, C., Tzedakis, T. and Kenis, P.J.A. (2005) *Journal of the American Chemical Society*, **127**, 10466.

426 Miyazaki, M., Kaneno, J., Yamaori, S., Honda, T., Briones, M.P.P., Uehara, M., Arima, K., Kanno, K., Yamashita, K., Yamaguchi, Y., Nakamura, H., Yonezawa, H., Fujii, M. and Maeda, H. (2005) *Protein and Peptide Letters*, **12**, 207.

427 Miyazaki, M., Kaneno, J., Kohama, R., Uehara, M., Kanno, K., Fujii, M., Shimizu, H. and Maeda, H. (2004) *Chemical Engineering Journal*, **101**, 277.

428 Müller, D.H., Liauw, M.A. and Greiner, L. (2005) *Chemical Engineering & Technology*, **28**, 1569.

429 Koch, K., Van den berg, R.J.F., Nieuwland, P.J., Wijtmans, R., Schoemaker, H.E., Van

Hest, J.C.M. and Rutjes, F.P.J.T. (2007) *Biotechnology and Bioengineering*, doi:10.1002/bit.21649.

430 Kawakami, K., Sera, Y., Sakai, S., Ono, T. and Ijima, H. (2005) *Industrial & Engineering Chemistry Research*, **44**, 236.

431 Maruyama, T., Uchida, J., Ohkawa, T., Futami, T., Katayama, K., Nishizawa, K., Sotowa, K., Kubota, F., Kamiya, N. and Goto, M. (2003) *Lab on a Chip*, **3**, 308.

432 Srinivasan, A., Wu, X.Q., Lee, M.Y. and Dordick, J.S. (2003) *Biotechnology and Bioengineering*, **81**, 563.

433 Srinivasan, A., Bach, H., Sherman, D.H. and Dordick, J.S. (2004) *Biotechnology and Bioengineering*, **88**, 528.

434 Ku, B.S., Cha, J.H., Srinivasan, A., Kwon, S.J., Jeong, J.C., Sherman, D.H. and Dordick, J.S. (2006) *Biotechnology Progress*, **22**, 1102.

435 Lee, M.Y., Srinivasan, A., Ku, B. and Dordick, J.S. (2003) *Biotechnology and Bioengineering*, **83**, 20.

436 Honda, T., Miyazaki, M., Yamaguchi, Y., Nakamura, H. and Maeda, H. (2007) *Lab on a Chip*, **7**, 366.

437 Monod, J. (1950) *Annals de Institut Pasteur, Paris*, **79**, 390.

438 Novick, A. (1955) *Annual Review of Microbiology*, **9**, 97.

439 Zanzotto, A., Szita, N., Boccazzi, P., Lessard, P., Sinskey, A.J. and Jensen, K.F. (2004) *Biotechnology and Bioengineering*, **87**, 243.

440 Kostov, Y., Harms, P., Randers-Eichhorn, L. and Rao, G. (2001) *Biotechnology and Bioengineering*, **72**, 346.

441 Kim, J.W. and Lee, Y.H. (1998) *Journal of the Korean Physical Society*, **33**, S462.

442 Maharbiz, M.M., Holtz, W.J., Howe, R.T. and Keasling, J.D. (2004) *Biotechnology and Bioengineering*, **86**, 485.

443 He, M.Y., Edgar, J.S., Jeffries, G.D.M., Lorenz, R.M., Shelby, J.P. and Chiu, D.T. (2005) *Analytical Chemistry*, **77**, 1539.

444 Dittrich, P.S., Jahnz, M. and Schwille, P. (2005) *ChemBioChem: A European Journal of Chemical Biology*, **6**, 811.

445 El-Ali, J., Sorger, P.K. and Jensen, K.F. (2006) *Nature*, **442**, 403.

5

Industrial Microreactor Process Development up to Production

Volker Hessel, Patrick Löb, Holger Löwe

This chapter is about industrial applications of microprocess technology. Industry is not very active in publishing not only because of confidentiality reasons but also because of the absence of direct need for a profile. Thus, one cannot expect that the ongoing developments can be shown with a similar degree of completeness, clarity and detailedness, as this is possible for reviewing scientific contributions from academy with a wealth of peer-reviewed papers. Especially, the information of major interest of what has been transferred to production and how do the companies make profit with the new technology is usually kept secret. Thus, it has to be accepted that some information is missing among the chapters and that this is only the tip of the iceberg. Also, the sources of information are not as validated as refereed literature because in some cases one has to rely on interviews and other information given in a magazine format.

Knowing these shortcomings, the contents were grouped in sections on industrial mission statements, laboratory process development, pilots and chemical production. For the process development part of this chapter, papers were also considered with industrial coauthors, assuming that the work was (partly) done under industrial perspective and reflects fields of interest and activity of industry.

5.1
Mission Statement from Industry on Impact and Hurdles

Fine chemistry is the present major application area of microprocess technology. Degussa in Hanau, Germany, a leading fine-chemical company, has tested the new approach for various liquid and gas processes including even transfer to plasma reactions [1]. Dominique Roberge, head of a project to evaluate microreactors at Lonza in Visp, Switzerland, states 'The question of whether microreactors are going to be used in the future, I think this is already answered "yes" [2]. The open question is what per cent of the market in fine chemicals they will take.' Lonza is a Swiss company that manufactures intermediates for the drug industry.

> *"The question of whether microreactors are going to be used in the future, I think this is already answered "yes".*" Dominique Roberge/Lonza

Microreactors in Organic Synthesis and Catalysis. Edited by Thomas Wirth
Copyright © 2008 WILEY-VCH Verlag GmbH & Co. KGaA. All rights reserved.
ISBN: 978-3-527-31869-8

Georg Markowz, senior process engineer at Degussa, says 'At Degussa, we believe in the potential of microchannel process technology both as a development tool in the lab and for production technology.' Ralf Pfirmann from Clariant Pharmaceuticals, global pharmaceutical market management director, summarizes some of the primary effects of microreaction engineering and adds 'Use of microreactors in pharmaceutical synthesis ... much higher yields, with higher selectivities, and with economics therefore not possible.' [3]. Fabian Wahl, manager of R&D for Europe at Sigma–Aldrich, emphasizes the secondary effects and adds '... allowing far more precise reaction control and far better product quality control than can be achieved with conventional reactors' [4]. Wahl adds that about 800 from the 2000 reactions in Sigma–Aldrich's portfolio are suitable for microreactor processing, with or without minor process modifications. For the microreactor cases studied so far, reduction of reaction time and cost were major drivers. An application is especially seen in the Sigma–Aldrich's custom synthesis business with fast process development as the prime goal. Wahl expects a 40% reduction in process development time by using microreaction technology. Tony Wood, head of discovery chemistry for Pfizer in Sandwich, UK, adds 'What's interesting to me is the opportunity to pursue fields such as electrochemistry or photochemistry. That would enable us to functionalize molecules in a quite different way from mainstream transformations.' [2].

Besides the technical challenges one still has to take care of the soft factors, which is the human personnel that requires a change in the mindset. 'You are in a small, innovative team in an established company that has more than 100 years' experience in chemical production and you want to change things – there are some barriers beyond the technical', says Dominique Roberge from Lonza [2].

John Brophy, former general manager of corporate research of BP Chemicals in Sunbury-on-Thames, UK, states 'Microchannel process technology is being hailed as the next big thing for the process technologies.' He adds 'To date, industry has been sceptical, adopting a "show us" approach to such radical change in plant technology ... But now the pace is heating up with several companies developing and scaling up this brand-new technology' [1].

Concerning the potential of microprocess technology for large-scale and bulk chemicals production, Dow corporate leader Jon Siddall points out at the example of olefins production, which is the chemical produced in largest volume: 'Microchannel technology may allow us to reduce costs or otherwise improve the process for making an essential commodity of world commerce ... the irony of making large-volume chemicals in small-volume reactors is unmistakable.' [1].

5.2
Screening Studies in Laboratory

5.2.1
Peptide Synthesis

The Nervous System Research branch of Novartis Pharma Ltd in Basel, Switzerland, and the University of Hull investigated peptide synthesis in chip-based microreactors

[5]. β-Amino acids were chosen for demonstrating feasibility of microreactor processing, as there are no chiral centers that may complicate the analysis of the products [6].

Peptides are typically synthesized by solid-phase chemistry on polymer beads, a route discovered by and named after Merrifield [7,8]. These polymer supports are expensive. Additional steps for linkage to and cleavage from the polymer are required. Hence, the motivation is to test solution chemistries as an alternative to the Merrifield approach.

The impact of mixing in a microreactor was demonstrated through the experiment of β-dipeptide synthesis by carbodiimide coupling using Dmab O-protection. Boc-β-alanine was O-protected (carboxylic moiety) by DMAP (4-dimethylamino pyridine) coupling with DmabOH (4-{N-[1-(4,4-dimethyl-2,6-dioxocyclohexyliden)-3-methylbutyl]amino}benzyl alcohol) yielding Dmab-β-alanine, whereas the Fmoc group was used for N-protection of β-alanine [9]. Thereby, orthogonal protecting groups were established. By carbodiimide coupling, Dmab-β-alanine and Fmoc-β-alanine reacted and the synthesis of the corresponding β-dipeptide was realized.

Electroosmotic flow (EOF) conditions were applied and yielded only 10% conversion with constant reactant movement [9]. The use of stopped-flow techniques, which periodically push and mix the flow, led to a 50% increase in yield. A change in the coupling agent from 1-(3-dimethylaminopropyl)-3-ethyl-carbodiimide hydrochloride (EDCI) to dicyclohexylcarbodiimide (DCC) for reasons of limited solubility resulted in a 93% yield of the dipeptide. Batch β-dipeptide synthesis using EDCI gave a yield of 50% [6].

A reduction in reaction time by virtue of the improved transport properties in electroosmotic-driven microreactors was demonstrated for the β-dipeptide synthesis using pentafluorophenyl O-activation. Fmoc-β-alanine was preactivated by introducing the pentafluorophenyl function as ester group [9]. Dmab-β-alanine and the pentafluorophenyl ester of Fmoc-β-alanine reacted, and the synthesis of the corresponding β-dipeptide was realized.

Quantitative yield of the dipeptide in only 20 min was achieved when using electroosmotic-driven microreactor, whereas batch synthesis under the same conditions gave only 46% yield needing 24 h [6,9].

The same findings with respect to reaction time reduction were made for the β-dipeptide synthesis using pentafluorophenyl *O*-activation. Boc-β-alanine was preactivated by introducing the pentafluorophenyl function as ester group [9]. Dmab-β-alanine and the pentafluorophenyl ester of Boc-β-alanine reacted, and the synthesis of the corresponding β-dipeptide was realized.

Electroosmotic-driven microreactor processing gave quantitative yield of the dipeptide in only 20 min, whereas batch synthesis under the same conditions gave only 57% yield needing 24 h [6,9].

The formation of more complex dipeptides was also demonstrated by microchannel processing through the experiment of converting the amino acid *N*-ε-Boc-ʟ-lysine with the additional amino function [5,6]. Although the batch yield is poor with only 9% yield, microreactor synthesis is quantitative in 20 min.

In the next step, the preparation of longer chain peptides was done in electroosmotic microreactors. For this purpose, deprotection and peptide bond forming reactions had to be developed resulting in a yield of 30% [6,9]. In this way, the formation of tripeptides was achieved. Dmab-β-alanine and Fmoc-β-alanine were reacted in the first step to a dipeptide [6]. After cleavage of the Fmoc function, Fmoc-β-alanine was added to such a dipeptide that resulted in tripeptide formation in 30% yield.

The degree of racemization was also monitored through the experiment of a simple carboxylic acid used in peptide synthesis, 2-phenylbutyric acid. The penta-fluorophenyl ester of (R)-2-phenylbutyric acid and (S)-1-phenylethylamine reacts with the corresponding amino acid via an EDCI coupling [5]. In a control experiment, (R)-2-phenylbutyric acid and (R)-1-phenylethylamine were reacted as well. For the amino acid formation of (R)-2-phenylbutyric acid and (S)-1-phenylethylamine, a 4.2% racemization was found [6]. At higher concentration (0.5 M instead of 0.1 M), a higher degree of racemization was observed (7.8%).

5.2.2
Hantzsch Synthesis

GlaxoSmithKline Pharmaceuticals in Harlow, UK, performed the Hantzsch synthe-sis of 2-bromo-4′-methylacetophenone and 1-acetyl-2-thiourea in NMP (N-methyl-2-pyrrolidone) using a microchip reactor under EOF conditions [10] (for EOF see [11]) [10]. This is claimed to be the first example of a heated organic reaction performed on a glass chip reactor under electroosmotic flow control, whereas only room tempera-ture reactions were made earlier. In a wider scope, the Hantzsch synthesis is a further example to evaluate the potential of microfluidic systems for high-throughput screening.

Yields from 42 to 99% were reported. Comparative and better yields were achieved when using a microchip reactor than the conventional lab batch technology. In case of improvement, the increase in yield amounted to about 10–20% [10,12].

5.2.3
Knorr Synthesis

A Knorr synthesis of pyrazoles using a microchip reactor under electroosmotic flow conditions was developed by GlaxoSmithKline Pharmaceuticals in Harlow, UK [13]. The target was a high-diversity screening campaign (7 × 32 libraries in the long run) by parallel operation of microchannels after the feasibility of EOF processing had been demonstrated through the experiment of the Hantzsch synthesis. The Knorr synthesis is of interest for drug applications as products with a wide range of biological activity can be generated this way.

In the Knorr route, 1,3-dicarbonyl compounds react with hydrazines under ring closure to pyrazoles [13].

The following library consisting of seven (A1–A7) 1,3-dicarbonyl compounds and three hydrazines (B1–B3) was synthesized.

The 3 × 7 library was made in a sequential and automated way with conversions between 35 and 99% (quantitative; for 16 reactions). The corresponding chip is a commercial product of Caliper Technologies Company (110 Caliper chip), which was originally designated for μTAS applications. The chips were constructed from two glass plates by means of standard photolithography. The etched microchannels have different widths for more stable flow, for example, to avoid dependence of capillary forces in the reservoirs. The glass chip is glued to a polymer caddy for interfacing with a multiport control device, the Caliper 42 Workstation. This automated system consists of an autosampler (CTC-HTS Pal system) that introduces the reactant solutions in the chip via capillaries. A pumping system (μ-HPLC-CEC system) serves for fluid motion by hydrodynamic-driven flow. A dilution system (Jasco PU-15(5)) for slug dilution on-chip, a detection system (Jasco UV-1575) for detection and an analysis system (LC–MS, Agilent 1100 series capLC-Waters micromass ZQ) were used. All components were online and self-configured.

The results obtained were compared for consistency to those of single-reaction processing on the same chip. No cross-contamination was found during preparation of the library [13]. Neither products, by-products, nor the hydrazine in excess were intermixed.

5.2.4
Enamine Synthesis

GlaxoSmithKline Pharmaceuticals in Harlow, UK, performed an enamine synthesis using a microchip reactor under electroosmotic flow conditions [13]. The aim was on eliminating the need of using Lewis acid catalysts for enamine formation by use of microreactors [14]. In addition, operation under mild conditions such as room temperature processing was favored.

The microreactor yield (up to 42%) is comparable to that of batch Stork enamine reactions using *p*-toluene sulfonic acid in methanol under Dean and Stark conditions, that is, under water separation in a water trap (Dean–Stark apparatus).

5.2.5
Aldol Reaction

The Nervous Systems Research branch of Novartis Pharma Ltd in Basel, Switzerland, carried out the aldol reaction using a microchip reactor under electroosmotic flow conditions. Aldol reactions are well-established routes for C–C bond formation in organic chemistry.

The reaction requires the formation of enolates that by themselves are one of the most profound species enabling C–C bond formation [15]. Reducing processing time is a driver for microchannel processing of aldol reactions [15] that can be accomplished using reactive reactants such as silyl enol ethers. For example, the reaction between 4-bromobenzaldehyde and the silyl enol ether of acetophenone was performed in a microreactor [15].

For this reaction, 100% conversion with respect to the silyl enol ether was achieved in 20 min [15]. The corresponding time for batch synthesis amounted to about 1 day.

5.2.6
Wittig Reaction

SmithKline Beecham Pharmaceuticals in Harlow, UK, carried out Wittig reactions using a microchip reactor under electroosmotic flow conditions. 2-Nitrobenzyltriphenylphosphonium bromide was reacted with methyl 4-formylbenzoate and four other aldehydes, 3-benzyloxybenzaldehyde, 2-naphthaldehyde, 5-nitrothiophene-2-carboxaldehyde and 3-dimethylamino-4-propoxybenzaldehyde [16,17].

With the use of optimized reaction conditions, the Wittig reactions with four of the five aldehydes indeed resulted in improvement of their yields. The ratio of (*E*)- and (*Z*)-alkenes could be changed by simply adjusting the voltages in the electroosmotic-flow-driven chip [18]. For a 1 : 1 ratio of the reactants, the *Z/E* ratio changed from 2.35–3.0 (premixed) to 0.82–1.09 (not premixed, separate movement) [17].

5.2.7
Polyethylene Formation

DOW in Midland, USA, performed metallocene-catalyzed polymerization of ethylene using a homebuilt tube reactor setup with advanced microflow tailored plant peripherals for heating, temperature monitoring, pressure control and dosing via smart valves and injectors. Screening of process conditions was a driver [19]. Also, flexibility with regard to temperature and pressure at low sample consumption was an issue. Quality of the information is another motivation due to the advanced process control and sensing.

Ethylene is handled at 60 °C, well above the critical temperature [19]. Various combinations of precatalysts and activators were sampled and loaded by an autoinjector.

Temperature profiles versus time were taken for different positions at the reactor tube [19]. The maximum rise in temperature was about 23 °C. Improved pressure control was exerted by using advanced pressure control electronics [19]. In the regions of large temperature increase, pressure was slightly fluctuating; this effect diminished downstream. By deliberately changing pressure (in a loop), the temperature response followed immediately [19]. This proved that control of pressure is crucial for obtaining stable temperature baselines.

Catalyst-plug-induced microchannel ethylene polymerization allows to process about 10 runs per hour [19]. This is considerably more than achievable with conventional equipment (Parr reactors) processing only 4–6 runs per day.

5.2.8
Diastereoselective Alkylation

The Novartis Institute for BioMedical Research in Basel, Switzerland, and the University of Hull, UK, performed the diastereoselective alkylation of metal-stabilized enolates using a pressure-driven microreactor at −100 °C, whereby increased conversions and diastereoselectivity were observed compared to the batch process [20].

The diastereoselective synthesis of (2′S,4R,5S)-3-(2′-methyl-3′-phenylpropionyl)-4-methyl-5-phenyloxazolidin-2-one **3** was demonstrated in a chip microreactor, whereby diastereoselectivities of >91 : 9 (**3** : **4**) were obtained compared to 85 : 15 in the batch mode.

5.2.9
Multistep Synthesis of a Radiolabeled Imaging Probe

Several academic partners and Siemens Medical Solutions USA Inc. (Molecular Imaging) in Culver City, USA, made the synthesis of an [18F]fluoride-radiolabeled molecular imaging probe, 2-deoxy-2-[18F]fluoro-D-glucose in an integrated micro-fluidic device (see Figure 5.1) [21]. Five sequential processes were made, and they are [18F]fluoride concentration, water evaporation, radiofluorination, solvent exchange and hydrolytic deprotection. The half-life of [18F]fluorine ($t_{1/2} = 110$ min) makes rapid synthesis of doses essential. This is one of the first examples of an automated multistep synthesis in microflow fashion.

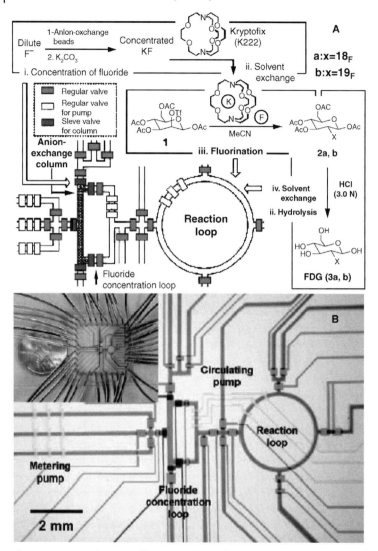

Figure 5.1 Top: a schematic of how to translate the chemical reaction steps of the synthesis of 2-deoxy-2-fluoro-ᴅ-glucose into a microfluidic design with microchannels and valves for consecutive fluid transport und manipulation (by courtesy of AAAS) [21]. Bottom: central area of the microfluidic circuit with channels filled with food dyes for visualization. Inset: photograph of the device (by courtesy of AAAS) [21].

In comparison to the conventional automated synthesis, the radiochemical yield and purity of the compound obtained by microreactor processing was higher and also had shorter synthesis time [21]. Multiple doses of 2-deoxy-2-[^{18}F]fluoro-ᴅ-glucose for positron emission tomography imaging studies in mice were prepared. Today, 2-deoxy-2-[^{18}F]fluoro-ᴅ-glucose is routinely produced in about 50 min with the use of expensive commercial synthesizers and the radiolabeled compound from ~10 to 100 doses is produced in a single run.

5.3
Process Development at Laboratory Scale

5.3.1
Nitration of Substituted Benzene Derivatives

BASF in Ludwigshafen, Germany, and university partners reported the nitration of several disubstituted benzene derivatives using a capillary-flow microreactor [22]. The exact nature of these species, however, was not disclosed.

Because the benzene derivative and nitric acid are immiscible, the impact of mixing/ distribution on slug formation was investigated. Uniform slugs of the aromatic compound/nitric acid were formed in a Y-piece [22]. The capillary attached has a stabilizing effect on the slug flow. The deviation of slug size distribution is very small (about 5%). Hence, interfacial area is nearly constant for this type of capillary flow.

For the nitration of benzene derivatives, experiments were performed at two temperature levels, 60 and 120 °C [22]. For the 60 °C experiment, very small amounts of by-products such as phenol derivatives and dinitrated species were formed, not exceeding 50 ppm each. At 120 °C, high levels of by-products were found, with 300 ppm dinitrated species and 200 ppm phenol derivatives. The mechanism of by-product formation was also investigated [22]. Dinitrated products were generated by nitration of the mononitrated product. It was concluded that phenolic by-products were formed directly from the aromatic starting material rather than from the mononitrated product. This proposed reaction mechanism could be confirmed by performing selective nitration of the mononitrated product.

The influence of interphase mass transfer between liquid–liquid slugs was investigated for nitration of aromatic compounds in a capillary-flow reactor (see Figure 5.2) [22]. This was achieved by changing flow velocity via volume flow setting, while residence time was kept constant by increasing the capillary length.

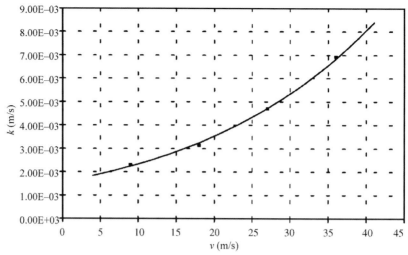

Figure 5.2 Interphase mass transfer coefficient obtained from reaction engineering model (by courtesy of Elsevier) [22].

Conversion to the mononitrated benzene derivative increased linearly with increasing flow velocity because of enhanced mass transfer. The formation of phenol by-products increased in the same manner for similar reasons. In turn, consecutive by-products, dinitrated aromatics, were formed in a linear decreasing fashion. This was explained by a mass-transfer-induced removal of the mononitrated product from the reacting slug.

In the last step, a reactor model was developed taking into account of both the mass transfer of organic components between the two phases and the homogeneous reaction within the aqueous phase, the latter relying on literature data [22]. From there, an extended kinetic model was developed and applied, considering the kinetics of the homogeneous side reactions as well [22]. With these efforts, the activation energies of these processes could be derived.

5.3.2
Phenyl Boronic Acid Synthesis

Clariant GmbH in Frankfurt, Germany, performed the synthesis of phenyl boronic acid from phenyl magnesium bromide and boronic acid trimethyl ester on a pilot-scale level [23].

This reaction suffers, like many organometallic reactions, from insufficient mixing because its reaction speed (at room temperature and below) is faster than the mixing times of many conventional mixer equipments. Thus, the reaction proceeds under nonstoichiometric conditions with timely and spatially changing concentration profiles, which promotes consecutive reactions. In addition, too long processing times, quite typical for batch reactions, favor side reactions such as oxidations and hydrolysis. This is solved in conventional processing by slowing down the reaction speed by virtue of cryogenic conditions, which reduces the impact of mixing and expands the mixing time scale so that selectivity is sufficient. For this purpose, cooling utilities that will add to the capital investment and make the process energy consumptive need to be installed.

For the microreactor process, a micromixer–tube rig was used with an interdigital mixer as a laboratory tool and a caterpillar mixer as a pilot tool [23]. The superior mixing of micromixers allows to perform the reaction at much higher reaction speed so that high selectivity of 90% could be demonstrated even at room temperature and above. This saves energy costs and also reduces the respective CAPEX (capital-related) investments. The crude yield was about 25% higher than that for the industrial batch production. The purity of the crude product was increased by about 10%. This had an impact downstream the reaction processing. Purification was simplified and could be achieved by means of crystallization only, whereas in the current industrial process energy-consumptive distillation is required (see Figure 5.3).

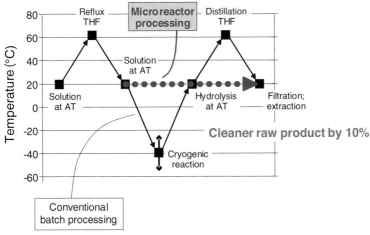

Figure 5.3 Temperature profile of the phenyl boronic acid synthesis along the major steps of the process flow scheme (AT: ambient temperature; THF: tetrahydrofuran). The difference in temperatures of the conventional batch and the microreactor processes stand for the reduction in energy consumption and respective heat transfer equipment when using the latter (by courtesy of ACS) [23].

5.3.3
Azo Pigment Yellow 12 Manufacture

Trust Chem Company in Hangzhou, China, was involved in the synthesis of the commercial azo pigment Yellow 12 [24].

Yellow 12

Particle synthesis, in general, forms products that differ in particle average size and distribution, shape, morphology, selectivity and many more properties. Mixing is the key step for particle seed formation and growth and can determine the above-mentioned product qualities. Owing to uniform concentration profiles and a good correlation between experiment and theory, product properties of particles made in microdevices are often superior and process development and upscaling can be faster and more predictable as compared to conventional technologies. In the ideal case of numbering up, the laboratory-scale performance is kept during piloting and production because the fluid dynamics are not changed or at least shift by analogy in a known manner.

The synthesis of the azo pigment Yellow 12 involves a very fast precipitation and, consequently, is largely impacted by mixing in micromixers. Target of an investigation of the Trust Chem Company was to obtain narrow-sized crystals by microprocess technology. Using a laboratory-scale slit-type interdigital micromixer–reactor [24], smaller particles with a uniform size distribution were obtained for the commercial azo pigment Yellow 12 (see Figure 5.4) [24].

Related product properties are improved, for example, optical properties such as the glossiness and transparency (see Table 5.1). There are no negative consequences for dye manufacturing with the new microreactor made crystals, as the tinctorial power is the same as for conventional synthesis. Tinctorial power is a measure for the adhesion of the pigment to wool stuff or similar material. The intensification in coloration properties means that the same amount of material can be treated now with less amounts of the Yellow 12 azo pigment that reduces materials costs and increases the profitability of the pigment manufacture.

5.3.4
Desymmetrization of Thioureas

At the early stage of microreactor development, Merck AG in Darmstadt, Germany, and the University of Chemnitz, Germany, performed an industrial study on the general applicability of microreactors towards organic synthesis for pharmaceutical applications, aiming at performing combinatorial chemistry by microflow processing

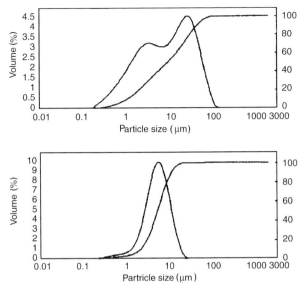

Figure 5.4 Particle size distribution of the batch (top) and the micromixer-based synthesis (bottom) of Yellow 12 (by courtesy of ACS) [24].

in the long run [25]. The scouting studies focused on determining suitable reaction parameters and to show basic feasibility, that is ability to reach yields similar to batch processes.

With the use of a micromixer/commercial tube reactor, the synthesis of a thiourea from phenyl isothiocyanate and cyclohexylamine at $0\,^{\circ}C$ was carried out [25].

A single mixing device connected to a stainless steel tube of about 10 m length and 0.25 mm diameter was used [25]. The feasibility of performing a nearly spontaneous reaction could be shown. Further studies on the desymmetrization of thioureas showed that for the diphenyl thiourea/cyclohexylamine system reasonable reaction

Table 5.1 Glossiness of the imprinted color (GU = glossiness units) for micromixer-based and conventional Yellow 12 pigment manufacture.

No.	Micromixer (10 ml/min)/GU	Micromixer (30 ml/min)/GU	Micromixer (50 ml/min)/GU	Yellow 12 standard/GU
Mean	40.9	47.1	51.0	29.4
σ	1.8	1.4	1.3	1.8

(From [24]). σ: standard deviation.

rates and conversions could be achieved [25,26] as the short use of high-temperature operation up to 91 °C, exceeding the boiling point of the solvent (acetonitrile), was the key for this reaction that can be easily accomplished in microreactors.

5.3.5
Vitamin Precursor Synthesis

BASF in Ludwigshafen, Germany, carried out process development for a reaction as part of a multistep process finally yielding a vitamin [27,28]. This concerned short-temperature processing (<4 s) that was simply not possible using macroscopic bench-scale apparatus. In the latter case, almost 50% of the reaction heat was released already at the mixer unit rather than after entering the subsequent heat exchanger. The temperature rise led to side reactions reducing the yield.

The reactants and the product were not disclosed in the open literature [27,28]. Concentrated sulfuric acid is present in quantitative amounts besides the organic solvent so that a liquid/liquid process results. The reactant forms quickly an intermediate that again quickly reacts to the product. Thermally induced side reactions occur at each stage.

A maximum yield of 80–85% was obtained at 4 s residence time and a temperature of 50 °C by microreaction system processing [27,28]. The use of ordinary lab processing with standard lab glassware yielded only 25%. The continuous industrial process had a yield of 80–85%; the former employed with semibatch industrial process gave 70% yield. The temperature and the residence time of industrial and microreactor continuous processing were identical.

5.3.6
Ester Hydrolysis to Produce an Alcohol

Sigma–Aldrich performed an ester hydrolysis to produce an alcohol that decomposes quickly. None of these compounds could be disclosed [4]. Scale-up could not satisfy the commercial interest in the labile product, as the yield decreased strongly with batch vessel size. Seventy percent yield was found at 5 l scale, 35% at 20 l and 10% at 100 l.

The investigation in a microreactor was initially hampered by the presence of an insoluble compound for the ester process [4]. Because this reactant had to be changed, a process development study had to be carried out. A model reaction, an ester hydrolysis yielding a stable alcohol, was used instead of the real one in order to facile the process development. Twelve different conditions were run. Having acquired the process know-how, the same kind of process development was done for the real reaction in only 2 h with success.

5.3.7
Synthesis of Methylenecyclopentane

Methylenecyclopentane was synthesized by Sigma–Aldrich using an undisclosed route [4].

The reaction is highly exothermic, which makes scale-up difficult [4]. In addition, 30% of the yield is the thermodynamically stable side product 1-methylcyclopentene. Product and side product are difficult to separate. Both unsolved issues led to a stop in process development using conventional technology.

With the use of microreactor processing, 70% conversion was achieved that gave only product and no side product [4]. Actually, the microreactor conversion is lower than the conversion for batch, but the former is now the preferred route as the separation of the product from the reactant can be accomplished, whereas product, as mentioned, can be hardly purified from the side product. A throughput of 300 g/h was achieved.

5.3.8
Condensation of 2-Trimethylsilylethanol

Sigma–Aldrich performed the condensation of 2-trimethylsilylethanol and *p*-nitrophenyl chloroformate to give 2-(trimethylsilyl)ethyl 4-nitrophenyl carbonate [4]. Reduction of residence time is the issue to high selectivity, as the product degrades and side products also form. In contrast to the conventional technology that needs 14 h to accomplish the reaction, a microreactor that enhances selectivity accomplishes the same in about 18 min. The reason for the better microreactor performance is not detailed and if the process parameters (e.g. temperature) were changed.

5.3.9
(*S*)-2-Acetyl Tetrahydrofuran Synthesis

SK Corporation in Daejeon, Korea, performed the synthesis of (*S*)-2-acetyl tetrahydrofuran with an alkylating step using a Grignard reaction and a hydrolysis step finally yielding a ketone moiety in the side chain starting with a cyano group [29].

The methylating agent MeMgCl is very reactive, even when compared to other Grignard reagents, and thus not easy to handle at large scale; this can also substantially cause safety and hazardous issues [29]. The microreactor allowed the minimization of moisture and oxidizing effects due to encased processing at low degree of contamination of decomposing species for Grignard reactants.

Overalkylation can lead to tertiary alcohol formation by consecutive reaction [29]. Product quality demands to keep this impurity level <0.2%. Microreactor operation yielded the overalkylated alcohol follow-up product at 0.18%, whereas level of impurity for the batch process was 1.56% [29]. The reason is probably the lower back-mixing in the microflow system, with concentration profiles being less deteriorated from ideal; that is, no excess of alkylating agent is generated locally to promote the follow-up reaction.

The α-hydrogen of the reactant is unstable under the basic reaction conditions applied, leading to a small degree of racemization [29]. Conservation of stereochemistry was largely achieved by microreactor operation; 98.4% enantiomeric excess (ee) was found as compared to 97.9% ee at batch level (see Table 5.2).

5.3.10
Synthesis of Intermediate for Quinolone Antibiotic Drug

LG Chem in Daejeon, Korea, performed the multistep synthesis of Gemifloxacin (FACTIVE), a quinolone antibiotic drug with enhanced activity against G(+) bacteria [30]. This drug has high activity against G(−) bacteria, atypical strains and major respiratory pathogens.

Gemifloxacin (FACTIVE)

Table 5.2 Impurity and optical purity of the batch and microreactor
processes in the (S)-2-acetyl tetrahydrofuran synthesis [29].

	Impurity (%)	Optical purity (%)
Batch	1.56	97.7
Microreactor process	0.18	98.4

In one reaction step, the enamine moiety is protected by the *t*-Boc group in a fast
and highly exothermic ($\Delta H = -213$ kJ/mol) reaction [30].

A consecutive reaction occurs with higher rate constant and higher activation
energy, thus getting more dominant at higher temperatures [30]. The impurity level
generated by the consecutive step becomes too high at temperatures $>25\,^{\circ}$C. This can
be minimized by effective heat removal of the exothermic reaction. In addition, the
product can react with KOH so that efficient mixing is required that avoids spatially
strong varying concentrations of reactants. For similar reasons, it was speculated that
narrower residence time distribution could increase selectivity.

A slit-type interdigital micromixer was used for fast mixing [30] and compared to a
tubular reactor and five Kenics mixers connected in series. Details on further
components of the micromixer rig were not given, but most likely a capillary reactor
was added for efficient heat exchange.

Mixing is essential to achieve dispersion of the two-phase mixture. For the tube
reactor (without mixing function), the flow had to be very high to achieve turbulence
($Re > 2000$) and in this way to serve for dispersion [30]. As a result of the high flow rate,
residence time was too short to complete reaction and only 27% conversion was
achieved. At lower flow rate, phase separation occurs. An extrapolation of the reactor
length for complete conversion comes to an impractical tube length of about 2 km. The
combination of five Kenics static mixers guaranteed good mixing over an extended
length so that a conversion of 97% was achieved as a result of the micromixer. The
microreactor used has good mixing and was also able to remove the heat of reaction.
No by-products were formed at conversions as high as 96%. Thus, although the
microreactor performance in terms of selectivity is equal to that of the static mixers, it
has the advantage of operation at ambient reaction temperature of 15 $^{\circ}$C, whereas the
latter need outside cooling of 0 or $-20\,^{\circ}$C because of the necessary heat removal.

5.3.11
Domino Cycloadditions in Parallel Fashion

GlaxoSmithKline Pharmaceuticals in Harlow, UK, investigated domino cycloaddi-
tions in a commercial chip and extended their process development by operation in

parallel fashion in a three-member array, which was one of the first examples of a parallel multireaction in microreactors [31].

(A)   (B)   (C)

The domino reaction consists of a Knoevenagel condensation giving an intermediate that immediately undergoes an intramolecular hetero-Diels–Alder reaction with inverse electron demand [31]. As aldehydes, *rac*-citronellal, an aromatic aldehyde, and two commercially available 1,3-diketones, 1,3-dimethylbarbituric acid and Meldrum's acid, were selected. By combinations of these reactants, different cycloadducts were generated.

Process development was accomplished in single-run reactions [31]. The conversion of the microchannel processing typically amounted to about 50–75%, depending on the nature of cycloadduct and the residence time chosen, and was comparable to batch syntheses results [31]. By combinations of aldehydes and 1,3-diketones, different cycloadducts were generated simultaneously in one run on one chip, that is, an undesired transfer of solutions from one channel to another by imperfect sealing between these channels. The conversions were comparable to the single runs, with one exception. Also, cross-contamination was observed. It ranged from a few percent to about 50%.

5.3.12
Ciprofloxazin Multistep Synthesis

The synthesis of ciprofloxazin was one among several syntheses being performed in contract research by a microreactor developer for pharmaceutical industry and feasibility was demonstrated [32]. In this multistep synthesis, alkylamino-defluorinations were the essential part of the chemistry. Ciprofloxazin, the final product, is an antibiotic compound with a high sales volume.

Ciprofloxazin

This reaction scheme involves two substitutions of fluorine moieties at the aromatic ring by amines, yielding the final product for pharmaceutical applications [32,33]. All in all, five synthesis steps are actually required and performed subsequently in a microreactor system to get the target molecule.

5.3.13
Methyl Carbamate Synthesis

The Chemical Development & Drug Evaluation branch of Johnson & Johnson Pharmaceutical Research & Development LLC in Raritan, USA, performed the exothermic reaction of methyl chloroformate with amines to methyl carbamates [34]. Owing to large heat release, hot spots occur. For the reaction of N-methoxycarbonyl-L-*tert*-leucine with methyl chloroformate to the amino acid derivative, it is even observed at laboratory scale.

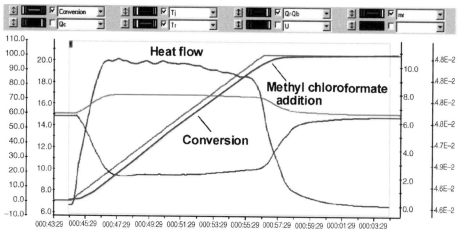

Figure 5.5 Reaction calorimetry results from the addition of methyl chloroformate (slight excess) to L-*tert*-leucine (by courtesy of ACS) [34].

Calorimetric measurements show that the addition is exothermic [34]. The heat release rate is mainly feed controlled, as the square shape of the heat flow curve demonstrates (see Figure 5.5). Whenever feed is added, the heat flow responds without delay.

In case of complete malfunction of cooling and stirring systems, the temperature may exceed the solvent reflux temperature [34]. Accordingly, a slow dosing of the methyl chloroformate is necessary to have control over the heat release. After having determined the reaction parameters at 1 g scale, the reaction was carried out in a microreactor with 91% yield at 7 min residence time. More than 1 kg of *N*-methoxycarbonyl-L-*tert*-leucine was prepared within 12 h.

Scale-up by using a microreactor was also done for the amidation of *p*-tolyl chlorodithionoformate with dimethylamine to *p*-tolyl dimethyldithiocarbamate without further safety precautions at 96% yield that is comparable to the batch process [34]. At 1.4 min residence time, a capacity of 155 g/h was achieved.

5.3.14
Newman–Kuart Rearrangement

The Chemical Development & Drug Evaluation branch of Johnson & Johnson Pharmaceutical Research & Development LLC in Raritan, USA, tested microreactors for processes at elevated temperatures above the limit of most multipurpose conventional reactors, which is above $\sim 140°C$ [34]. Operation above this limit is only possible by means of special reactors equipped with heat transfer units.

The Newman–Kuart rearrangement is an example of a high-temperature reaction [34]. With the use of a microreactor, the reaction temperature could be extended up to 200 °C. *O*-(2-Nitrophenyl)-*N,N*-dimethylthiocarbamate was converted to *S*-(2-nitrophenyl)-*N,N*-dimethylcarbamothioate at 170 °C in 14 min at 90% yield. Quantitative conversion with a throughput of 34 g/h was achieved with sulfolane as solvent at the same temperature and reaction time.

5.3.15
Ring-Expansion Reaction of *N*-Boc-4-Piperidone

The Chemical Development & Drug Evaluation branch of Johnson & Johnson Pharmaceutical Research & Development LLC in Raritan, USA, investigated the ring-expansion reaction of *N*-Boc-4-piperidone with ethyl diazoacetate in a microreactor system as an example of processing hazardous substances [34].

A crude yield of 90% was obtained in ether at −25 °C. When performed in a batch mode on 70 mg scale, no safety issues were taken and an 81% yield was obtained [34]. Reaction calorimetry reveals a very exothermic reaction after feeding and subsequent mixing with an initiation period; that is, the response of the heat flow curve is delayed by about 1 min as compared to that of the feed curve (see Figure 5.6). Therefore, the heat release rate with this mode of addition is not feed controlled. The reaction is very sluggish because the reaction occurs at a single blow as soon as 60% of the material has been added. Calculating the worst-case temperature rise shows that the reaction would rapidly increase with temperature and can approach the solvent reflux temperature, possibly throwing out the reaction mixture in case of cooling or stirring malfunction. This would particularly include the full accidental release of all the $BF_3 \cdot Et_2O$.

The calculated worst-case temperature of 45.6 °C also approaches the reflux temperature of the solvent (diethyl ether) but does not reach the decomposition temperature of ethyl diazoacetate [34]. Furthermore, the evolution of large amounts of nitrogen gas during the reaction could lead to an overpressurization of the reaction vessel. All these reasons together prevented a scaling of the ring-expansion reaction to kilogram scales as a result of the related safety issues. Operating the microreactor

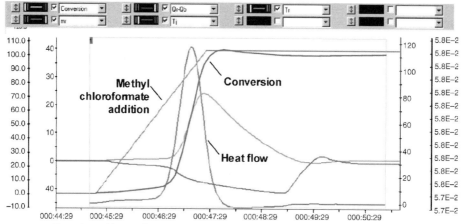

Figure 5.6 Addition of $BF_3 \cdot Et_2O$ to N-Boc-4-piperidone and ethyl diazoacetate (by courtesy of ACS) [34].

system, however, allowed without any further optimization a precise control of the reaction, and 89% yield was obtained. Reaction time was only 1.8 min with a throughput of 91 g/h.

5.3.16
Grignard and Organolithium Reagents

Lonza Ltd in Visp, Switzerland, performed the reaction of an acid chloride with phenylethyl magnesium bromide in THF in a multiple-injection microreactor [35]. Many injection points were chosen to split and delocalize the reaction and heat release so that several small hot spots arise instead of having one large hot spot ($\Delta H = -260$ kJ/mol, $\Delta T_{adiabatic} =$ about 70 °C). The reaction is quenched at the reactor outlet with water.

The reaction performance was given for mixers with one to six injection points (see Figure 5.7) [35]. A considerable increase in yield from 21 to 38% can be seen.

In another process optimization step, the cooling was further improved (from 'sufficient' to 'efficient') by using a lower temperature of the cooling liquid, with 0 °C instead of 20 °C [35]. In this way, a yield of 39% was obtained with only three injection points, thus reducing the complexity of the microdevice and pressure drop (see Figure 5.8).

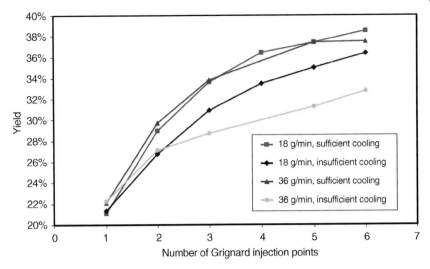

Figure 5.7 The impact of number of injection points and cooling on the yield of the reaction of an acid chloride with a phenylethyl Grignard reagent (by courtesy of D. Roberge/Lonza) [35].

The Chemical Development & Drug Evaluation branch of Johnson & Johnson Pharmaceutical Research & Development LLC in Raritan, USA, used the reaction of the Grignard reagent 3-methoxyphenylmagnesium bromide and a ketone to obtain the product Tramadol. *Cis-* and *trans-*isomers are formed in about 4 : 1 ratio [34]. The

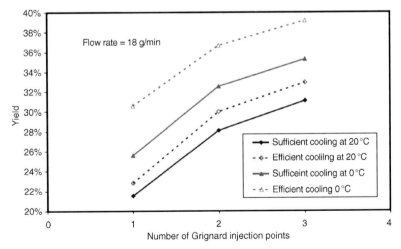

Figure 5.8 The impact of improved cooling on the yield of the reaction of an acid chloride with a phenylethyl Grignard reagent (by courtesy of D. Roberge/Lonza) [35].

organometallic compounds formed are unstable intermediates and are sensitive to moisture.

The use of aryllithium instead of the Grignard reagent resulted in a higher ratio of *cis*-isomer formation [34]. In reaction calorimetric studies, it was found that both steps, the formation of 3-methoxyphenyllithium and its addition to ketone, are pretty exothermic with worst-case temperature rises of up to 62 and 133 °C, respectively. The lithium intermediate has to be kept at very low reaction temperatures to prevent decomposition. We concluded that a continuous reaction may be a good alternative to batch synthesis to improve the reaction yield and to minimize the safety concerns because of the exothermicity of the reaction sequence.

3-Methoxyphenyllithium and cyclohexanone were reacted in a batch mode at -10 and -65 °C to give yields of 32 and 80%, the expected tertiary alcohol, respectively [34]. Such temperature effect was also planned to be used in the microreactor. The metal–halogen exchange step could be performed at -14 °C with 17 s residence time and the lithium intermediate further reacted with cyclohexanone in batch mode at -40 °C. Lower temperatures were not possible because of chiller limitations, and the availability of only one microreactor accounted for the combined continuous flow–batch processing. In this way, a yield of 87% yield at a throughput of 54 g/h was achieved.

5.4
Pilots Plants and Production

5.4.1
Hydrogen Peroxide Synthesis

$$H_2 + O_2 \rightarrow H_2O_2$$

The process developing Company UOP LLC in Des Plaines, USA, searched for direct routes for hydrogen peroxide manufacture from elementary hydrogen and oxygen by microprocess technology that almost under all process conditions inevitably involves processing in the explosive area [36,37]. The better use of raw materials and less technical expenditure owing to process simplification are the advantages of the direct over the indirect routes. The few process windows out of the explosive range need large dilution and/or multistep processing with the addition of intermediates as carriers for oxygen, such as the anthraquinone route.

On the contrary, operation in the explosive area ('New Process Windows') benefits from higher space–time yield because elevated temperatures, pressures and reactant concentrations are used [36,37]. This is enabled by the higher safety potential of microreactors, which are less accident-prone by thermal runaway because of enhanced heat transfer and also do not allow uncontrolled, ever increasing radical chain propagation by providing an even high rate of radical quenching at the large surface area of the channel walls [38]. Owing to the latter flame-arrestor effect that is intrinsic to the devices, the term 'inherently safe' was assigned. However, this only refers to the device inner volume and is a theoretical term. For example, walls between microchannels may break under very high pressure shock waves or other similar incidents may occur so that a dimensional increase above the safety threshold may occur. Still, a practical safety evaluation needs to be done, and safety issues are different in the interconnections and fluid distributions to the device, which holds even more in the plant peripherals. Thus, although the intrinsic safety of the microchannel on a device level can be assumed in a theoretical manner, the same feature of the microprocess technology as a whole has to be verified and authorized for each chemical process. Despite these issues, advantages remain because of faster safety approval.

On the basis of this information, a process flow based on a mini-trickle-bed operation was developed where a premixed gas mixture of hydrogen and oxygen (without inert gas) was dispersed with an aqueous solution flowing through the interstices of a catalyst bed [36,37]. Experimental investigations were first done at a laboratory level using slit-type high-pressure interdigital micromixer connected to a cartridge filled with catalyst. Special procedures and plant details were developed and added to have safety on a process level and to encounter typical failure situations such as total oxidation taking place outside the catalyst cartridge with water as heat absorbing medium.

The direct contact of hydrogen and oxygen offers other advantages besides having high concentrations and high reaction rates [36,37]. Explosive regime mixing allows to generate hydrogen/oxygen mixtures in a stable and reliable manner. This will transfer the noble metal catalyst into a partially oxidized state that will then generate hydrogen peroxide with high selectivity. In the fully oxidized or fully reduced state of the catalyst, either only water is formed or no reaction is achieved, respectively.

Operation of this process at only 20 bar and the use of oxygen/hydrogen ratios close to 1 : 1 were considerably improved [36,37] compared to processes from the patent literature [39,40]. For oxygen/hydrogen ratios of 1.5–3, a selectivity as high as 85 at 90% conversion and a space–time yield of 2 g hydrogen peroxide per gram of catalyst per hour was achieved, exceeding industrial benchmarks from patent literature at that point of time. These laboratory experiments were followed by pilot tests that resulted in a basic engineering design for the production of about 150 000 t hydrogen peroxide per year.

FMC Corporation in Princeton, USA, one of the largest producers of hydrogen peroxide, works with Stevens Institute in Hoboken, USA, in a publicly funded project for the on-demand production of hydrogen peroxide in a microreactor at end-user sites [1,41]. Although today's hydrogen peroxide is mainly produced as a concentrated solution (70% in the anthraquinone process), many user applications demand solutions <15% so that dilution is common, and this overrides the efforts made by the energy-consumptive distillation to yield concentrated solutions. FMC summarizes the prospects of a direct microreactor route for on-site production to reduce transportation and storage costs as well as 'concentration–dilution' costs. To target the safe operation in the explosive regime, a low-pressure and energy-efficient process is designed. Solvent processing, gas recycling and treatment, and product purification are avoided as compared to the commercial anthraquinone process. The initial goal is to achieve the production of a 1% solution at laboratory scale and to transfer this to an on-site production scale within 5 years.

Hydrogen peroxide was generated in a microreactor in a gas–liquid–solid process with direct mixing of hydrogen and oxygen (using air), a liquid solvent and a solid catalyst [41], which included operation in the explosive regime. The particles of the self-developed platinum group catalyst supported on an oxide were packed in the microchannel. In addition, sol–gel wall coatings were adhered to the microchannels, and cellular structures were formed by different deposition routes involving closed-channel flow coating, open-channel surface-selective and dip coating methods. Using moderate pressure and temperature, the residence time could be decreased by almost two orders of magnitude as compared to the conventional reactors. The hydrogen peroxide concentrations were within the industrial relevant range. On the basis of the identification of the elemental reactions involved, a study of the reaction mechanism was done to determine the kinetics. Following the laboratory studies, an optimized process analysis, design and simulation was developed for a pilot plant comprising a multichannel, multilayered microreactor system built by Chart Energy and Chemicals in The Woodlands, USA (Figure 5.9). First runs proved the production of hydrogen peroxide in an industrial relevant range.

5.4.2
Diverse Case Studies at Lonza

Lonza Ltd in Visp, Switzerland, reported several further process developments [35]. For an organolithium exchange reaction up to 250 kg of product was obtained in a few weeks of operations using a Corning multi-injection reactor. At the end of the campaign, the reactor was cleaned by standard procedure and used again for a new project. For an organolithium coupling reaction, several kilograms of product were obtained by continuous operation for 1 week. The Corning multi-injection reactor as well as the Corning single-injection reactor was used for validation. In a nitration reaction, a few kilograms of product were obtained in 24 h operation using the Corning multi-injection reactor.

With the use of dedicated reactor technology, chlorination was carried out in a falling-film reactor that has a highly exothermic reaction ($\Delta T_{adiabatic} > 250\,^\circ C$) [35]. Dehydrogenations were done using a heterogeneous catalyst filled in a tubular packed bed at high temperatures above $500\,^\circ C$. An organolithium coupling reaction was made in a small CSTR reactor to avoid reactor plugging.

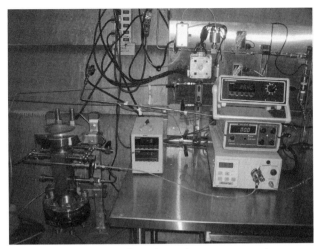

Figure 5.9 Hydrogen peroxide synthesis pilot microprocess plant by courtesy of FMC Corporation in Princeton, USA, and Niyi Lawal/Stevens Institute) [41].

A so-called continuous launch production unit R-01 was built with a capacity in the range of 150 kg/h and designed as a multipurpose system (see Figure 5.10) [35]. Conventional technology such as static mixers, mini-heat exchangers and so on was used following experience with microprocess technology. Such devices are the mesoscale extension of microdevices, albeit not always using the same mixing and heat exchange principles. Several campaigns were performed with in-between cleaning for the following reactions. Tons of products were made using the

Figure 5.10 Continuous launch production unit R-01 at Lonza (by courtesy of D. Roberge/Lonza) [35].

Simmons–Smith reaction, and the production ran over weeks. A static mixer with conventional minitube heat exchangers was used. A similar amount of material was produced by an organolithium coupling reaction using a static mixer in an adiabatic regime.

5.4.3
Polyacrylate Formation

Axiva in Frankfurt, Germany (now Siemens-Axiva), performed the radical solution polymerization of acrylate resins using micromixer–tube reactors [42].

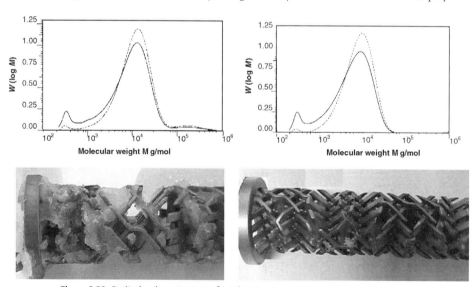

This reaction includes modified acrylates with or without the addition of styrene derivatives in combination with one or more initiators in a solvent [43]. In the molecular weight distribution of the polymer obtained in the micromixer, no high molecular weights above a mass of $>10^5$ that are usually insoluble were found [42,43]. As a result, no precipitates are formed on the surface of the microchannel despite the large surface-to-volume ratio (see Figure 5.11). In contrast to this result, polymer

Figure 5.11 Radical polymerization of acrylates using a static mixer–reactor without (left) and with (right) a micromixer as premixer. Top: molecular weight distribution; down: appearance of the static mixer to illustrate the degree of fouling (by courtesy of Springer) [42].

samples taken from conventional processes without micromixer displayed a small but significant fraction of high molecular weight polymer with a mass $>10^5$ (see Figure 5.11) [43]. Here, in some cases heavy precipitation occurred, resulting even in plugging of the static-mixer internals of the tube reactors [42].

A micromixer-based laboratory plant would give 50 t/a, when assuming an annual operation time of 8000 h. On the basis of these laboratory experiments, a prebasic design comprising numbering up of 28 micromixers and 4 tube reactors was proposed [42]. Accordingly, the production of such plant was calculated to be in the order of 2000 t/a, when assuming 8000 h operation.

5.4.4
Butyl Lithium-Based Alkylation Reactions

Lonza Ltd in Visp, Switzerland, developed at the beginning of the 1990s a process for the halogen exchange of a brominated aromatic compound with butyl lithium followed by the C−C coupling to a ketone [44].

Ar–Br + ∼∼∼Li → Ar–Li + ∼∼∼Br

Intermediate

Ar–Li + R^1–CO–R^2 → R^1–C(O–Li)(R^2)–Ar

Intermediate

The lithium intermediate is unstable, even at temperatures as low as −60 °C. Only a continuous process with a short residence time between the two reactions was allowed to avoid decomposition and have sufficient selectivity for an economical process [44]. Clogging is a major issue for the first process. Owing to heat release issues, high dilution is applied, and the recycling of the solvent was an important issue that needed to be considered.

The first reaction is highly exothermic with an adiabatic temperature rise >100 °C (classified as Type A in [45]), whereas the second reaction is only slightly exothermic with an adiabatic temperature rise <30 °C (this follows a classification given in [45]: Types A and B, respectively) [44]. Despite the latter, the second reaction is accompanied by the formation of large amounts of side products at long residence times >1 min, which almost excludes batch processing here. A conventional continuous system, a static mixer with much shorter residence times, was successfully used. The first reaction can be performed batchwise because the lithium intermediate is stable for a few hours. Still, a stop of production for several hours would lead to a significant loss of product.

Thus, it is more flexible to run both reactions continuously [44]. A microreactor is used for the more demanding, highly exothermic reaction, whereas a static mixer is sufficient for the second reaction. After laboratory process development, a pilot phase

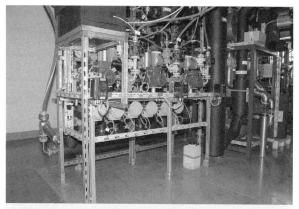

Figure 5.12 Lonza c-SSP with combined microreactor and conventional technologies (static mixers), process view (top) and service view (bottom) (by courtesy of PharmaChem/B5Srl) [44].

in the so-called continuous small-scale production (c-SSP) comprising microreactor technology followed (see Figure 5.12). The c-SSP plant is a multipurpose and modular approach and can operate from cryogenic to high temperature.

The c-SSP plant typically operates at a total flow rate of 100 g/min that relates to 70 kg product/week of product [44]. This gives sufficient material for preclinical studies or phase I clinical trials during a typical pilot campaign cycle time, which is 1–2 weeks (see Figure 5.13). The c-SSP plant qualified for c-GMP (GMP: Good Manufacturing Practice; guidelines for quality management for production processes and environment not only in the fields of pharmaceutical and medicinal products but also in food and animal feed industry) and meets the ATEX ('Atmosphère Explosible'; stands for two guidelines given by European Union in the field of explosion prevention) standards. The microreactor for cryogenic operation is insulated in the black box unit, as shown in the upper image of Figure 5.12. The alkyl lithium process was performed at multiple kilogram scale, was safe and led to a yield increase.

Methodologies for scale-out to phase II–III clinical trials are also given in Figure 5.12 [44]. One approach would be to scale up the c-SSP using static mixers and mini-heat exchangers, which are already in operation at Lonza. The possible

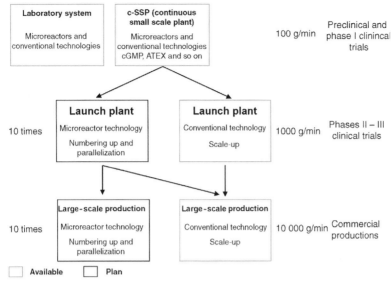

Figure 5.13 A schematic representation with typical process steps in the fine chemical and pharmaceutical industries and recommendation when to use microreactor technology or continuous processes based on a multipurpose approach (by courtesy of PharmaChem/B5Srl) [44].

flow rate of 1500 g/min is 10–15 higher than that of the c-SSP. Another approach is the replication of the laboratory system in a few number of identical base units (numbering up). A maximum of 10 base units is estimated for sufficient throughput to supply phase II–III clinical trials or quantities in the range of a few tons. The technical transfer would be more rapid than done conventionally, and scale-up issues would be avoided especially with exothermic reactions.

For commercial production, other issues arise [44]. Among these, the production cost is the key driver, as this last step in process development increasingly consumes large resources because of a number of additional tasks (scale-up, supply chain, waste management, ecology, location, etc.). The specific microreactor-related advantages such as speed in process R&D and avoidance of scale-up issues have not the same importance here as in the clinical productions. Thus, microprocess technology is here in competition to other technologies and one advantage may not be enough, as a highly complex scenario has to be considered.

5.4.5
German Project Cluster 2005

In 2004, the Federal Ministry of Education and Research (BMBF) gave microprocess engineering a research priority under its frame program 'Microsystems' [46]. In 2005, six industrial cooperation projects were launched, supervised by the funding agency VDI/VDE-IT, which are described in detail in the following. The target is the development of pilot plants for different reactions in several application fields for demonstration of the technical and economic feasibility. A further comprehensive project was started to enable small- and medium-sized enterprises to judge if their processes are technically feasible and economic by microprocess engineering. This is done by means of compilation of a compendium. This research project cluster is accompanied by a similar education and training project cluster that aims to educate researchers with the specific skills and know-how necessary in the field. The education cluster is jointly initiated and funded by the Federation of the Chemical Industry (VCI), the German Federal Foundation for the Environment (DBU) and the Ministry of Education and Research. Eight new projects have been started to develop educational courses and to equip universities and advanced technical institutes with educational devices.

5.4.6
Development for OLED Materials Production

Within the German public funded project POKOMI a microreactor-based pilot plant for the synthesis of polymeric, light-emitting semiconductors for use in displays (OLED materials) is developed [47]. COVION GmbH in Frankfurt, Germany, is a producer of these materials and the other partners are mikroglas chemtech GmbH, hte AG, IMM GmbH, Stiftung caesar and JUMO GmbH.

The microreactor plant will be equipped with the process control and online analysis [47]. The Suzuki coupling reaction is investigated as a common synthetic route for polymeric semiconductors. The transfer of the developed microflow process into chemical production is accompanied by economic and ecologic evaluations.

5.4.7
Development for Liquid/Liquid and Gas/Liquid Fine-Chemicals Production

Within the German public funded project μ.PRO.CHEM, a concept for a continuously operated modularly assembled flexible pilot plants for highly exothermic two-phase liquid–liquid or gas–liquid reactions will be developed and validated [48]. The plant features process intensifying microprocess technologies. A goal of the project is the demonstration of the technical and economic feasibility of the plant concept on pilot scale with selected model processes.

The use of microprocess technology is to overcome limitations in mass transfer through large specific interfaces and in heat transfer that so far prevented using higher conversion rates and selectivity by process intensifying measures (e.g. higher temperatures or pressures). Retrofit of existing pilot plants and classical production plants with the newly developed microstructured reactors will be done, and rules for their numbering up are developed as 'scale-up free' methodologies. Process safety and reliability are issues beyond the demonstration of process advantages.

The new modular, decentralized plant concept for on-site production will be opposed to traditional chargewise processing in multiproduct and multipurpose plants. As potential process examples, the synthesis of organic peroxides (liquid–liquid) and the oxidation of cyclohexane with air (gas–liquid) were initially selected, embracing fine- and bulk-chemical applications. Because this is an open collection, changes and additions in the chosen reactions during the project course are likely.

5.4.8
Development of Pharmaceutical Intermediates Production by Ozonolysis and Halogenation

Bayer Health Care (Bayer Schering Pharma) together with mikroglas chemtech GmbH, Mainz, and the Leibniz-Institut für Katalyse e. V. an der Universität Rostock (LIKAT), branch Berlin, develops the preparation of pharmaceutical intermediates by microprocess engineering in the German public funded project ZOHIR [49,50]. These intermediates are required for the development of pharmaceuticals against skin diseases or for hormone replacement therapy and are typically produced at 1–10 t/a scale. The main focus of the work is the development and construction of equipment consisting of connectable components for safe and continuous ozonolysis, fluorination and bromination/hydrobromination, which are standard reactions in organic chemistry. Some of these reactions are highly exothermic so that heat transfer limitations are given that can be overcome by process intensifying measures. The heat may even be released in sudden bursts leading to explosions. As a result, large-scale synthesis in stirred batch reactors is here run under suboptimal synthesis efficiency conditions with large expenditure, and some syntheses are even totally excluded from scale-up. The reactants and products are also safety hazards, being either highly reactive or thermally instable. The goal is to achieve a productivity of 10–100 kg/week.

Miniaturized near-infrared sensors were developed and implemented for online analysis and automated process control to also meet the safety requirements for handling of ozone and halogenating agents [49,50]. A target is to reduce the time from process idea to production (time-to-market) as well as development costs and costs for installation of the production unit. As pharmaceutical industry relies on the manufacture of many different products on smaller scale, and intermediates in quantities ranging from some kilograms to tons per year a modular approach toward a multipurpose microreactor plant is demanded.

A multistep gas–liquid process using a falling-film microreactor was carried out for the ozonolysis of a steroid that is a fast and highly exothermic reaction [50].

Intermediate II   Aldehyde   Hydroperoxide   Alcohol

The ozonolytic cleavage of the double bond leads to the formation of hydroperoxide and aldehyde species, which are subsequently reduced to the alcohol [50]. Semibatch

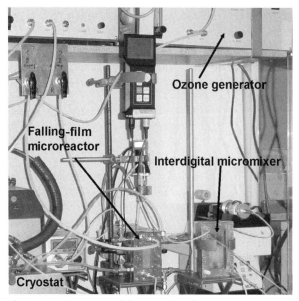

Figure 5.14 Microprocess laboratory plant for ozonolysis of steroids using a falling-film microreactor (by courtesy of K. Jähnisch/LIKAT) [50].

conditions are −60 °C, 7 h for dosing of ozone and 1 h for quenching. The processed volume is restricted to 10 l because of the danger of peroxide accumulation.

A laboratory-scale microprocess plant was built using a falling-film microreactor for the first step and an interdigital mixer–microchannel reactor for the second one (see Figure 5.14) [50].

The aldehyde and the hydroperoxide concentrations were detected by means of FT-IR in self-made small online flow cells (see Figure 5.15) [50].

As a result of microreactor processing, the reaction temperature could be increased up to −20 °C at a residence time of only 15 s [50]. The reduction was done at 10–25 °C within several minutes. The quality of the product and the throughput were comparable to that derived from the minibatch processing. A throughput of about 100–200 g/day was achieved.

As another example, a continuous two-step liquid phase process was developed involving the geminal difluorination of a 17-keto steroid using diethyl amino sulfur trifluoride (DAST) and subsequent quenching in alkaline solution to consume the excess reactants [50].

DAST

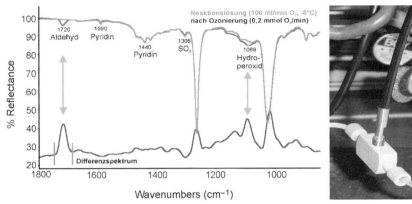

Figure 5.15 Online FT-IR monitoring of the microprocessing for the ozonolysis of steroids (left) and corresponding mini online flow cell (right) (by courtesy of K. Jähnisch/LIKAT) [50].

The corresponding semibatch process is a rather slow reaction at 90 °C with simultaneous exothermic decomposition of DAST. Thus, the processed volume is restricted to laboratory scale (<1 l) [50]. The transfer to production in a stirred tank is prohibited because of these safety reasons. A micromixer–tube reactor approach was chosen using the convective-flow-driven bas-relief caterpillar micromixer and tubes with diameters of 1–5 mm and lengths up to 20 m, respectively, and tube reactor volumes up to 500 ml (see Figure 5.16).

Again, process monitoring was made by means of FT-IR in self-made small online flow cells (see Figure 5.17) [50].

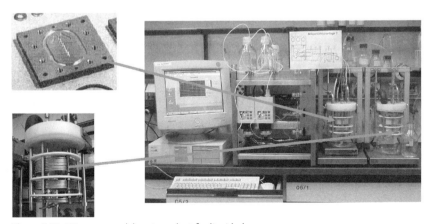

Figure 5.16 Microprocess laboratory plant for liquid phase fluorination of a 17-keto steroid with DAST (by courtesy of U. Budde/Bayer Schering Pharma) [50].

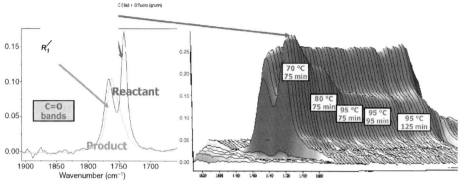

Figure 5.17 Online FT-IR monitoring of the microprocessing for the liquid-phase fluorination of a 17-keto steroid with DAST (by courtesy of U. Budde/Bayer Schering Pharma) [50].

With the use of microprocess technology, the fluorination with DAST can be performed under decomposition conditions in continuous-flow mode [50]. Temperatures of 90–100 °C and reaction times of 60–120 min are necessary for high conversions in order to compensate for the slow intrinsic reaction rate. A continuous quality control allows regulation of the process parameters (PAT, process analytical technology). A throughput of 5–10 kg/day using three parallel modules was achieved.

5.4.9
Industrial Photochemistry

The German public funded project μ-PR aims at the development, supply and testing of a microphotoreactor system in production experiments for chemical, pharmaceutical and biotechnological industry [51]. The reactor will be tested for exemplary applications of fine chemicals and photobiology.

Photoreactors are not largely used on industrial scale basically because the supply of light energy is not on the same standard or cannot be done with the same expenditure as for thermal energy. The radiation sources have high costs, emit widebanded radiation, have limited lifetime and are energy consumptive, that is, produce large heats, which also decreases selectivity by thermal side reactions. Radiation sources have the largest share on capital costs for industrial photochemical plants. Standard Hg-based radiators have a lifetime of about 1000 operating hours, with recognizable deterioration right from the start. This aging changes the emitted radiation power and also lead to changes in the spectrum. Sensitizers are often needed to transfer the correct spectral energy to the molecules for reaction. An alternative here is the spectral filters; still this suffers from inefficient energy use. The main innovation of microreactors here is to provide defined channel architectures that can be tailored so that the light energy can be homogeneously distributed and optimally used. High-performance LED arrays offer a technological and economic perspective and have lifetimes of 10 000 to 50 000 operating hours. In this way, the big advantage of photochemical reactions as compared to thermally induced reactions can be used. The former principally can introduce energy in a sparing and selective manner. This is especially valid for natural product synthesis with large molecules and numerous functional groups that easily can lead to side reactions when treated thermally.

The applications investigated focus on two key market sectors with high added value, pharmaceutical industry with its high purity demands (cGMP) and biotechnological industry using microorganisms such as algae and bacteria. After the demonstration of feasibility, the photomicroreactors will be tested on-site by the partners.

5.4.10
Development of Ionic Liquid Production

The German public funded project NEMESIS focuses on the design and development of microreactors for the synthesis of ionic liquids at pilot scale [52]. Scientific objectives are to increase the yield of the corresponding ionic liquid as well as to decrease reaction time from hours up to days currently. Ionic liquids, a new innovative class of materials, are synthesized using microreaction technology. Possible application fields are their use as electrolytes for the elaborate deposition of metals. A concept for regeneration of the electrolyte is also considered.

In particular, the expectation is that by numbering up the same product quality can be assured for the production and laboratory level; obviously, this is not the case for the current practice, at least in a number of cases. Thus, increase in purity is one goal of the investigations. Online analysis with temporal and spatial resolution is used for process control. An economic and ecologic evaluation of the results is made. Applications are the sale of the ionic liquids as laboratory chemicals and a special use of these for electrochemical deposition for surface refinement of mass products.

5.4.11
Japanese Project Cluster 2002

The known industrial implementation of microprocess technology in Japan so far is done in a consorted action [53]. The Ministry of Economy, Trade and Industry (METI) launched the first project cluster in 2002, entitled 'High Efficiency Micro-Chemical Process Technology Project'. One year later, the focus of the associated works was more shifted into early industrial implementation, and the project cluster name was changed to 'Production, Analysis and Measurement System for Micro-Chemical Process Technologies'. The management of this project was done by the Association of Micro-Chemical Process Technology (MCPT), which is an umbrella for 30 chemical and instrument companies. The projects were organized in three groups covering different facets, which are the development of microchemical plant technology, the development of microchip technology and systematization of microchemical process technology. In the following chapters, developments done under the first issue will be reported with focus on the transfer of process developments into pilot plants. It is mentioned that researchers from 7 companies and 11 research groups at Kyoto University and 4 other universities are engaged in the latter action. The schedule is to have 3-year fundamental development and then transfer to pilot plants. Another project cluster was started in July 2006, named 'Development of Microspace and Nanospace Reaction Environment Technology for Functional Materials'. This program is planned for 5 years and besides MCPT, the National Institute of Advanced Industrial Science and Technology (AIST) is involved. Furthermore, seven universities carry out research, with focal points at Kyoto University and the Tsukuba Intensive Research Center.

One major project refers to the isolation of activation and reaction spaces [53]. The idea is to create highly reactive intermediates in an 'activation space' by the use of energy and short reaction time and then to react these with another reactant in a 'reaction space' to the product. The conventional path is characterized by having activation and reaction space not separated and thus exposing all the reactants, intermediates and products to the harsh reaction conditions. The claim is that the new path is more selective and produces much less by-products. In addition, it allows to transfer multi-step synthesis to a one-step reaction, but now with several steps of activation that also saves purification steps. For this goal, a new type of hardware has to be developed that allows precisely to control the reaction conditions so that these activation and reaction steps can be carried really one after the other and do not partly merge. In particular, the setting of residence time is challenging.

The energy provision will not only be done conventionally, for example, by electrical heating. Alternative energy sources such as microwaves and light are tested as well [53]. The goal is to accelerate reaction rates by choosing much harsher reaction conditions, for example, in terms of temperature and pressure, than normal existing conditions in chemistry. The small spaces in microreactors help to conduct such extreme processing in a safe manner.

Because problems with fouling and erosion were encountered in previous projects, a fault detection and diagnosis system is developed as well [53]. This should detect blockage in an early stage. The aim is to also have new designs of microsystems with fewer blockages. Finally, a surface finishing method is one research target for protection of the microsystems from erosion. This is detected by an erosion monitoring method.

5.4.12
Pilot Plant for MMA Manufacture

Idemitsu Kosan in Chiba, Japan, operates a pilot plant for the free radical polymerization of methylmethacrylate (MMA) [54], following prior process development at Kyoto University [55].

The plant has a capacity of 10 t/a (see Figure 5.18) [54]. Eight microreactor blocks form the reactor core and each comprises three tube reactors with micron inner dimensions (500 μm internal diameter and 2 m length) in series. It could be shown that the numbering-up principle is valid; that is, relevant process figures such as the polydispersity index, yield and average number-based molecular weight were similar for single and parallel tube operation.

5.4.13
Grignard Exchange Reaction

Researchers at Kyoto University performed the Grignard exchange reaction of ethylmagnesium bromide and bromopentafluorobenzene to give pentafluorophenylmagnesium bromide [56]. Scale-out was made using small- and medium-scale microflow systems consisting of a micromixer and a microheat exchanger. The shell and tube microheat exchanger used for the medium-scale microflow systems could be operated at a high flow rate of 6 l/h. The heat exchanger consists of 55 microtubes (i.d. 490 μm × 200 mm) embedded in a shell (i.d. 16.7 mm × 200 mm). Water as coolant is circulated through the shell.

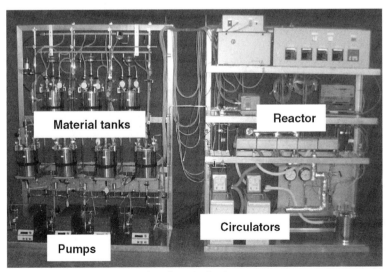

Figure 5.18 Microprocess pilot plant for radical polymerization reaction (by courtesy of ACS) [54].

In a detailed process optimization study, the impact of the type of micromixers and process parameters was determined [56]. As a result, a pilot with a Toray Hi-mixer connected to a shell and tube microheat exchanger was constructed. Continuous operation for 24 h was carried out to obtain pentafluorobenzene (PFB) after protonation (92% yield). In this time, 14.7 kg of the product was produced, that is, about 5 t/a. Thus, the industrial-scale production carried out using a batch reactor ($10\,\text{m}^3$) can be replaced by adding only four microflow systems of the scale investigated. The pilot plant produces 0.5 kg in 6 h continuous operation, thus about 730 kg/a (see Figure 5.19). The name of the industrial company was not disclosed.

5.4.14
Halogen–Lithium Exchange Pilot Plant

The University of Kyoto, Japan, reports about a halogen–lithium exchange reaction of aryl bromides with butyl lithium (see Figure 5.20) [53,57]. The intermediate was trapped with an electrophile. The whole process was done under noncryogenic conditions at $0\,^\circ$C.

The pilot plant produces 0.5 kg in 6 h continuous operation, thus about 730 kg/a [53,57]. The name of the industrial company was not disclosed (see Figure 5.21).

Figure 5.19 Microprocess pilot plant for Grignard exchange reaction (by courtesy of ACS) [56].

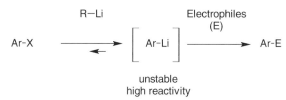

Figure 5.20 Halogen–lithium exchange reactions (by courtesy of Wiley-VCH Verlag GmbH) [57].

Figure 5.21 Microprocess pilot plant for halogen–lithium exchange process (by courtesy of Wiley-VCH Verlag GmbH) [57].

5.4.15
Swern–Moffat Oxidation Pilot Plant

Ube Industries Ltd in Yamaguchi, Japan, and Kyoto University investigated the Swern oxidation for pharmaceutical intermediates [57,58]. In this reaction, alcohols are oxidized to carbonyl compounds using dimethyl sulfoxide. The reaction variant using dimethyl sulfoxide activated by trifluoroacetic anhydride (shown below) has found industrial application, but is limited to low-temperature operation ($-50\,°C$ or below) to avoid decomposition of an intermediate.

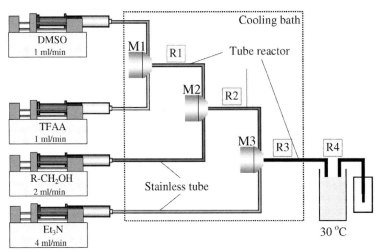

In a microscale tubular reactor, Swern oxidations were performed between -20 and $20\,°C$. Mixing was performed stagewise with a series of rapid mixing functions (see Figure 5.22) [57,58]. First, dimethyl sulfoxide and trifluoroacetic anhydride were contacted in an interdigital micromixer followed by a stainless steel tube reactor R1. After addition of the alcohol and reaction in reactor R2, the mixture was then contacted with a triethylamine solution and passed through two more reactors (R3 and R4) to complete the reaction.

This microreactor system thus allows fast mixing, good temperature control and changing process parameters at short residence times – the overall time to pass all

Figure 5.22 Schematic diagram of the microscale flow system for the Swern oxidation (by courtesy of Wiley-VCH Verlag GmbH) [58].

Figure 5.23 Microprocess pilot plant for Swern–Moffat reaction (by courtesy of J.-I. Yoshida/University of Kyoto).

reactors is between 8 and 11 s [57,58]. The timescale between end and start of a new operation is 0.01 s in order to avoid significant decomposition. Thereby, conversions and yields were determined for primary, secondary, cyclic and benzylic alcohols in the temperature range of -20 and $20\,°C$, which were equal or better than lower temperature batch reactions. The oxidation of cyclohexanol was run for 3 h at $20\,°C$, and a stable process in terms of conversion and selectivity was observed.

Then, a pilot plant with a capacity of 10 t/a was build for the Swern–Moffat oxidation [57,58]. The yield of the batch process is 83% at $-70\,°C$, whereas the microchemical process achieved a yield of 88% at $20\,°C$. The pilot plant was operated under stable reaction conditions for a long run with similar product yields as in the laboratory experiment (see Figure 5.23).

5.4.16
Yellow Nano Pigment Plant

The University of Kyoto, Japan, and Fuji in Tokyo, Japan, developed a pilot plant for the production of a yellow nano pigment with a capacity of 70 t/a, which was also developed and operated by Kyoto University and Fuji Company (see Figure 5.24) [59]. The particle size spectra show a clear impact of the flow conditions with smaller particles at higher flow rates.

5.4.17
Polycondensation

The MCPT research group investigated polycondensation reactions for the synthesis of highly thermal resistant polymers such as polyamide and polyimide [57]. These reactions are exothermic, which negatively impacts average molecular weights and the corresponding distributions when using conventional technology. A two-step reaction with polycondensation and terminal modification with NA was conducted in a microreactor as shown in the following scheme.

NA =

NMP: *N*-Methyl pyrrolidone
GBL: γ-Butyrolactone

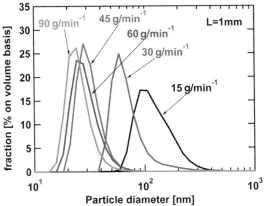

Figure 5.24 Top: yellow nano pigment pilot microprocess plant. Bottom: particle size spectra at various volume flows (by courtesy H. Maeta, Fuji) [59].

The reaction was faster in the microreactor than in the batch system, which is assigned to be a mixing effect [57]. The molecular weight distribution obtained in the microreactor is slightly narrower than that of the batch reactor, which is attributed to the good thermal management of the microreactor.

5.4.18
Friedel–Crafts Alkylation

The MCPT research group investigated Friedel–Crafts alkylations, a widely used reaction path in organic chemistry [57]. As an electron-donating substituent is introduced in the first reaction, the monoalkylated product is more reactive than the starting material and the second alkylation takes place more readily. This leads to by-products by dialkylation or even polyalkylation. This can be avoided by using a large excess of the starting material. However, a smart solution would be the achievement of a selective monoalkylation using only one equivalent of an aromatic compound. The key to this may be micromixing, as respective impact on selectivity has been found in earlier investigations.

monoalkylation   dialkylation   polyalkylation

As industrial relevant Friedel–Crafts reaction, the synthesis of Bisphenol-F, a material for epoxy resin, from phenol and formaldehyde was chosen [57]. This reaction involves formation of higher order condensates such as tris-phenols. To minimize the latter, the molar ratio of phenol to formaldehyde is set to a very high value (30–40), which is more than 15 times larger than the amount theoretically necessary. Three types of micromixers were used. These are a T-shaped mixer with 500 µm inner diameter, a multilaminating interdigital micro-mixer with 40 µm channels and a so-called self-made K-M micromixer with center collision mixing.

Bis-phenols   Tris-phenols

The selectivity increased with increasing phenol/hydroxybenzylalcohol ratio (see Figure 5.25) [57]. Best performance was given by the K-M mixer. The phenol/hydroxybenzylalcohol ratio was decreased to half in this way.

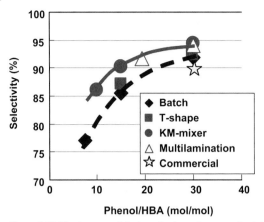

Figure 5.25 The impact of choice of micromixer on selectivity when varying the phenol/hydroxybenzylalcohol molar ratio (by courtesy of Wiley-VCH Verlag GmbH) [57].

5.4.19
H₂O₂ Based Oxidation to 2-Methyl-1,4-Naphthoquinone

The MCPT research team investigated the synthesis of 2-methyl-1,4-naphthoquinone (vitamin K3) by the oxidation of 2-methylnaphthalene [57].

oxidation
in AcOH

2MN   VK3

(2-Methylnaphthalene)   (2-Methyl-1,4-naphthoquinone)

This is done in the commercial production with chromium trioxide, and the aim was to test for the possibility of substitution by using aqueous hydrogen peroxide (60%) as oxidant [57]. Hydrogen peroxide (H_2O_2) is a mild and environmentally friendly oxidant. For better performance, the oxidation could be made faster by using higher temperatures and higher concentrations of the oxidant. This, however, is prohibited by side reactions such as overoxidation, side-chain oxidation and decomposition of the oxidant. Therefore, dropwise addition of hydrogen peroxide is common to achieve high selectivities, usually at low temperatures. This results in long reaction times.

With the use of a microreactor and a feed comprising also palladium acetate and sulfuric acid, the reaction time was shortened to 10 min at 70 °C as compared to the

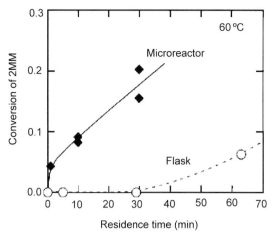

Figure 5.26 Speeding up the oxidation of 2-methylnaphthalene (2MN) in a microreactor by virtue of faster hydrogen peroxide addition and higher temperatures (by courtesy of Wiley-VCH Verlag GmbH) [57].

industrial semibatch process (see Figure 5.26) [57]. However, the selectivity was lower. This could be solved by process modification using peracetic acid, generated *in situ* from acetic acid and hydrogen peroxide.

5.4.20
Direct Fluorination of Ethyl 3-Oxobutanoate

Following extensive laboratory studies at the University of Durham, UK, on the direct fluorination of ethyl 3-oxobutanoate in formic acid, scale-out was made from a three- to a nine-channel microstructured reactor, in cooperation with Asahi Glass Co. in Yokohama, Japan [60].

The process was performed for many months yielding 700 g of monofluorinated product with a nine-channel microstructured reactor [60]. A continuous 150 h operation was performed without decline of yield or conversion. Even in the scale-out to a 30-channel reactor (see Figure 5.27), no loss in performance was noticed. A single feed system distributed the reactants and reagents to the various microchannels.

If the results of these first pilot studies are extrapolated to large-scale synthesis, a two-sided 30-channel device with 60 channels would be capable of synthesizing about

Figure 5.27 A 30-channel microstructured reactor for the direct fluorination of ethyl 3-oxobutanoate (by courtesy of the Royal Society of Chemistry) [60].

300 g product per day [60]. By external numbering to 10 reactors about 3 kg of product per day would result. Such pilot plant benefits from low expenditure and having exactly the same operating conditions as given for the laboratory processing. In addition, convenient maintenance and purification by distillation at improved safety features are predicted as major advantages.

5.4.21
Propene Oxide Formation

The propene oxidation process is globally conducted on a 5 million t/a scale. A new gas-phase oxidation process of propene to propene oxide using hydrogen peroxide would save the use of solvents and is expected to have higher selectivity.

$$\text{\Large $\diagup\!\!\diagdown\!\!\diagdown$} \xrightarrow[\text{$-H_2O$}]{\text{$H_2O_2$ (vap)}} \text{\Large $\diagup\!\!\diagdown\!\!\bigtriangledown$}_O$$

Microstructured reactors have the potential to perform a key step here to thermally treat concentrated hydrogen peroxide solutions for vaporization that would be otherwise dangerous when using conventional technology because of the handling of the combustible hydrogen peroxide. Thus, large microreactor units have to be built that are capable of safe operation of the hydrogen peroxide route. In addition, these units need to transfer the large reaction heats produced in the exothermic propylene oxide reaction.

Following these ideas, Degussa in Hanau, Germany, developed and built, together with the large-scale plant manufacturer Uhde in Dortmund, Germany, and other university and institute partners in a publicly funded project (DEMIS), a 6-m long, two-story high pilot-scale microstructured reactor by numbering up of meter-long and meter-wide plates brought to micron distance (see Figure 5.28) [61]. The lower part of the reactor comprises the vaporizer and in the longer upper part the reaction plates are positioned. The construction uses a simplified, robust approach with no real walls for laterally defining microchannels, but the few fins only have mechanical function, and the flow dimension in lateral direction is virtually infinite compared to that of typical microchannels. This simplification in reactor design has to be paid by the development of a catalyst coating technique for such extended dimensions. Typically, the small dimensions in microchannels help in having good mechanical adhesion to the substrate and also promote uniform coating because of surface

Figure 5.28 Left: DEMIS pilot reactor for propylene oxide formation at Degussa site, superposed by schematic on the reactor construction. Right: assembly of the reactor internals (by courtesy of Wiley-VCH Verlag GmbH) [61].

forces. Initial results are very stable and reproducible, and Degussa regards this as encouraging.

5.4.22
Diverse Industrial Pilot-Oriented Involvements

Bayer Company operates several microprocess plants. Clariant GmbH has formed a competence center on microreaction technology in early 2004 to offer the technology in its custom synthesis business. Degussa AG started a new Project House on Process Intensification encompassing the microreaction technology. Lonza has also used microreaction technology over years for pharmaceutical customers. The pharmaceutical company Brystol-Meyers-Squibb, Lucent Technologies and the Stevens Institute have a public funded project on hydrogenation reactions [62]. They started with a model reaction, the hydrogenation of *o*-nitroanisole to *o*-anisidine, and will move on to proprietary reactions.

$$H_3C{-}O{-}C_6H_4{-}NO_2 \xrightarrow{\ H_2,\ Pd/C\ } H_3C{-}O{-}C_6H_4{-}NH_2$$

The reaction engineering uses the innovative approach of fixing single Pd/C catalyst particles in thousands of small microstructured traps [62]. This is a mixed approach relying on traditional catalysts and using the innovative flow features in microchannels. In this way, higher specific surfaces between gas, liquid and catalyst are ensured than possible by using fixed beds. In comparison to wall coatings, it is expected that the trapped catalyst provide a more reliable and potentially higher catalyst coating.

Brystol-Meyers-Squibb has filed patents for making glycosides using noncryogenic processes that involve the lithiation of aromatic compounds [62]. These aryllithium derivatives react with carbonyl groups of other reactants to form a glycoside. Both steps are highly exothermic and rapid, which are tailored to reveal the advantages of microprocess technology in terms of selectivity and reduced cost of manufacture.

The Syrris Company, a system manufacturer, and GlaxoSmithKline developed an automated flow system AFRICA (Automated Flow Reaction Incubation and Control Apparatus) where a sequence of reactions can be run under several process conditions using the advantages of miniature-scale flow [62]. Attached to the system is an online high-pressure liquid chromatography for immediate product separation to analyze the performance of process optimization. As further scientific support, Syrris has entered into a cooperation with S.V. Ley at Cambridge University who studies flow-through liquid reactions with immobilized reagents. Pfizer has some high-throughput oriented research and development work with a microreactor combined to an automated multiple reactant feed upstream and UV detection downstream.

Dow Chemical in Midland, USA, the microprocess technologist Velocys in Plain City, USA, and PNNL in Richland, USA, as research institute in microreactor technology have a public funded project on high-intensity production of ethylene and other olefins by oxidation such as the formation of ethylene from ethane [1]. A two-step reactor engineering is performed, starting with a bench-scale reactor with microchannel dimensions equal to the latter commercial unit and followed by numbering to the latter. An economic analysis with focus on reactor costs and energy consumption completes the project.

5.4.23
Production of Polymer Intermediates

DSM Fine Chemicals GmbH in Linz, Austria, inserted a microstructured reactor into an existing production plant for the Ritter reaction in a retrofit manner (see Figure 5.29). A high-value intermediate, not disclosed by chemical formula, for the polymer industry is produced [63]. This approach, under involvement of the product team right from the start, stands for a plant philosophy which may best paraphrased by 'minimal invasive plant surgery', different from some holistic approaches on a total change in plant design. This enabled the replacement of a central reaction route in a very large reactor tank encasing several tons of explosive and corrosive chemicals. Design and fabrication of this reactor was done at the Institut für Mikroverfahrenstechnik (IMVT) in the Forschungszentrum Karlsruhe (FZK), Germany.

During a 10-week production campaign, over 300 t of the polymer product was produced [63]. The microstructured reactor (65 cm long, 290 kg heavy, special Nickel alloy, several ten thousands of microchannels) was operated at a throughput of 1700 kg liquid chemicals per hour. A crucial issue was the removal of the reaction heat

Figure 5.29 Production-type microstructured reactor for throughput at 1700 kg/h and transfer of a power of 100 kW. This apparatus was used for the manufacture of a high-value product for plastics industry at DSM in Linz, Austria (by courtesy of Wiley-VCH Verlag GmbH; from [64]).

that was accomplished within seconds. The yield exceeds that of the former route, albeit it was not detailed. It was also found that the process safety for handling the corrosive chemicals was higher for the microreactor process. The use of raw materials and the waste streams were reduced, improving the cost and efficiency of the process.

5.4.24
Synthesis of Diazo Pigments

Clariant carried out diazo-coupling for pigment synthesis. Laboratory developments were transferred to pilot scale, reported in the literature, and a further transfer to production scale was announced at conferences [65].

$$\text{Ar-N}^{+}_{\overset{\displaystyle |}{N}}\ \ \text{Y}^{-}\ \ +\ \ \text{RH}\ \ \longrightarrow\ \ \text{Ar-N}_{\overset{\displaystyle \diagdown}{N-R}}\ \ +\ \ \text{HY}$$

Particle synthesis is known to be highly sensitive to mixing, as this controls seed formation and crystal growth. With the Clariant and later works, it became evident that by the use of micromixing technology particles with more uniform size, defined morphology, or chemical selectivity can be prepared, that is with considerably improved product qualities, which is the main driver for such investigations. The motivation to use microreactors also stems from the benefits of developing a continuous process, for example, the production of flexible quantities and eliminating the need of refining, such as milling the pigment in the end of the production line, as typically given for batch processes. A further argument for microreaction technology in the case of diazonium salt-based synthesis comes from the hazardous potential of that intermediate. A major hurdle is to find processing solutions that allow to handle solids generated in microchannels.

Two commercial azo pigments, one yellow and one red colored, were made on the basis of azo coupling; the nature of the substituents was not disclosed [65]. Diazotation was performed in batchwise manner. A CPC microreactor with multi-lamination mixer (lamellae $<100\,\mu m$) was used. The two azo pigments had a color strength of 119 and 139%, a five and six times glossier brightness, and a five and six steps higher transparency, respectively, than the same products made by batch processing (see Table 5.3) [65]. This originated from the formation of smaller particles with more narrow size distribution (microreactor: $D_{50} = 250\,nm$, $s = 1.5$; batch: $D_{50} = 600\,nm$, $s = 2.0$).

The same features were found for pilot-size microreactor operation (see Figure 5.30). Brightness and transparency were the same, and the color strength could even be increased to 149% [65]. The mean particle size was even smaller than the lab-scale microreactor processing (microreactor: $D_{50} = 90\,nm$, $s = 1.5$; batch: $D_{50} = 600\,nm$, $s = 2.0$), probably because of process optimization.

Pilot-size microreactor operation was done using a flow of 500 ml/h, which is equivalent to a production in the range of 10 t/a that was estimated when accounting

Table 5.3 Coloristic properties of pigments synthesized in two different microreactors compared to the batch standard.

	Microreactor pigment 1	Microreactor pigment 2
Color strength	119%	139%
Brightness	Five steps glossier	Six steps glossier
Transparency	Five steps more transparent	Six steps more transparent

From [65].

for 8000 h annual running time [65]. The increase in throughput compared to laboratory-scale microreactors used prior (1 t/a; 20–80 ml/h) was achieved by both internal and external numbering up accompanied by a slight scale-up of internal dimensions. More reaction plates were assembled in parallel within one device; in addition, three such devices were connected in parallel. Furthermore, slightly larger microchannels were used, still ensuring laminar flow.

To test for fouling, a 24 h run of a pilot-scale microreactor for azo pigment production was performed using a diazo suspension [65]. At the end of this period, the pressure loss of the microreactor increased exponentially. Special means were developed to prevent clogging and instable operation. By partial removal of the deposits, the pressure loss was brought back to normal.

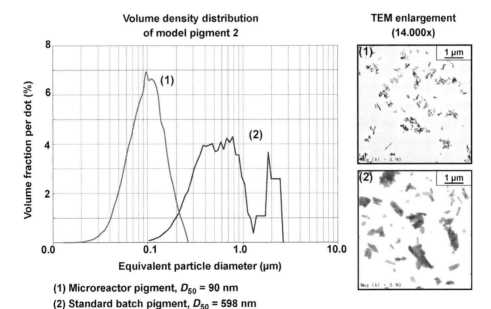

(1) Microreactor pigment, D_{50} = 90 nm
(2) Standard batch pigment, D_{50} = 598 nm

Figure 5.30 Size characterization of an azo pigment, yielded in the pilot plant microreactor, compared to the batch standard. Left: volume density distribution; right: TEM magnification (by courtesy of Wiley-VCH Verlag GmbH; from [66]).

5.4.25
Nitroglycerine Production

The Xi'an Huian Industrial Group in Xi'an, China, operates a nitroglycerin plant at 15 kg/h production that has been developed and installed by IMM (see Figure 5.31) [62].

The nitroglycerine is of pharmaceutical grade and used as medicine for acute cardiac infarction. This demands high selectivity and low levels of impurity. First manual plant start-up tests demonstrated that this can be achieved by performing the reaction in a microreactor. Safe operation was found during the first runs. In the second step, full automation of the plant is planned. The extension of the process chain is another future issue, adding a purification unit for washing and drying and finally a unit for formulation and packaging to encapsulate the nitroglycerine drug safely in tablets. Advanced wastewater treatment and a closed water cycle should lead to an environmentally clean process.

Figure 5.31 Microprocess production plant for pharmaceutical nitroglycerine (by courtesy of H. Löwe/IMM).

5.4.26
Fine-Chemical Production Process

Microinnova KEG, Graz, Austria, made process development and the installation of a StarLam 3000 microstructured mixer in an existing production plant of an undisclosed customer and for an undisclosed chemical process (see Figure 5.32) [67]. The aim was to double the capacity of a running two-step batch process. This was achieved by installing the microreactor for the first reaction step. A higher reaction rate made it possible to reach overall throughputs of 3.6 t/h. Additionally energy savings were achieved. The microstructured mixer is running in production for more than a year now.

The application refers to the production of fine chemicals [67]. A 10 m^3 batch vessel was used to perform the two-step chemical process. The first strongly exothermic reaction step needed cooling because of the volatility of one of the starting materials. The reaction was finished when all of the volatile reactant had reacted. The second step was endothermic and the batch vessel had to be heated for some hours to complete the reaction. It took several hours to perform these two steps and a throughput of about 1800 kg/h was achieved.

In laboratory-scale investigations, the reactor and attached tube reactor were kept at temperatures of about 150 °C [67]. The experiments showed that in the microprocess lab plant the first reaction step can be finished in less then 60 s, whereas it takes about 4 h to perform it in the cooled batch vessel of the production plant.

Figure 5.32 StarLam 3000 microstructured mixer retrofitted to existing plant peripherals and tank reactor (by courtesy of Wiley-VCH Verlag GmbH) [67].

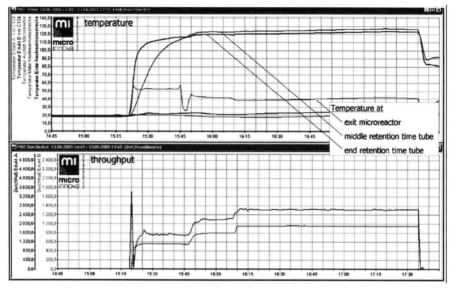

Figure 5.33 Temperature diagram of the microreactor assisted processing (by courtesy of Wiley-VCH Verlag GmbH) [67].

The inlet pipes of the two starting reactants to the batch vessel were simply connected to the StarLam mixer [67]. The only difference to the previous feed lines was the installation of filter cartridges before the entries to the microstructured mixer, necessary to avoid blocking of the reactor. The pressure drop in the lines was lower than 3 bar so that it was possible to keep the pumps used before in the plant. At the outlet of the reactor, a tube reactor was installed. During optimization it was found that it is sufficient to insulate this tube to reach the temperature needed to finish the reaction. The pipe ended directly in the batch vessel where the second endothermic reaction step was carried out as before.

The first start-up of the plant was in June 2005 [67]. A temperature diagram shows that most of the heat is released in the retention time tube (see Figure 5.33). The temperature measured directly at the outlet of the microstructured mixer–reactor StarLam 3000 was below 50 °C even at higher throughputs. During the retention time, tube temperatures up to about 130 °C were reached. The throughputs for this first test run were increased in three steps to up to 3600 kg/h.

5.4.27
Grignard-Based Enolate Formation

Merck KGaA in Darmstadt, Germany, investigated a reaction of a Grignard reagent with a high reaction enthalpy of 300 kJ/mol and high reaction speed so that heat transfer limitations result for conventional technology [68]. The Grignard reagent having a long alkyl chain was added to a keto compound, the substituents remain disclosed. Thereby, an enolate is formed that is further reacted in the frame of a multistage fine-chemical industrial process.

$$X = Br, Cl$$

A yield of 95% was obtained by a micromixer-based process (<10 s, at $-10\,°C$), whereas the industrial batch process (6 m^3 stirred vessel) had only 72% yield (5 h, at $-20\,°C$) [68]. The lab-scale batch process (0.5 l flask; 0.5 h, at $-40\,°C$) had 88% yield (see Table 5.4). Pilot-scale studies followed with a homebuilt minimixer for reasons of clogging, which was not decisive at the lab scale. With one minimixer at the pilot scale, a yield of 92% was obtained (<10 s, at $-10\,°C$). The validity of the numbering-up concept was proven by operating also five minimixers of the same type at a yield of 92% (<10 s, at $-10\,°C$). This was the central part of the actual production process, running for more than 3 years until the life cycle of the commercial product of the corresponding multistage process run out (see Figure 5.34).

Meanwhile, Merck has reported to have 20 microreactor plants under operation for diverse reactions [65]. The production costs are typically reduced by 20% as compared to prior conventional technology. The throughputs range from 50 g/h to 4 kg/h, which corresponds to 146 kg/a and 11.7 t/a, respectively.

5.5
Challenges and Concerns

Concerns about an industrial use of microprocess technology are still existing [62]. Process chemists need to be familiarized with the new tool. Often it seems that these soft factors are even more relevant than the hard factors. Nonetheless, the performance of microprocess technology must show up a clear driver in the interplay of operating and capital costs of existing equipments and respective costs on the microflow processing side.

Processes that work best with microreactors are fast and generate high-value materials. This restricts the use currently to such niches [62]. However, it is also more

Table 5.4 Comparison of reaction time and yield for different reactor types.

Reactor type	Temperature (°C)	Residence time	Yield (%)
Flask 0.5 l	-40	0.5 h	88
Production (stirred vessel, 6 m^3)	-20	5 h	72
Microreactor (lab setup)	-10	<10 s	95
Minireactor (pilot scale)	-10	<10 s	92
Five minireactors (production)	-10	<10 s	92

From [68].

Figure 5.34 Merck production plant for Grignard-based enolate formation (by courtesy of Springer) [68].

and more recognized that speed-up of reactions can be achieved by changes in the chemical processing and synthesis in view of what is tailored for microreactors. This reorientation has now been started with an initiative on new process windows (DBU project cluster in Germany, DBU German Environmental Agency), but will take many years though.

Microprocess technology is still regarded as a rather radical change in chemical engineering. It will need probably another decade until it has become more routine business. On this way, intermediate solutions such as the use of mesoscale equipment, still of advanced nature, will complement the choice for smart continuous manufacturing. Also, other modern technologies such as microwave organic synthesis, ionic liquids and supercritical processing will probably be used jointly with microreactors in selected cases. In this way, some current limitations of microprocess technology may be overcome (e.g. concerning solubility, upper operating temperature and heat supply).

Microprocess technology is strongly knowledge driven. Education and training will have a major role [46].

References

1 Freemantle, M. (2004) *Chemical & Engineering News*, **82**, 39.
2 Hogan, J. (2006) *Nature*, **442**, 351.
3 Reisch, M. (2004) *Chemical & Engineering News*, **82**, 9.
4 Rouhi, A.M. (2004) *Chemical & Engineering News*, **82**, 18.

5 Watts, P., Wiles, C., Haswell, S.J. and Pombo-Villar, E. (2002) *Lab on a Chip*, **2**, 141.

6 Watts, P., Wiles, C., Haswell, S.J. and Pombo-Villar, E. (2002) *Tetrahedron*, **58**, 5427.

7 Merrifield, R.B. (1963) *Journal of the American Chemical Society*, **85**, 2149.

8 Erickson, B.W. and Merrifield, R.B. (1976) *The Proteins*, 3rd edn, vol. 2 (eds H. Neurath and R.L. Hill), Academic Press, New York, p. 256.

9 Watts, P., Wiles, C., Haswell, S.J., Pombo-Villar, E. and Styring, P. (2001) Microreaction Technology – IMRET 5: Proceedings of the 5th International Conference on Microreaction Technology (eds M. Matlosz, W. Ehrfeld and J.P. Baselt), Springer-Verlag, Berlin, p. 508.

10 Garcia-Egido, E., Wong, S.Y.F. and Warrington, B.H. (2002) *Lab on a Chip*, **2**, 31.

11 Fletcher, P.D.I., Haswell, S.J., Pombo-Villar, E., Warrington, B.H., Watts, P., Wong, S.Y.F. and Zhang, X. (2002) *Tetrahedron*, **58**, 4735.

12 Garcia-Egido, E. and Wong, S.Y.F. (2001) *Micro Total Analysis Systems* (eds J.M. Ramsey and A. van den Berg), Kluwer Academic Publishers, Dordrecht, p. 517.

13 Garcia-Egido, E., Spikmans, V., Wong, S.Y.F. and Warrington, B.H. (2003) *Lab on a Chip*, **3**, 67.

14 Sands, M., Haswell, S.J., Kelly, S.M., Skelton, V., Morgan, D., Styring, P. and Warrington, B. (2001) *Lab on a Chip*, **1**, 64.

15 Wiles, C., Watts, P., Haswell, S.J. and Pombo-Villar, E. (2001) *Lab on a Chip*, **1**, 100.

16 Skelton, V., Grenway, G.M., Haswell, S.J., Styring, P., Morgan, D.O., Warrington, B.H. and Wong, S.Y.F. (2001) *Analyst*, **126**, 11.

17 Skelton, V., Greenway, G.M., Haswell, S.J., Styring, P., Morgan, D.O., Warrington, B.H. and Wong, S. (2000)4th International Conference on Microreaction Technology, IMRET 4, AIChE Topical Conference Proceedings, Atlanta, USA, p. 78.

18 Haswell, S.J. (2001) *Micro Total Analysis System* (eds A. van den Berg and J.M. Ramsay), Kluwer Academic Publishers, Dordrecht, p. 637.

19 Nielsen, C.A., Chrisman, R.W., LaPointe, E. and Miller, T.E. (2002) *Analytical Chemistry*, **74**, 3112.

20 Wiles, C., Watts, P., Haswell, S.J. and Pombo-Villar, E. (2004) *Lab on a Chip*, **4**, 171.

21 Lee, C.-C., Sui, G., Elizarov, A., Shu, C.J., Shin, Y.-S., Dooley, A.N., Huang, J., Daridon, A., Wyatt, P., Stout, D., Kolb, H.C., Witte, O.N., Satyamurthy, N., Heath, J.R., Phelps, M.E., Quake, S.R. and Tseng, H.-R. (2005) *Science*, **310**, 1793.

22 Dummann, G., Quitmann, U., Gröschel, L., Agar, D.W., Wörz, O. and Morgenschweis, K. (2002) Catalysis Today, Special Edition – 4th International Symposium on Catalysis in Multiphase Reactors, CAMURE IV, vols 78–79, 433.

23 Hessel, V., Hofmann, C., Löwe, H., Meudt, A., Scherer, S., Schönfeld, F. and Werner, B. (2004) *Organic Process Research & Development*, **8**, 511.

24 Pennemann, H., Hessel, V., Löwe, H., Forster, S. and Kinkel, J. (2005) *Organic Process Research & Development*, **9**, 188.

25 Schwesinger, N., Marufke, O., Qiao, F., Devant, R. and Wurziger, H. (1998) Process Miniaturization: 2nd International Conference on Microreaction Technology, IMRET 2; Topical Conference Preprints (eds W. Ehrfeld, I.H. Rinard and R.S. Wegeng), AIChE, New Orleans, USA, p. 124.

26 Ehrfeld, W., Hessel, V. and Löwe, H. (2000) *Microreactors*, Wiley-VCH Verlag GmbH, Weinheim.

27 Wörz, O., Jäckel, K.-P., Richter, T. and Wolf, A. (2000) *Chemie Ingenieur Technik*, **72**, 460.

28 Wörz, O., Jäckel, K.-P., Richter, T. and Wolf, A. (2001) *Chemical Engineering & Technology*, **24**, 138.

29 Kim, J., Park, J.-K., and Kwak, B.-S. (2005) 4th Asia-Pacific Chemical Reaction

Engineering Symposium, APCRE05, Gyeongju, Korea, p. 441.

30 Choe, J., Song, K.-H. and Kwon, Y. (2005) 4th Asia-Pacific Chemical Reaction Engineering Symposium, APCRE05, Gyeongju, Korea, p. 435.

31 Fernandez-Suarez, M., Wong, S.Y.F. and Warrington, B.H. (2002) *Lab on a Chip*, **2**, 170.

32 Schwalbe, T., Autze, V. and Wille, G. (2002) *Chimia*, **56**, 636.

33 Taghavi-Moghadam, S., Kleemann, A. and Overbeck, S. (2000) VDE World Microtechnologies Congress, MICRO.tec 2000, vol. 2, VDE Verlag, Berlin, EXPO Hannover, p. 489.

34 Zhang, X., Stefanick, S. and Villani, F.J. (2004) *Organic Process Research & Development*, **8**, 455.

35 Roberge, D.M. (2007) AIChE Spring National Meeting, Houston, TX, published on CD.

36 Pennemann, H., Hessel, V. and Löwe, H. (2004) *Chemical Engineering Science*, **59**, 4789.

37 Vanden Bussche, K.M. (2005) Symposium on Modeling of Complex Processes, Austin, TX.

38 Veser, G. (2001) *Chemical Engineering Science*, **56**, 1265.

39 Brownstein, A. (1999) *European Chemical News*, November, 36.

40 Huckins, H.A. (1995) US Patent 5,641,467, Princeton Advanced Technology, Inc., Hilton Head, USA.

41 Lawal, A., Voloshin, Y. and Dada, D. (2007) AIChE Spring National Meeting, Houston, TX, published on CD.

42 Bayer, T., Pysall, D. and Wachsen, O. (2000) Microreaction Technology: 3rd International Conference on Microreaction Technology, Proceedings of IMRET 3 (ed. W. Ehrfeld), Springer-Verlag, Berlin, p. 165.

43 Pysall, D., Wachsen, O., Bayer, T. and Wulf, S. (1998) DE 19816886, Aventis Research & Technologies GmbH & Co. KG, Germany.

44 Roberge, D.M., Bieler, N. and Thalmann, M. (2006) *PharmaChem*, **28**, 14.

45 Roberge, D.M., Ducry, L., Bieler, N., Cretton, P. and Zimmermann, B. (2005) *Chemical Engineering and Technology*, **28**, 318.

46 Ackermann, U. (2007) *MST News 1*, 7.

47 www.mstonline.de/foerderung/ projektliste/detail_html? vb_nr=V3MVT016 (2005).

48 www.mstonline.de/foerderung/ projektliste/printable_pdf? vb_nr=V3MVT021 (2007).

49 http://www.mstonline.de/foerderung/ projektliste/printable_pdf? vb_nr=V3MVT001(2007).

50 Budde, U. (2007) *Statuskolloquium Mikroverfahrenstechnik, Deutsche Bundesstiftung Umwelt*, Osnabrück, Germany.

51 www.mstonline.de/foerderung/ projektliste/printable_pdf? vb_nr=V3MVT009 (2006).

52 http://www.mstonline.de/foerderung/ projektliste/detail_html? vb_nr=V3MVT004 (2006).

53 Hasebe, S. (2007) *MST News 1*, 10.

54 Iwasaki, T., Kawano, N. and Yoshida, J.-I. (2006) *Organic Process Research & Development*, **10**, 1126.

55 Iwasaki, T. and Yoshida, J.-I. (2005) *Macromolecules*, **38**, 1159.

56 Wakami, H. and Yoshida, J.-I. (2005) *Organic Process Research & Development*, **9**, 787.

57 Yoshida, J.-I. and Okamoto, H. (2006) Micro Process Engineering – Fundamentals, Devices, Fabrication, and Applications (eds O. von Brand, G.K. Fedder, C. Hierold, J.G. Korvink and O. Tabata), Book of the Series Advanced Micro and Nanosystems, vol. 5 (ed. N. Kockmann).

58 Kawaguchi, T., Miyata, H., Ataka, K., Mae, K. and Yoshida, J.-I. (2005) *Angewandte Chemie-International Edition*, **44**, 2413.

59 Maeta, H., Sato, T., Nagasawa, H. and Mae, K. (2006) AIChE Spring National Meeting, Orlando, FL, published on CD.

60 Chambers, R.D., Fox, M.A., Holling, D., Nakano, T., Okazoe, T. and Sandford, G. (2005) *Lab on a Chip*, **5**, 191.

61 Markowz, G., Schirrmeister, S., Albrecht, J., Becker, F., Schütte, R., Caspary, K.J. and Klemm, E. (2005) *Chemical Engineering and Technology*, **28**, 459.

62 Thayer, A.M. (2005) *Chemical & Engineering News*, **83**, 43.

63 www.fzk.de/fzk/idcplg?IdcService =FZK&node=2374&document= ID_050927, Karlsruhe, Germany (2005).

64 Hessel, V., Serra, C., Löwe, H. and Hadziioannou, G. (2005) *Chemie Ingenieur Technik*, **77**, 1693.

65 Wille, C., Autze, V., Kim, H., Nickel, U., Oberbeck, S., Schwalbe, T. and Unverdorben, L. (2002) 6th International Conference on Microreaction Technology, IMRET 6, AIChE Pub. No. 164, New Orleans, USA, p. 7.

66 Hessel, V., Hardt, S. and Löwe, H. (2004) *Chemical Micro Process Engineering – Fundamentals, Modelling and Reactions*, Wiley-VCH Verlag GmbH, Weinheim.

67 Kirschneck, D. and Tekautz, G. (2007) *Chemical Engineering and Technology*, **30**, 305.

68 Krummradt, H., Kopp, U. and Stoldt, J. (2000) Microreaction Technology: 3rd International Conference on Microreaction Technology, Proceedings of IMRET 3 (ed. W. Ehrfeld), Springer-Verlag, Berlin, p. 181.

Index

Microreactors in Organic Synthesis and Catalysis. Edited by Thomas Wirth
Copyright © 2008 WILEY-VCH Verlag GmbH & Co. KGaA. All rights reserved.
ISBN: 978-3-527-31869-8